SpringerWienNewYork

CISM COURSES AND LECTURES

The series presents lecture notes, monographs, edited works and proceedings in the field of Mechanics, Engineering, Computer Science and Applied Mathematics.
Purpose of the series is to make known in the international scientific and technical community results obtained in some of the activities organized by CISM, the International Centre for Mechanical Sciences.

INTERNATIONAL CENTRE FOR MECHANICAL SCIENCES

COURSES AND LECTURES - No.507

SIMULATION TECHIQUES FOR APPLIED DYNAMICS

EDITED BY

MARTIN ARNOLD
MARTIN LUTHER UNIVERSITY HALLE-WITTENBERG, GERMANY

WERNER SCHIEHLEN
UNIVERSITY OF STUTTGART, GERMANY

SpringerWienNewYork

This volume contains 210 illustrations

Printed in Italy
SPIN 12576876

All contributions have been typeset by the authors.

ISBN 978-3-211-89547-4 SpringerWienNewYork

PREFACE

In general, dynamical systems consist of mechanical, electrical, hydraulic and biological components as well as control devices often based on computer hardware and software. The design of such systems requires advanced modelling and simulation techniques that are tailored to multidisciplinary problems. The coupling of models from different physical domains and the efficient and reliable simulation of coupled physical phenomena in engineering applications are topics of active research in various fields of engineering, in simulation technology and in the development and analysis of numerical solvers.

The methods and software tools of multibody dynamics illustrate the large potential of advanced monodisciplinary simulation techniques in the integrated design of heterogeneous engineering systems. Non-mechanical system components are either considered by appropriate extensions of classical multibody system simulation tools or by the coupling with standard tools from other disciplines like hydraulics or computer aided control engineering in a co-simulation framework.

Alternatively, the heterogeneous engineering system may be described in a unified modelling framework for multidisciplinary applications. Bond graph modelling or, more general, port-based modelling techniques as well as general modelling languages like VHDL-AMS and Modelica are typical representatives of this unified approach.

The Advanced School "Simulation Techniques for Applied Dynamics" was held at CISM, International Centre for Mechanical Sciences, Udine (Italy), from the 17th to the 21st of September, 2007. The Course was addressed to engineers, mathematicians and physicists from industry and research institutes, who are concerned with system dynamics, control and computer simulation of mechanical and mechatronic systems, and to research scientists and students with interests in the theoretical background and in practical applications of computer simulation in applied dynamics.

The lecturers aimed at providing detailed knowledge on modelling and simulation of mechatronic systems with applications to dynamical analysis and model based controller design. It is a pleasure to see that all seven lecturers have contributed to this proceedings volume.

The book starts with some basics in the classical theory of rigid multibody systems and a typical extension to biomechanic system sim-

ulation (W. Schiehlen and P. Eberhard).

The dynamic analysis of flexible multibody systems results in challenging problems for modelling, simulation and control that are discussed in the contribution of O. Brüls et al. The methods are applied to simulation and control of a vehicle with semi-active suspension and to the motion and vibration control of a large flexible manipulator.

Modelling, simulation and control of mechatronic systems are studied in great detail by M. Valášek who starts with a general introduction to mechatronic systems and considers methods for model setup, simulation and control that are applied in suspension and engine studies illustrating the benefits of model predictive control strategies.

Port-based techniques for the unified modelling of multidomain physical systems in engineering applications are introduced in the contribution of P. Breedveld. The underlying bond graph notation is domain independent and combines physical and computational structure which contributes to a substantial speed-up of the modelling process.

The efficient and reliable simulation of complex engineering systems is a demanding task for numerical solution methods. The extension of highly developed numerical solvers from multibody dynamics to multidisciplinary coupled problems is discussed in the chapter on numerical methods for simulation in applied dynamics (M. Arnold).

Hardware- and human-in-the-loop applications require dynamical simulations in real time. J. Cuadrado et al. investigate a real-time capable combination of optimized multibody formalisms with appropriate time integration methods. Their approach is validated by experiments with a prototype car and may be used as well for controller design through human-in-the-loop simulations.

In a final contribution, P. Eberhard et al. present advanced applications of the methods of multibody dynamics in technical design processes. A flatness-based control concept and multi-criteria optimization techniques are applied to highly nonlinear parallel kinematic machines. The dynamics of gearwheels is analysed taking into account the details of impact-like contacts in geartrains for Diesel engines.

It was the common goal of all authors to provide a comprehensive introduction to state-of-the-art and recent developments in the dynamical simulation and control of heterogeneous systems.

Martin Arnold and Werner Schiehlen

CONTENTS

Multibody Systems and Applied Dynamics

Werner Schiehlen and Peter Eberhard
Institute of Engineering and Computational Mechanics,
University of Stuttgart, Stuttgart, Germany

Abstract Modern applied dynamics is characterized by systems consisting of mechanical, electrical, hydraulical and biological components subject to control. For the modelling and design of controlled systems the method of multibody systems is most appropriate due to the comparatively low number of degrees of freedom. Therefore, the method of multibody systems is briefly introduced in the following.

Human locomotion is used as a typical example for a mechanical system controlled by redundant muscle actuators requiring optimization approaches for control design. The simulation techniques applied are based on inverse dynamics considering measurement results, too.

1 Multibody Dynamics

Multibody dynamics is based on analytical mechanics dealing with rigid bodies. Early applications of the dynamics of rigid bodies are related to gyrodynamics, mechanism theory and biomechanics as reviewed by Schiehlen (1997). However, the requirements for more complex models of satellites and spacecrafts, and the fast development of more and more powerful computers led to a new branch of mechanics: multibody system dynamics. The results of classical mechanics had to be extended to computer algorithms, the multibody formalisms. One of the first formalisms is due to Hooker and Margulies (1965). This approach was developed for satellites consisting of an arbitrary number of rigid bodies interconnected by spherical joints. Another formalism was published by Roberson and Wittenburg (1967). In addition to these numerical formalisms, the progress in computer hardware and software allowed formula manipulation with the result of symbolical equations of motion, too. First contributions are due to Levinson (1977) and Schiehlen and Kreuzer (1977). In the 1980s complete software systems for the modeling, simulation and animation were offered on the market as described by Schwertassek and Roberson (1986). The state-of-the-art achieved by the end of the 20th century was documented in Schiehlen

(1990). Reviews on multibody dynamics including analysis methods and applications were presented by Kortüm and Schiehlen (1985) and Huston. The progress achieved in flexible multibody systems was documented by Shabana (2005, 2003) who established the Absolute Nodal Coordinate Formulation (ANCF). Bauchau (2006) considered computational multibody dynamics approaches including impact problems. Today, software packages for multibody dynamics analysis are widely used in academia and industry, see e.g. http://real.uwaterloo.ca/~mbody/#Software.

Recent research topics cover theoretical and computational methods, flexible multibody systems, large deformation phenomena, parameter identification and parameter models, model reduction, contact and impact problems, control and mechatronics, nonholonomic systems, multiphysics problems, algorithms, integration codes and software, simulation and virtual reality, experiments and numerical verification, optimization of mechanical systems with applications to machines and mechanisms, vehicle dynamics, robotics and walking machines, biomechanical problems as well as education in multibody dynamics as presented by Schiehlen (2007). The method of multibody systems is reviewed in this section as one of the most efficient modelling approaches in applied dynamics.

1.1 Mechanical Modelling and Kinematics

First of all the engineering system has to be replaced by the elements of the multibody system approach: rigid and/or flexible bodies, joints, gravity, springs, dampers and position and/or force actuators. A system constrained by bearings and joints is disassembled as free body system using an appropriate number of inertial, moving reference and body fixed frames for the mathematical description.

A system of p free rigid bodies holds $6p$ degrees of freedom characterized by translation vectors and rotation tensors with respect to the inertial frame as

$$\boldsymbol{r}_i = \begin{bmatrix} r_{i1} & r_{i2} & r_{i3} \end{bmatrix}^T, \quad \boldsymbol{S}_i = \boldsymbol{S}_i(\alpha_i, \beta_i, \gamma_i), \quad i = 1(1)p, \tag{1}$$

see, e.g., Eberhard and Schiehlen (2006). Thus, the position vector of the free system can be written as

$$\boldsymbol{x} = [r_{11}\ r_{12}\ r_{13}\ r_{21}\ \dots\ \alpha_p\ \beta_p\ \gamma_p]^T \quad \in I\!R^{6p}. \tag{2}$$

Then, the free system's translation and rotation remain as

$$\boldsymbol{r_i} = \boldsymbol{r_i}(\boldsymbol{x}), \quad \boldsymbol{S_i} = \boldsymbol{S_i}(\boldsymbol{x}). \tag{3}$$

Assembling the system by q holonomic, rheonomic constraints reduces the number of degrees of freedom to $f = 6p - q$. The corresponding constraint equations may be written in explicit or implicit form, respectively, as

$$\boldsymbol{x} = \boldsymbol{x}(\boldsymbol{y}, t) \qquad \text{or} \qquad \boldsymbol{\Phi}(\boldsymbol{x}, t) = \boldsymbol{0} \tag{4}$$

where the generalized position vector $\boldsymbol{y}$ summarizes the f generalized coordinates of the holonomic system

$$\boldsymbol{y}(t) = [y_1 \;\; y_2 \;\; y_3 \;\; \ldots \;\; y_f]^T. \tag{5}$$

Then, for the holonomic system's translation and rotation it remains

$$\boldsymbol{r}_i = \boldsymbol{r}_i(\boldsymbol{y}, t), \quad \boldsymbol{S}_i = \boldsymbol{S}_i(\boldsymbol{y}, t). \tag{6}$$

By differentiation the absolute translational and rotational velocity vectors are found

$$\boldsymbol{v}_i = \dot{\boldsymbol{r}}_i = \frac{\partial \boldsymbol{r}_i}{\partial \boldsymbol{y}^T}\dot{\boldsymbol{y}} + \frac{\partial \boldsymbol{r}_i}{\partial t} = \boldsymbol{J}_{Ti}(\boldsymbol{y}, t)\dot{\boldsymbol{y}} + \overline{\boldsymbol{v}}_i(\boldsymbol{y}, t), \tag{7}$$

$$\boldsymbol{\omega}_i = \dot{\boldsymbol{s}}_i = \frac{\partial \boldsymbol{s}_i}{\partial \boldsymbol{y}^T}\dot{\boldsymbol{y}} + \frac{\partial \boldsymbol{s}_i}{\partial t} = \boldsymbol{J}_{Ri}(\boldsymbol{y}, t)\dot{\boldsymbol{y}} + \overline{\boldsymbol{\omega}}_i(\boldsymbol{y}, t) \tag{8}$$

where $\boldsymbol{s}_i$ means a vector of infinitesimal rotations following from the corresponding rotation tensor, see, e.g., Eberhard and Schiehlen (2006), and $\overline{\boldsymbol{v}}_i, \overline{\boldsymbol{\omega}}_i$ are the local velocities. Further, the Jacobian matrices $\boldsymbol{J}_{Ti}$ and $\boldsymbol{J}_{Ri}$ for translation and rotation are defined by Eqs. (7) and (8).

The system may be subject to additional r nonholonomic constraints which do not affect the $f = 6p - q$ positional degrees of freedom, but they reduce the velocity dependent degrees of freedom to $g = f - r = 6p - q - r$. The corresponding constraint equations can be written explicitly or implicitly, too,

$$\dot{\boldsymbol{y}} = \dot{\boldsymbol{y}}(\boldsymbol{y}, \boldsymbol{z}, t) \quad \text{or} \quad \boldsymbol{\Psi}(\boldsymbol{y}, \boldsymbol{z}, t) = \boldsymbol{0} \tag{9}$$

where the generalized velocities are summarized by

$$\boldsymbol{z}(t) = [z_1 \; z_2 \; z_3 \; \ldots \; z_g]^T. \tag{10}$$

For the system's translational and rotational velocities it follows from Eqs. (7) to (9)

$$\boldsymbol{v}_i = \boldsymbol{v}_i(\boldsymbol{y}, \boldsymbol{z}, t) \quad \text{and} \quad \boldsymbol{\omega}_i = \boldsymbol{\omega}_i(\boldsymbol{y}, \boldsymbol{z}, t). \tag{11}$$

By differentiation the acceleration vectors are obtained, e.g., the translational acceleration follows as

$$\boldsymbol{a}_i = \frac{\partial \boldsymbol{v}_i}{\partial \boldsymbol{z}^T}\dot{\boldsymbol{z}} + \frac{\partial \boldsymbol{v}_i}{\partial \boldsymbol{y}^T}\dot{\boldsymbol{y}} + \frac{\partial \boldsymbol{v}_i}{\partial t} = \boldsymbol{L}_{Ti}(\boldsymbol{y}, \boldsymbol{z}, t)\dot{\boldsymbol{z}} + \overline{\boldsymbol{a}}_i(\boldsymbol{y}, \boldsymbol{z}, t) \tag{12}$$

where $\overline{\boldsymbol{a}}_i$ denotes the so called local accelerations. A similar equation yields the rotational acceleration. The Jacobian matrices $\boldsymbol{L}_{Ti}$ and $\boldsymbol{L}_{Ri}$, respectively, are related to the generalized velocities, for translations as well as for rotations.

1.2 Equations of Motion of Rigid Body Systems

Newton's equations and Euler's equations are based on the velocities and accelerations from section 1.1 as well as on the applied forces and torques, and the constraint forces and torques acting on all bodies. The reactions or constraint forces and torques, respectively, can be reduced to a minimal number of generalized constraint forces also known as Lagrange multipliers.

In matrix notation the following equations are obtained, see also Schiehlen (1997). One gets for free body system kinematics and holonomic constraint forces

$$\overline{\overline{\boldsymbol{M}}}\ddot{\boldsymbol{x}} + \overline{\boldsymbol{q}}^c(\boldsymbol{x}, \dot{\boldsymbol{x}}, t) = \overline{\boldsymbol{q}}^e(\boldsymbol{x}, \dot{\boldsymbol{x}}, t) + \overline{\boldsymbol{Q}}\boldsymbol{g}, \quad \overline{\boldsymbol{Q}} = -\boldsymbol{\Phi}_x^T, \tag{13}$$

for holonomic system kinematics and constraints

$$\overline{\overline{\boldsymbol{M}}}\,\overline{\boldsymbol{J}}\ddot{\boldsymbol{y}} + \overline{\boldsymbol{q}}^c(\boldsymbol{y}, \dot{\boldsymbol{y}}, t) = \overline{\boldsymbol{q}}^e(\boldsymbol{y}, \dot{\boldsymbol{y}}, t) + \overline{\boldsymbol{Q}}\boldsymbol{g}, \tag{14}$$

and for nonholonomic system kinematics and constraints

$$\overline{\overline{\boldsymbol{M}}}\,\overline{\boldsymbol{L}}\dot{\boldsymbol{z}} + \overline{\boldsymbol{q}}^c(\boldsymbol{y}, \boldsymbol{z}, t) = \overline{\boldsymbol{q}}^e(\boldsymbol{y}, \boldsymbol{z}, t) + \overline{\boldsymbol{Q}}\boldsymbol{g}. \tag{15}$$

On the left hand side of Eqs. (13) to (15) the inertia forces are characterized by the inertia matrix $\overline{\overline{\boldsymbol{M}}}$, the global Jacobian matrices $\overline{\boldsymbol{J}}, \overline{\boldsymbol{L}}$ and the vector $\overline{\boldsymbol{q}}^c$ of the Coriolis forces. On the right hand side the vector $\overline{\boldsymbol{q}}^e$ of the applied forces, which include control forces, and the constraint forces composed by a global distribution matrix $\overline{\boldsymbol{Q}}$ and the vector of the generalized constraint forces $\boldsymbol{g}$ are found.

Each of the Eqs. (13) to (15) represents $6p$ scalar equations. However, the number of unknowns is different. In Eq. (13) there are $6p + q$ unknowns resulting from the vectors $\boldsymbol{x}$ and $\boldsymbol{g}$. In Eq. (14) the number of unknowns is exactly $6p = f + q$ represented by the vectors $\boldsymbol{y}$ and $\boldsymbol{g}$, while in Eq. (15) the number of unknowns is $12p - q$ due to the additional velocity vector $\boldsymbol{z}$ and an extended constraint vector $\boldsymbol{g}$. Obviously, the Newton-Euler equations have to be supplemented for the simulation of motion.

The equations of motion are complete sets of equations to be solved by vibration analysis and/or numerical integration. There are two frequently used approaches resulting in differential-algebraic equations (DAE) or ordinary differential equations (ODE), respectively.

For the DAE approach the implicit constraint equation (4) is differentiated twice and added to the Newton-Euler equations (13) resulting in

$$\begin{bmatrix} \overline{\overline{\boldsymbol{M}}} & \boldsymbol{\Phi}_x^T \\ \boldsymbol{\Phi}_x & \boldsymbol{0} \end{bmatrix} \begin{bmatrix} \ddot{\boldsymbol{x}} \\ \boldsymbol{g} \end{bmatrix} = \begin{bmatrix} \overline{\boldsymbol{q}}^e - \overline{\boldsymbol{q}}^c \\ -\dot{\boldsymbol{\Phi}}_t - \dot{\boldsymbol{\Phi}}_x \dot{\boldsymbol{x}} \end{bmatrix}. \tag{16}$$

Equation (16) is numerically unstable due to a double zero eigenvalue originating from the differentiation of the constraints. During the last decade great progress was achieved in the stabilization of the solutions of Eq. (16). This is, e.g., documented in Eich-Soellner and Führer (1998), and also discussed by Arnold (2008).

The ODE approach is based on the elimination of the constraint forces using the orthogonality of generalized motions and constraints, $\overline{\boldsymbol{J}}^T \overline{\boldsymbol{Q}} = \boldsymbol{0}$, also known as d'Alembert's principle, see D'Alembert (1743), for holonomic systems. Then a minimal number of equations remains

$$\boldsymbol{M}(\boldsymbol{y}, t)\ddot{\boldsymbol{y}} + \boldsymbol{k}(\boldsymbol{y}, \dot{\boldsymbol{y}}, t) = \boldsymbol{q}(\boldsymbol{y}, \dot{\boldsymbol{y}}, t). \tag{17}$$

The orthogonality may also be used for nonholonomic systems, $\overline{\boldsymbol{L}}^T \overline{\boldsymbol{Q}} = \boldsymbol{0}$, corresponding to Jourdain's principle, see Jourdain (1909), and Kane's equations, see Kane and Levinson (1985). However, the explicit form of the nonholonomic constraints (9) has to be added,

$$\dot{\boldsymbol{y}} = \dot{\boldsymbol{y}}(\boldsymbol{y}, \boldsymbol{z}, t), \quad \boldsymbol{M}(\boldsymbol{y}, \boldsymbol{z}, t)\dot{\boldsymbol{z}} + \boldsymbol{k}(\boldsymbol{y}, \boldsymbol{z}, t) = \boldsymbol{q}(\boldsymbol{y}, \boldsymbol{z}, t). \tag{18}$$

Equations (17) and (18) can now be solved by any standard time integration code.

1.3 Equations of Motion for Flexible Systems

The equations presented can also be extended to flexible bodies as shown in Fig. 1. For the analysis of small structural vibrations often the relative nodal coordinate formulation (RNCF) with a floating frame of reference is used while for large deformations the absolute nodal coordinate formulation (ANCF) turned out to be very efficient, see, e.g., Melzer (1996) and Shabana (2005, 2003).

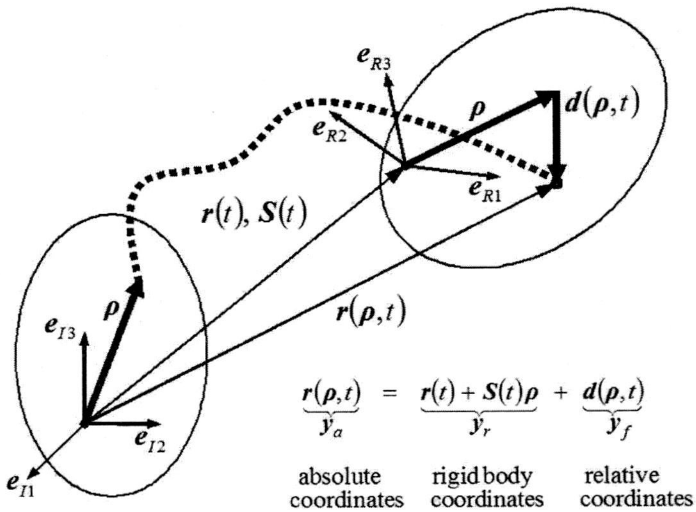

Figure 1. Reference systems for flexible multibody systems

Within the RNCF the small number of f_f relative coordinates describing the elastic deformations are added to the large number of f_r rigid body coordinates of the reference frame moving with translation $\boldsymbol{r}(t)$ and rotation $\boldsymbol{S}(t)$ resulting in an extended position vector

$$\boldsymbol{y}(t) = [\boldsymbol{y}_r^T \ \boldsymbol{y}_f^T]^T \tag{19}$$

where the subvectors $\boldsymbol{y}_r$, $\boldsymbol{y}_f$ summarize the corresponding coordinates. Then, the extended equations of motion read as

$$\boldsymbol{M}(\boldsymbol{y},t)\ddot{\boldsymbol{y}} + \boldsymbol{k}(\boldsymbol{y},\dot{\boldsymbol{y}},t) + \boldsymbol{k}_i(\boldsymbol{y},\dot{\boldsymbol{y}}) = \boldsymbol{q}(\boldsymbol{y},\dot{\boldsymbol{y}},t). \tag{20}$$

In comparison to Eq. (17) the additional term

$$\boldsymbol{k}_i(\boldsymbol{y},\dot{\boldsymbol{y}}) = \begin{bmatrix} 0 & 0 \\ 0 & \boldsymbol{K} \end{bmatrix} \boldsymbol{y} + \begin{bmatrix} 0 & 0 \\ 0 & \boldsymbol{D} \end{bmatrix} \dot{\boldsymbol{y}} \tag{21}$$

depends only on the stiffness and damping matrices $\boldsymbol{K}$ and $\boldsymbol{D}$ of the flexible bodies. Moreover, the inertia matrix shows the inertia coupling due to the relative coordinates

$$\boldsymbol{M} = \begin{bmatrix} \boldsymbol{M}_{rr} & \boldsymbol{M}_{rf} \\ \boldsymbol{M}_{rf}^T & \boldsymbol{M}_{ff} \end{bmatrix}. \tag{22}$$

Within the ANCF for highly flexible bodies f_a absolute coordinates are summarized in a vector $\boldsymbol{y}_a$ characterizing the material points of the bodies by an appropriate shape function. Then, the equations of motion read as

$$\boldsymbol{M}\ddot{\boldsymbol{y}}_a + \boldsymbol{K}_a(\boldsymbol{y}_a)\boldsymbol{y}_a = \boldsymbol{q}(\boldsymbol{y}_a, t) \tag{23}$$

where $\boldsymbol{M}$ is a constant mass matrix and the vector $\boldsymbol{k}$ of the generalized Coriolis forces is vanishing due to the absolute coordinates. This is true for standard finite elements like Euler beams or bricks. However, for Timoshenko beams with rotary inertia Eq. (20) may be found again as pointed out in von Dombrowski (2002). In any case, the stiffness matrix $\boldsymbol{K}_a$ is highly nonlinear and requires special evaluation as shown by Shabana (2005).

1.4 Linear Motion Analysis

The dynamical analysis of multibody systems is closely related to vibration theory. For engineering applications mechanical vibrations of holonomic, rheonomic systems are most important. The dynamical phenomena are classified according to the linear and nonlinear equations of vibrations.

For small vibrations the nonlinear Eq. (17) can be linearized resulting in

$$\boldsymbol{M}(t)\ddot{\boldsymbol{y}} + \boldsymbol{P}(t)\dot{\boldsymbol{y}} + \boldsymbol{Q}(t)\boldsymbol{y} = \boldsymbol{h}(t). \tag{24}$$

This system may feature parametrically excited vibrations due to the time-varying, often periodic matrices. In the case of time-invariant matrices one gets after decomposition into the symmetric and skew-symmetric parts

$$\boldsymbol{M}\ddot{\boldsymbol{y}} + (\boldsymbol{D} + \boldsymbol{G})\dot{\boldsymbol{y}} + (\boldsymbol{K} + \boldsymbol{N})\boldsymbol{y} = \boldsymbol{h}(t), \tag{25}$$

a system which performs forced vibrations due to the external excitation on the right hand side. In the case of $\boldsymbol{h}(t) = \boldsymbol{0}$ only free vibrations remain. Furthermore, if the damping matrix $\boldsymbol{D}$, the gyroscopic matrix $\boldsymbol{G}$, and the circulatory matrix $\boldsymbol{N}$ are missing, the conservative system

$$\boldsymbol{M}\ddot{\boldsymbol{y}} + \boldsymbol{K}\boldsymbol{y} = \boldsymbol{0} \tag{26}$$

with free undamped vibrations is found. On the other hand, nonlinear time-variant mechanical systems represented by Eq. (17), even with just one degree of freedom, may show chaotic vibrations.

The special structure of Eqs. (25) and (26) simplifies the analysis. Marginal stability of Eqs. (26) is guaranteed if the stiffness matrix $\boldsymbol{K}$ is positive definite. Free damped vibrations due to Eq. (25) with $\boldsymbol{G} = \boldsymbol{N} = \boldsymbol{0}$ are asymptotically stable if both, the stiffness matrix $\boldsymbol{K}$ is positive definite and the damping matrix $\boldsymbol{D}$ is positive definite or pervasively positive semidefinite, respectively, see Müller and Schiehlen (1985). Moreover, Eq. (25) is asymptotically stable if all eigenvalues have a negative real part. The general solution of Eqs. (26) reads as

$$\boldsymbol{y}(t) = \boldsymbol{\Psi}_1(t)\boldsymbol{y}_0 + \boldsymbol{\Psi}_2(t)\dot{\boldsymbol{y}}_0 \tag{27}$$

where the transition matrices $\boldsymbol{\Psi}_1(t)$, $\boldsymbol{\Psi}_2(t)$ are found from a real eigenvalue analysis of dimension f. The general solution of Eqs. (25) can be written in state space form with the state vector $\boldsymbol{x}(t)$ summarizing the system's state given by the generalized coordinates and their first time derivatives as

$$\boldsymbol{x}(t) = \begin{bmatrix} \boldsymbol{y}(t) \\ \dot{\boldsymbol{y}}(t) \end{bmatrix} . \tag{28}$$

Then, the general solution reads simply

$$\boldsymbol{x}(t) = \boldsymbol{\Phi}(t)\boldsymbol{x}_0 \tag{29}$$

where $\boldsymbol{\Phi}(t)$ denotes here the state transition matrix following from a complex eigenvalue problem of dimension $2f$. Matrix methods for linear systems with harmonic excitation $\boldsymbol{h}(t)$ lead to the concept of frequency response matrices while random excitation processes require spectral density matrices or covariance matrices, respectively. In the case of Eq. (24) with periodically time-varying coefficients Floquet's theory allows closed form solutions, see Müller and Schiehlen (1985).

2 Human Locomotion

Human locomotion research requires simulation techniques provided by applied dynamics. Therefore, human locomotion serves in this section as a typical application for applied dynamics.

Dynamic simulation of human motion driven by physiological muscle models is increasingly used to understand human motion. Indeed, recent research has provided remarkable insights into the coordination of human walking. Because the neural excitations to the muscles are not known a priori, optimization techniques are used based on the assumption that the central nervous system (CNS) excites muscles in such a way as to optimize some physiological performance criteria. In walking, the metabolic cost of transport was shown to be an essential performance criterion, see e.g. Ralston (1976).

The resulting neuromuscular optimal control problem is traditionally solved by transforming the problem in an optimization by means of discretizing the controls, i.e. the neural excitations. The approach consists of repeatedly solving the initial value problem using the reconstructed control histories while minimizing a user-defined cost function. This approach, called dynamic optimization in this contribution, is being increasingly used to study human walking, see e.g. Gerritsen et al. (1998), Neptune et al. (2001), Anderson and Pandy (2001), or Umberger et al. (2003). However, the prohibitive computational effort arising from the several integrations of the differential equations describing the dynamics of the musculoskeletal system prevents a broader use. Anderson and Pandy (2001), for instance, report 10000 hours of CPU time to find a solution for a 3-D musculoskeletal model. The successful application of dynamic simulation to the design of assistive devices, to the evaluation of rehabilitation therapies or to the planning of surgical interventions depends on the reduction of computational effort to acceptable levels.

One promising alternative to avoid several integrations of the system equations is the use of inverse-dynamics approaches to solve the optimal control problem. For instance, Nagurka and Yen (1990) proposed an inverse-dynamics approach that consists of approximating the trajectories of the generalized coordinates of a nonlinear dynamic system by Fourier series, whose coefficients are incorporated to the design parameters. This allows the conversion of the optimal control problem into a parameter optimization problem, and consequently the use of standard and efficient nonlinear programming algorithms such as sequential quadratic programming (SQP). Furthermore, the controls are computed by inverse dynamics by means of the equations of motion from the generalized coordinates and their first and second time derivatives, which are obtained by analytical differentiation. As a result, numerical integrations of the system differential equations are not required and, thus, the computational effort is significantly reduced.

Recently, Saidouni and Bessonnet (2003) and Bessonnet et al. (2005) applied a similar approach to determine near-optimal walking trajectories and joint moments for biped robots. Instead of Fourier series, Bessonnet et al. (2005) employed splines of class C3, i.e. approximating polynomials of fourth order, to approximate the time histories of the generalized coordinates. They divide the gait cycle into phases and model contact by means of kinematic constraints that are activated and deactivated along the walking cycle. The application of this approach to human locomotion requires, however, the introduction of physiological models of the muscles and their dynamics. Furthermore, the redundancy in the actuation resulting from the number of muscles being greater than the number of degrees of freedom has to be addressed. In this context, an inverse-dynamics approach is proposed, see Ackermann (2007a), that allows for the determination of near-optimal time histories of the generalized coordinates, muscle forces and neural excitations taking the contraction and activation dynamics of the muscles into account.

As for the inverse-dynamics methods mentioned above, the trajectories of the generalized coordinates describing the motion of the skeletal system are approximated by parameterized functions. In addition, to deal with the redundancy caused by the introduction of muscles, also the time histories of the muscle forces are parameterized. The consistency between the generalized coordinates and the muscle forces is guaranteed by ensuring the fulfillment of constraints given by the equations of motion. The neural excitations, i.e. the controls, are computed by inverting the muscle contraction and activation dynamics. In this way, near-optimal time histories of the generalized coordinates and of the muscle forces are searched that optimize a time-integral cost function such as minimal metabolic cost of transport or maximal proximity to measured data. The method is described in detail in Ackermann (2007b).

2.1 Biomechanical Modelling

The model of the skeletal system adopted is shown in Fig. 2. The free model, i.e. without considering the ground contact constraints, is composed by 7 bodies, the HAT, the two thighs, the two shanks and the two feet, and has nine degrees of freedom. The motion of the model is performed in the sagittal plane and is described by $f = 9$ generalized coordinates as depicted in Fig. 2. The pelvis, upper part of the trunk, arms and head are modeled as a single rigid body referred to as HAT. The anthropometric parameters of the segments are obtained using the tables in de Leva (1996). The heels, the positions of the metatarsophalangeal joints, and the ankles are denoted by the capital letter A, B and C, respectively, followed by r for the right or l for the left foot. These points are used to define the constraints active at different walking phases. Each lower limb is actuated by 8 three-element Hill-type muscle units, see Fig. 3.

During walking contacts occur between the feet and the ground. The contacts are modeled by means of kinematic constraints as implemented by Bessonnet et al. (2005). The constraints are activated and deactivated throughout the gait cycle leading to a time-varying kinematic topology. The kinematic constraints occur between the points Ar, Br, Al and Bl in Fig. 2 and the ground. The gait cycle is divided into eight phases based on the events observed during normal walking. The phases 1, 2, 3 and 4 correspond to the periods limited by the right heel strike, right toe contact (flat contact of right sole), left toe off, right heel off, and left heel strike, respectively. The phases 5, 6, 7 and 8 correspond to the periods limited by the left heel

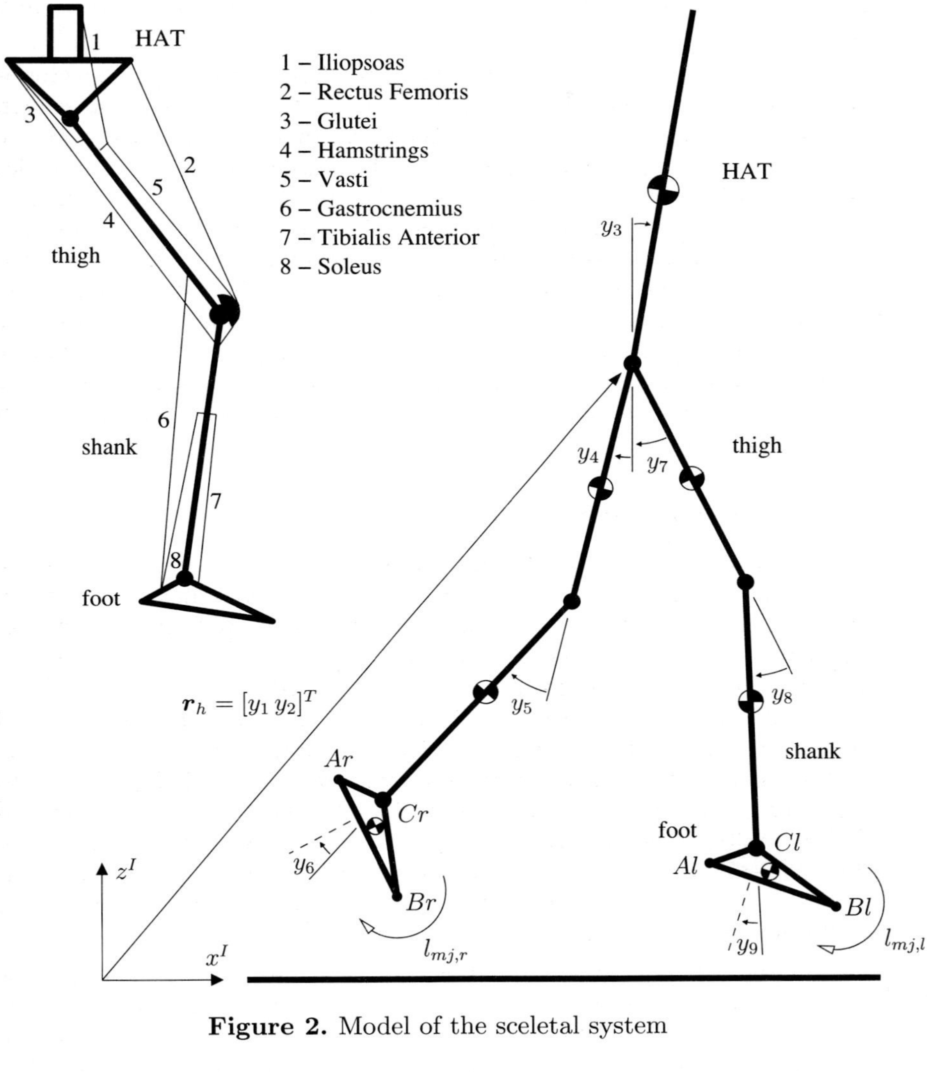

Figure 2. Model of the sceletal system

strike, left toe contact (flat contact of left sole), right toe off, left heel off, and right heel strike, respectively.

The equations of motion of the free model, which are a set of ordinary differential equations as described by Eq. (17), are modified to account for the additional kinematic constraints modeling the contacts occurring

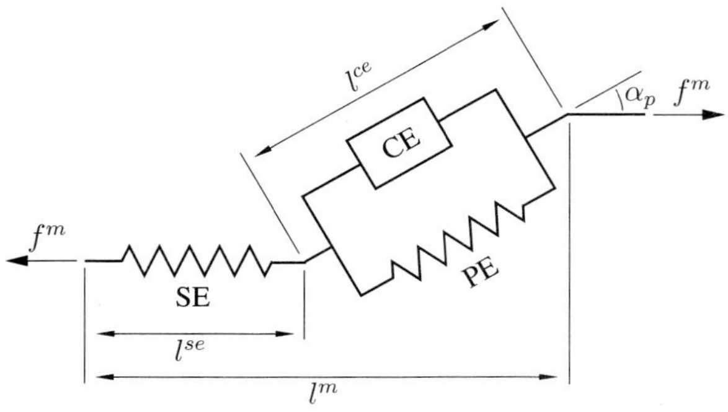

Figure 3. Hill-type muscle model

between the feet and the ground, see e.g. Schiehlen (2006), for each phase p as

$$\boldsymbol{M}(\boldsymbol{y})\ddot{\boldsymbol{y}} + \boldsymbol{k}(\boldsymbol{y}, \dot{\boldsymbol{y}}) = \boldsymbol{q}_r(\boldsymbol{y}, \dot{\boldsymbol{y}}) + \boldsymbol{R}(\boldsymbol{y})\boldsymbol{f}^m + \boldsymbol{C}_p^T(\boldsymbol{y})\boldsymbol{\lambda}_p, \tag{30}$$

$$\boldsymbol{c}_p = \boldsymbol{0}, \quad p = 1, ..., 8, \tag{31}$$

where $\boldsymbol{M}$ is the symmetric, positive definite mass matrix, $\boldsymbol{k}$ is the vector of generalized Coriolis forces, $\boldsymbol{q}_r$ is the vector of generalized forces other than the ones caused by the muscles, $\boldsymbol{f}^m$ is the vector of $2 * 8 = 16$ muscle forces, $\boldsymbol{R}$ is the matrix that transforms the muscle forces into generalized forces, $\boldsymbol{c}_p$ is the vector containing the kinematic constraints at the feet active in phase p, $\boldsymbol{\lambda}$ is the vector of Lagrangian multipliers, which may be interpreted as generalized constraint forces required to fulfill the kinematic constraints in $\boldsymbol{c}_p$, and $\boldsymbol{C}_p$ represents the Jacobian matrix of the constraints active in phase p. The muscles generate moments at the joints as $\boldsymbol{\tau} = \boldsymbol{A}\boldsymbol{f}^m$, where $\boldsymbol{A}$ is a matrix containing the muscles moment arms.

The eight muscle groups considered are shown in Fig. 2 for one leg. The Hill-type muscle model, see e.g. Zajac (1989), is composed by a contractile element CE that generates force and represents the muscle fibers, and a series elastic element SE, while the force of the parallel elastic element PE is set to zero, see Fig. 3. All the structures in parallel to the CE and the SE are represented by total passive moments at the joints, which include the moments generated by all other passive structures crossing the joints, like ligaments. The model for the passive moments at the joints is adopted from Riener and Edrich (1999). A linear viscous damping is added to the hip and knee joints, whose values are determined by pendulum experiments

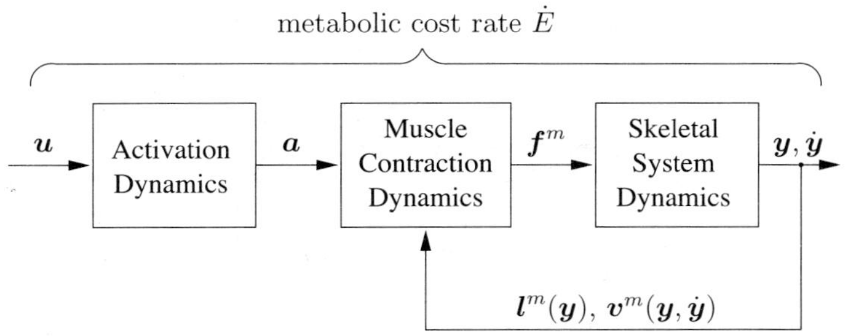

Figure 4. Muscoleskeletal system dynamics

performed by Stein et al. (1996). In order to account for the moments $l_{mj,r}$ and $l_{mj,l}$ applied at the metatarsophalangeal joints by the muscles *flexor halucis longus and flexor digitorum longus*, negative moments up to 10 Nm are allowed to act at points *Br* and *Bl.* Positive moments on the foot at the metatarsophalangeal joint pull the toes upwards having little or no influence on the dynamics of the model during walking. For this reason no positive moments are applied at the metatarsophalangeal joint in the model. Figure 4 shows a scheme of the dynamics of the complete musculoskeletal system having the neural excitations as controls. The vectors $\boldsymbol{u}$ and $\boldsymbol{a}$ are the vectors of the 16 neural excitations and muscle activations, respectively, of all muscle groups considered in the model. The vectors $\boldsymbol{l}^m$ and $\boldsymbol{\nu}^m$ contain the lengths and lengthening velocities of the muscles, respectively. The activation dynamics is modeled by a first order differential equation as in He et al. (1991). The first order contraction dynamics arises from the presence of the series elastic element (CE) and from the muscle force-length-velocity relations, refer e.g. to Ackermann (2007b), and is based on the model of Nagano and Gerritsen (2001).

In order to estimate the metabolic energy rate $\dot{E}$ consumed by the muscles, phenomenological muscle energy expenditure expressions recently proposed in Umberger et al. (2003) are used as

$$\dot{E} = \dot{E}(u, a, \nu^{ce}, l^{ce}, f^{ce}, \boldsymbol{p}^m) \tag{32}$$

where the muscle parameters are summarized in the vector $\boldsymbol{p}^m$. All muscle parameters are selected from Nagano and Gerritsen (2001), Menegaldo et al. (2003), and Umberger et al. (2003).

2.2 Optimization Approach

The approach proposed to solve the optimal neuromuscular problem is described in details in Ackermann (2007a,b). It consists in a conversion of the optimal control problem having neural excitations as controls into a parameter optimization based on the parameterization of states while the controls are determined by inverse dynamics. This dispenses with the need for numerically integrating the system differential equations, thus, reducing the computational cost. In order to deal with the fact that the musculoskeletal system is redundantly actuated, a parameterization of the muscle forces is proposed. The approach can be seen as a fusion of the approaches of Bessonnet et al. (2005) to determine near-optimal walking patterns for biped robots with the extended inverse dynamics (EID) approach of Ackermann and Schiehlen (2006) to determine optimal, physiological muscle forces and energetic requirements for a prescribed motion of the skeleton. The connection between both approaches occurs at the joint levels by imposing the agreement between joint moments applied by the muscles and joint moments required to generate the parameterized motion. The approach yields optimal motion patterns and joint moment histories that are compatible with muscle capabilities. Furthermore, more accurate estimations of metabolic cost can be assessed by using phenomenological muscle energy expenditure models from Eq. (32).

Figure 5 shows a schematic representation of the approach proposed. The time histories of the generalized coordinates or kinematics, respectively, and of the muscle forces are approximated by splines that interpolate nodal values, which are incorporated to the optimization variables. The consistency between muscle forces and motion is guaranteed by satisfying the equations of motion (30) at discrete control points. From the reconstructed muscle force histories, the neural excitations are computed by inverting the muscle contraction dynamics and activation dynamics. The contraction dynamics is inverted by solving for the activation $\boldsymbol{a}$ from the muscle length $\boldsymbol{l}^m(\boldsymbol{y})$ and lengthening velocity $\boldsymbol{\nu}^m(\boldsymbol{y}, \dot{\boldsymbol{y}})$, and from the muscle force $\boldsymbol{f}^m$ and its derivative $\dot{\boldsymbol{f}}^m$, obtained by numerical differentiation of $\boldsymbol{f}^m$. The activation dynamics is inverted subsequently by the corresponding expression for the neural excitation $\boldsymbol{u}$ from the muscle activation $\boldsymbol{a}$ and its derivative $\dot{\boldsymbol{a}}$, obtained by numerical differentiation of $\boldsymbol{a}$. The optimization problem is formulated as a search for optimal motion and muscle forces that minimize a cost function, e.g., the metabolic cost of transport and the deviation from normal measured patterns, subject mainly to constraints that ensure neural excitations bounded by 0 and 1, fulfillment of kinematic constraints in Eqs. (31), and consistency between muscle forces and kinematics, i.e. fulfillment

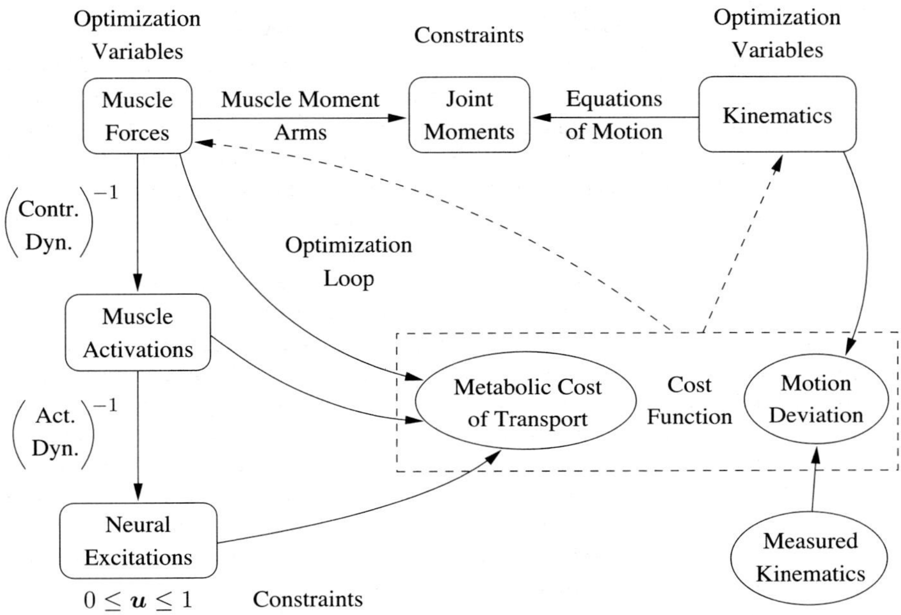

Figure 5. Optimization approach

of the equations of motion throughout the walking cycle, see Eq. (30).

The cost function adopted for the applications is characterized by the metabolic cost of transport E^t. The expression for the metabolic cost of transport reads as, see Ackermann (2007a),

$$E^t = \sum_{i=1}^{16} \int_{t_i}^{t_f} \dot{E}_i(t)dt \tag{33}$$

where $\dot{E}_i$ is the metabolic cost rate from Eq. (32) for muscle i.

2.3 Verification of Extended Inverse Dynamics Approach

One way to validate the extended inverse dynamics (EID) approach is to use the solution for the neural excitations obtained with the EID as controls for a numerical integration of the differential equations of the musculoskeletal model. If the kinematics obtained in this way agrees well with the measured one, used as input for the EID, it can be concluded that the approach works well and that the discretization adopted is sufficiently fine,

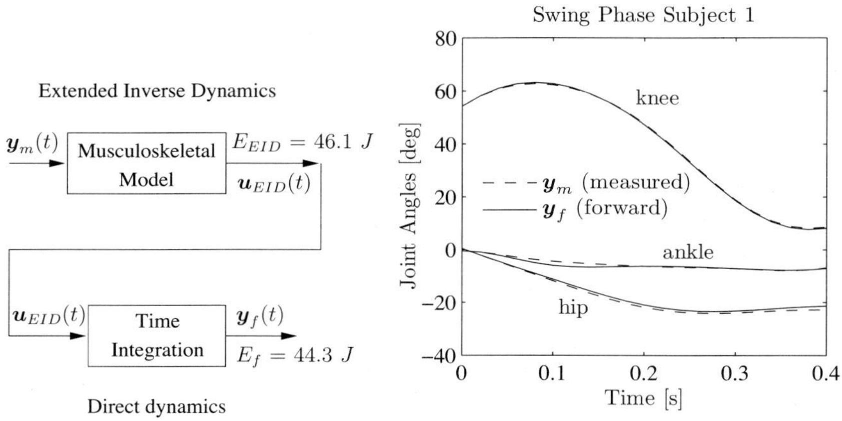

Figure 6. Schematic representation of the procedure to validate the EID approach on the left hand side.

i.e. the number n of nodes is sufficient to guarantee a proper modeling of the phenomenon. Fig. 6 shows a scheme of the described procedure on the left hand side. The right hand side shows a comparison between the measured kinematics $\boldsymbol{y}_m(t)$ of the swing phase of subject 1 walking normally with the kinematics $\boldsymbol{y}_f(t)$ computed by forward integrating the muscoleskeletal dynamics using the neural excitations obtained with the EID as controls.

The kinematics $\boldsymbol{y}_m(t)$ of the lower limb of a subject measured during the swing phase serves as input to the EID approach to compute the optimal neural excitations in $\boldsymbol{u}_{EID}$ and the total metabolic cost E_{EID} according to Eq. (33). The neural excitations at the nodes are then linearly interpolated and used as controls for a numerical forward integration of the differential equations with initial conditions for the states, muscle forces and muscle activations at $t = 0$, given by the values computed with the EID. The forward integration delivers the time history of the generalized coordinates $\boldsymbol{y}_f(t)$. A new estimation of the total metabolic cost E_f is then computed.

The results of the described procedure are shown in Fig. 6. On the left hand side, the metabolic cost estimations E^t obtained with the EID and with the forward integration are shown, $E_{EID} = 46.1J$ and $E_f = 44.3J$, respectively, representing a small difference of about 4%. The comparison of the kinematics on the right hand side of Fig. 6 features a good agreement. The results indicate, therefore, that the method works well and that the discretization is sufficiently fine by showing a good reconstruction of the original measured kinematics and a reasonable match of the metabolic cost estimations. In fact, this comparison would show an even better agreement if the states at all nodes computed with the EID were used for a numerical integration throughout the inter-node times.

3 Conclusion

In the first section the historical development of multibody dynamics is reviewed briefly. The fundamental approaches of analytical dynamics are summarized for rigid and flexible bodies. Multibody dynamics is an excellent foundation for applied dynamics and sophisticated control design. Multibody dynamics offers challenging problems for computational methods required advanced simulation techniques.

The second section illustrates the value of biomechanics based approaches to evaluate and analyze the potential of the inverse-dynamics based optimization framework. The results show how modelling and control of the sophisticated systems may be verified in applied dynamics by measurements and advanced simulation techniques.

Bibliography

M. Ackermann. A novel optimization approach to generate physiological human walking patterns. In *Proceedings of ASME, IDETC/CIE 2007, Las Vegas, USA*, 2007a. Paper: IDETC35014.

M. Ackermann. *Dynamics and energetics of walking with prostheses.* PhD Thesis, Shaker Verlag, Aachen, 2007b.

M. Ackermann and W. Schiehlen. Dynamic analysis of human gait disorder and metabolical cost estimation. *Archive of Applied Mechanics*, 75(10-12):569–594, 2006.

F.C. Anderson and M.G. Pandy. Dynamic optimization of human walking. *Journal of Biomechanical Engineering*, 123(5):381–390, 2001.

M. Arnold. Numerical methods for simulation in applied dynamics. In M. Arnold and W. Schiehlen, editors, *Simulation Techniques for Applied Dynamics.* Springer, Viena, 2008.

Bauchau, editor. *Comp. Methods Applied Mechanics Engineering.* volume 195(50-51): Special issue on multibody dynamics analysis, 2006.

G. Bessonnet, P. Seguin, and P. Sardain. A parametric optimization approach to walking pattern synthesis. *The International Journal of Robotics Research*, 24(7):523–536, 2005.

J. D'Alembert. *Traité de Dynamique.* David, Paris, 1743.

P. de Leva. Adjustments to Zatsiorsky - Seluyanov's segment inertia parameters. *Journal of Biomechanics*, 29(9):1223–1230, 1996.

P. Eberhard and W. Schiehlen. Computational dynamics of multibody systems: History, formalisms, and applications. *Journal of Computational and Nonlinear Dynamics*, 1(1):3–12, 2006.

E. Eich-Soellner and C. Führer. *Numerical Methods in Multibody Dynamics.* Teubner, Stuttgart, 1998.

K.G.M. Gerritsen, A.J. van den Bogert, M. Hulliger, and R.F. Zernicke. Intrinsic muscle properties facilitate locomotor control – a computer simulation study. *Motor Control*, 2:206–220, 1998.

J. He, W.S. Levine, and G.E. Loeb. Feedback gains for correcting small perturbations to standing posture. *IEEE Transactions on Autonomic Control*, 36:322–332, 1991.

W. Hooker and G. Margulies. The dynamical attitude equations for an n-body satellite. *J. Astronom. Science*, 12:123–128, 1965.

R.L. Huston. Multibody dynamics - modeling and analysis methods. *Applied Mechanics Review*, 44:109–117, 1991 and 49:35–40, 1996.

P.E.B. Jourdain. Note on an analogue at Gauss principle of least constraint. *Quarterly Journal on Pure Applied Mathematics*, 40:153–197, 1909.

T.R. Kane and D.A. Levinson. *Dynamics: Theory and Applications.* McGraw-Hill, New York, 1985.

W. Kortüm and W. Schiehlen. General purpose vehicle system dynamics software based on multibody formalisms. *Vehicle System Dynamics*, 14: 229–263, 1985.

D.A. Levinson. Equations of motion for multi-rigid-body systems via symbolical manipulation. *J. Spacecraft Rockets*, 14:479–487, 1977.

F. Melzer. Symbolic computations in flexible multibody systems. *Nonlinear Dynamics*, 9:147–163, 1996.

L.L. Menegaldo, A.T. Fleury, and H.I. Weber. Biomechanical modeling and optimal control of human posture. *Journal of Biomechanics*, 36(11): 1701–1712, 2003.

P.C. Müller and W. Schiehlen. *Linear Vibrations.* Martinus Nijhoff Publ., Dordrecht., 1985.

A. Nagano and K.G.M. Gerritsen. Effects of neuromuscular strength training on vertical jumping performance - a computer simulation study. *Journal of Applied Biomechanics*, 17(2):113–128, 2001.

M.L. Nagurka and V. Yen. Fourier-based optimal control of nonlinear dynamic systems. *Journal of Dynamic Systems, Measurement, and Control*, 112:17–26, 1990.

R.R. Neptune, S.A. Kautz, and F.E. Zajac. Contributions of the individual ankle plantar flexors to support, forward progression and swing initiation during walking. *Journal of Biomechanics*, 34(11):1387–1398, 2001.

H.J. Ralston. Energetics of human walking. In R.M. Herman et al., editor, *Neural Control of Locomotion*, pages 77–98. Plenum Press, New York, 1976.

R. Riener and T. Edrich. Identification of passive elastic joint moments in the lower extremities. *Journal of Biomechanics*, 32(5):539–544, 1999.

R.E. Roberson and J. Wittenburg. A dynamical formalism for an arbitrary number of interconnected rigid bodies, with reference to the problem of satellite attitude control. In *Proceedings 3rd Congr. Int. Fed. Autom. Control*, volume 1 number 3. Butterworth, London, 1967.

T. Saidouni and G. Bessonnet. Generating globally optimised sagittal gait cycles of a biped robot. *Robotica*, 21:199–210, 2003.

W. Schiehlen. *Multibody Systems Handbook*. Springer Berlin, 1990.

W. Schiehlen. Multibody system dynamics: Roots and perspectives. *Multibody System Dynamics*, 1(2):149–188, 1997.

W. Schiehlen. Computational dynamics: theory and applications of multibody systems. *European Journal of Mechanics A/Solids*, 25(4):566–594, 2006.

W. Schiehlen. Research trends in multibody system dynamics. *Multibody System Dynamics*, 18(1):3–13, 2007.

W. Schiehlen and E. Kreuzer. Aufstellen der Bewegungsgleichungen gewöhnlicher Mehrkörpersysteme. *Ingenieur-Archiv*, 46:185–194, 1977. (in German).

R. Schwertassek and R.E. Roberson. A perspective on computer-oriented multibody dynamical formalisms and their implementations. In G. Bianchi and W. Schiehlen, editors, *Dynamics of Multibody Systems*, pages 263–273. Springer, Berlin, 1986.

A.A. Shabana, editor. *Nonlinear Dynamics*. 4(1/2): Special Issue on Flexible Multibody Dynamics, 2003.

A.A. Shabana. *Dynamics of Multibody Systems*. Cambridge University Press, Cambridge, 2005.

R.B. Stein, M.K. Lebiedowska, D.B. Popovic, A. Scheiner, and H.J. Chizeck. Estimating mechanical parameters of leg segments in individuals with and without physical disabilities. *IEEE Transactions on Rehabilitation Engineering*, 4(3):201–211, 1996.

B.R. Umberger, K.G.M. Gerritsen, and P.E. Martin. A model of human muscle energy expenditure. *Computer Methods in Biomechanics & Biomedical Engineering*, 6(2):99–111, 2003.

S. von Dombrowski. Analysis of large flexible body deformation in multibody systems using absolute coordinates. *Multibody System Dynamics*, 8:409–432, 2002.

E.F. Zajac. Muscle and tendon: properties, models, scaling, and application to biomechanics and motor control. *CRC Critical Reviews in Biomedical Engineering*, 19(4):359–411, 1989.

Modelling, simulation and control of flexible multibody systems

Olivier Brüls*, Alberto Cardona† and Michel Géradin‡
* University of Liège, Liège, Belgium
† Universidad Nacional del Litoral - CONICET, Santa Fe, Argentina
‡ European Commission, Joint Research Centre, IPSC, Elsa Unit, Ispra, Italy

Abstract This chapter concerns the dynamic analysis of flexible multibody systems. After a brief review of the inertial frame, the corotational frame and the floating frame approaches, a general simulation framework is presented in detail. Based on the finite element concept, the proposed approach allows the coupled analysis of dynamic systems composed of rigid and flexible bodies, kinematic joints and control elements. The text is illustrated with some didactic examples and industrial applications.

1 Introduction

Multibody system simulations and structural analyses are often considered as complementary engineering tools. On the one hand, multibody system simulations can exploit the rigid body assumption to evaluate global performances of the multibody system and to compute the loads exerted on its mechanical components. On the other hand, structural analyses allow to study locally the behaviour of each isolated structural component and to check if the deformations and stresses are kept sufficiently small. This two-level approach is certainly relevant when structural components are designed to behave as rigid bodies and when deformations have minor influence on the global behaviour.

However, these quasi-rigidity requirements are quite restrictive and they should be reconsidered in many modern applications. Indeed, if the engineer allows more flexibility in the components of the system, a lighter mechanical design is possible with the advantages of reduced inertia forces, lower power consumption and better transportability. Typical examples of flexible mechanisms are lightweight manipulators, large space structures, high-speed machines, landing gears and wind turbines. For such systems, it is no longer possible to neglect the deformations for the analysis of the global motion nor to use pre-computed equivalent loads for a local structural analysis. Instead, an integrated approach is necessary for the dynamic analysis of the flexible multibody system.

A flexible multibody system can be defined as a collection of rigid and flexible bodies interconnected by rigid and flexible kinematic joints (e.g. revolute, prismatic or universal joints) and by force elements (e.g. springs or dampers). The global motion and the deformations are strongly coupled in the system. Several methods have been proposed in the literature to analyse both the finite motion and the deformations in a fully integrated way, see Wasfy and Noor (2003) for a detailed review.

Let us briefly discuss the application of inertial frame, corotational frame and floating frame approaches for the modelling of an isolated flexible body, as illustrated in Figure 1.

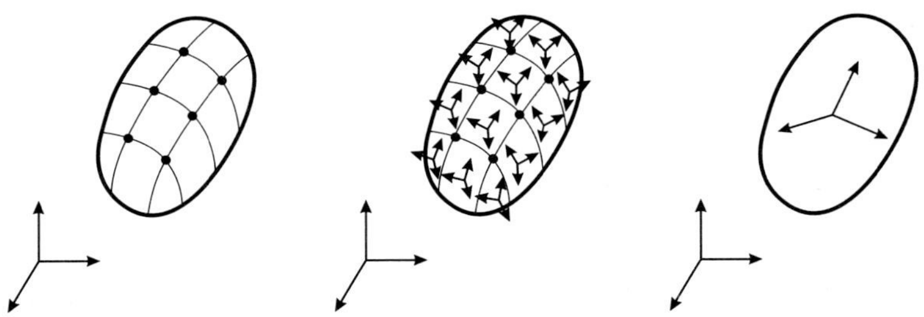

Figure 1. Inertial frame (left), corotational frame (middle) and floating frame of reference (right) approaches.

The *inertial frame approach* is closely related to the nonlinear finite element method. Absolute nodal translation and rotation coordinates are defined with respect to a unique inertial frame. Hence, there is no distinction between rigid and elastic coordinates. Due to the combined large amplitude motions and deformations, the relation between the strains and the generalized coordinates is nonlinear, with the consequence that the computation of elastic forces requires a nonlinear theory of elasticity.

The *corotational frame approach* has been proposed to simplify the computation of elastic forces in a nonlinear finite element model, see Belytschko and Hsieh (1973); Cardona and Géradin (1991); Crisfield et al. (1997). The flexible body is still described in terms of absolute nodal coordinates, but an intermediate corotational frame follows the gross motion of each element of the finite element mesh. The method relies on the assumption that the local displacements in the element are small with respect to this frame. In other words, the global motion of the element is decomposed into a large amplitude rigid motion of the corotational frame and a small amplitude elastic deformation. Hence, using the assumption of infinitesimal motion in the moving frame, the elastic forces can be computed using the standard linear theory of elasticity. Let us note that the corotational computation of the inertia forces requires special care, see Cardona and Géradin (1991) and Crisfield et al. (1997).

Both the inertial frame and the corotational approaches originate from the finite element community. In contrast, the *floating frame of reference approach* may be seen as an extension of rigid multibody formalisms to account for flexible bodies, see Fraeijs de Veubeke (1976); Shabana and Wehage (1983); Meirovitch and Kwak (1991); Lai et al. (1991). It relies on the definition of an intermediate frame at the body level, whose motion is described using rigid-body coordinates (e.g. Cartesian or relative coordinates). Additional flexible coordinates are then introduced to represent the deformations of the body. As in the corotational approach, the elastic forces can be efficiently evaluated in the floating frame using a linear model, but the computation of inertia forces is more involved since the dynamic equations for the rigid and the elastic coordinates are strongly coupled by complex nonlinear inertia terms.

The assumption of a linear elastic behaviour with respect to a floating frame is valid for dynamic systems characterized by small deformations and small velocities. For more general situations, higher-order terms may be included in the formulation in order to capture the geometrically nonlinear behaviour of the system, see Wallrapp and Schwertassek (1991). For example, the centrifugal stiffening effect arises in slender beams with large rotation speeds, where axial stresses induced by the inertia forces tend to increase the bending stiffness. In the floating frame of reference formulation, such phenomena may be captured through the implementation of a geometric stiffness matrix, whose coefficients are not constant but depend on the stresses in the body in a non-trivial way.

Compared to the corotational frame and to the floating frame of reference approaches, the nonlinear finite element method allows to account in a natural way for many nonlinear effects that are missed by linear elastic models (e.g. the geometric stiffening effect) and it leads to fairly simple expressions of the inertia forces. On the other hand, it usually leads to sparse but rather large sets of equations and it requires sophisticated algorithms for the computation of elastic forces. Hence, each method seems to have its own advantages and drawbacks. We believe that a suitable environment for the analysis of flexible multibody systems should allow *a combination of these approaches.*

One alternative is to combine the finite element approach with a model reduction strategy. In linear structural dynamics, component-mode techniques allow to simplify a detailed finite element model of a complex body under the small displacements assumption. The reduction procedure results in a *superelement*, which encapsulates in a compact way the dynamic behaviour of the whole flexible body or substructure. Following the work in Cardona and Géradin (1991), we will show that the corotational frame method can be used for the nonlinear formulation of a superelement with large amplitude motions. Compared to the floating frame of reference approach, the most significant difference is that the superelement is expressed in terms of interface nodal coordinates and internal modal coordinates, whereas the floating frame of reference method relies on an explicit separation between rigid and elastic coordinates.

Hence, a formulation based on absolute nodal coordinates (or finite element coordinates) offers the possibility to combine finite element models and superelements in order to reach the best compromise between accuracy and computational efficiency. The first part of this paper presents a unifying finite element approach for the simulation of dynamic systems composed of rigid bodies, flexible bodies and kinematic joints. The interested reader will be able to find more details about this approach in the book by Géradin and Cardona (2001). Let us note that this method is implemented in the general purpose MECANO software[1], which has been successfully exploited over the last 20 years to solve industrial problems in numerous fields of applications.

An extension of this finite element approach to deal with mechatronic systems will also be discussed. Using the block diagram language, the strongly coupled dynamic equations are obtained from a modular description of the system. A monolithic time integration scheme is then proposed to solve the coupled dynamic problem.

This chapter is organized as follows. Section 2 concerns the kinematic description of multibody systems using the finite element approach. The dynamic analysis is addressed in Section 3. Two formulations for the dynamics of flexible bodies are then presented: the nonlinear finite element method in Section 4 and the superelement technique in Section 5. Section 6 presents an extension of the finite element method for the strongly coupled analysis of mechatronic systems. Numerical time integration algorithms are discussed in Section 7. Finally, three applications are described in Section 8 and some conclusions are drawn in Section 9.

[1] for more information: `www.samcef.com`

2 Kinematics of multibody systems

In this section, the kinematics of a multibody system is described using a finite element mesh composed of elements and nodes. As in structural dynamics, beam, shell or volume elements can be defined from a classical finite element discretization of continuous flexible bodies. In a wider sense, an element can also represent any discrete subsystem of a multibody system, such as a rigid body, a kinematic joint or a force element.

The equations of motion are thus formulated using absolute nodal coordinates, which are collected in the vector of generalized coordinates $\mathbf{q}$. Due to the presence of joints and rigid-body conditions, these coordinates are not independent but they have to satisfy a set of kinematic constraints

$$\boldsymbol{\Phi}(\mathbf{q}) = \mathbf{0}. \tag{1}$$

In the following, the kinematics of the rigid body, of the hinge and of a particular rigid multibody system are described. Some important characteristics of the finite element approach will thus be illustrated but the description of flexible bodies is postponed until Section 4.

2.1 Kinematics of the rigid body

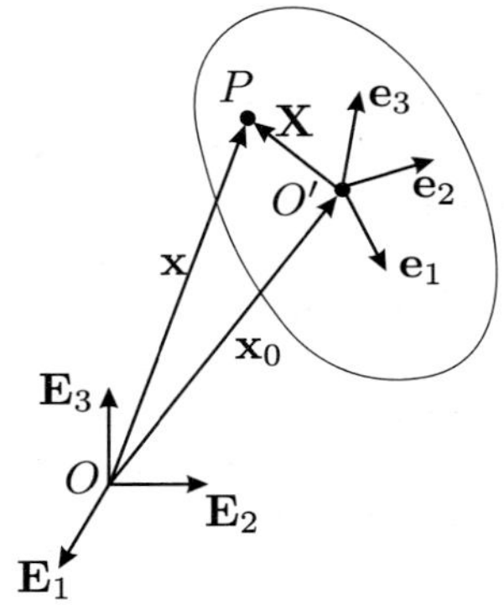

Figure 2. Kinematics of the rigid body.

Let us consider an arbitrary rigid body which is moving with respect to the inertial frame $\{\mathbf{E}_1, \mathbf{E}_2, \mathbf{E}_3\}$, see Figure 2. Its spatial motion can be represented using the three Cartesian coordinates of a reference point O' and the orientation of a reference frame $\{\mathbf{e}_1, \mathbf{e}_2, \mathbf{e}_3\}$ attached to the body. The position of an arbitrary point P of the rigid body is represented by the constant *material position* vector $\mathbf{X}$ in the body-attached frame, and by the time-varying *spatial position* vector $\mathbf{x}$ in the inertial frame. The relation between both vectors is given by

$$\mathbf{x} = \mathbf{x}_0 + \mathbf{R}\mathbf{X} \tag{2}$$

where $\mathbf{x}_0$ is the position vector of O' in the inertial frame and $\mathbf{R}$ is the 3×3 rotation matrix. The spherical motion is a special case of the rigid body motion when $\mathbf{x}_0 = \mathbf{0}$.

The rotation matrix is orthogonal, i.e.

$$\mathbf{R}^{-1} = \mathbf{R}^T. \tag{3}$$

Therefore, its nine components are not independent but they have to satisfy 6 orthogonality conditions. In principle, three independent parameters α_1, α_2 and α_3 can be used to describe a rotation and we can write

$$\mathbf{R}(\alpha_1, \alpha_2, \alpha_3). \tag{4}$$

Various choices are possible for the rotation parameters, e.g. Euler angles, Bryant angles, the conformal rotation vector or the Cartesian rotation vector. We refer to Géradin and Cardona (2001) and to Bauchau and Trainelli (2003) for a discussion on the parameterization of rotations.

For example, the Cartesian rotation vector $\boldsymbol{\psi}$ is the vector parallel to the rotation axis whose amplitude is the rotation angle ψ. The rotation matrix can be computed from $\boldsymbol{\psi}$ using the Rodrigues formula

$$\mathbf{R}(\boldsymbol{\psi}) = \mathbf{I} + \frac{\sin\psi}{\psi}\widetilde{\boldsymbol{\psi}} + \frac{1-\cos\psi}{\psi^2}\widetilde{\boldsymbol{\psi}}\widetilde{\boldsymbol{\psi}} \tag{5}$$

where $\widetilde{\boldsymbol{\psi}}$ is the skew-symmetric matrix formed by the components of $\boldsymbol{\psi}$

$$\widetilde{\boldsymbol{\psi}} = \begin{bmatrix} 0 & -\psi_3 & \psi_2 \\ \psi_3 & 0 & -\psi_1 \\ -\psi_2 & \psi_1 & 0 \end{bmatrix}. \tag{6}$$

Note that, by construction, $\widetilde{\boldsymbol{\psi}}\mathbf{v} = \boldsymbol{\psi} \times \mathbf{v}$ for any vector $\mathbf{v}$.

Another consequence of the orthogonality of the rotation matrix is that the matrix $\widetilde{\boldsymbol{\omega}}$ defined by

$$\widetilde{\boldsymbol{\omega}} = \dot{\mathbf{R}}\mathbf{R}^T \tag{7}$$

is skew-symmetric. It can be verified that this matrix is formed by the components of the vector $\boldsymbol{\omega}$, which represents the angular velocity in the inertial frame. Likewise, the skew-symmetric matrix

$$\widetilde{\boldsymbol{\Omega}} = \mathbf{R}^T\dot{\mathbf{R}} \tag{8}$$

is formed by the components of the vector $\boldsymbol{\Omega}$, which represents the angular velocity in the body-attached frame. The coordinate transformation from the body-attached frame to the inertial frame yields

$$\boldsymbol{\omega} = \mathbf{R}\boldsymbol{\Omega}. \tag{9}$$

The angular velocities can be computed from the time derivatives of the rotation parameters. For example, using the Cartesian rotation vector, we have

$$\boldsymbol{\Omega} = \mathbf{T}(\boldsymbol{\psi})\dot{\boldsymbol{\psi}} \tag{10}$$

where $\mathbf{T}(\boldsymbol{\psi})$ is the tangent operator of the parameterization defined by

$$\mathbf{T}(\boldsymbol{\psi}) = \mathbf{I} + \frac{\cos\psi - 1}{\psi^2}\widetilde{\boldsymbol{\psi}} + \left(1 - \frac{\sin\psi}{\psi}\right)\frac{\widetilde{\boldsymbol{\psi}}\widetilde{\boldsymbol{\psi}}}{\psi^2}. \tag{11}$$

One difficulty when dealing with arbitrary large rotations comes from the absence of global parameterization based on three independent parameters. Indeed, any such parameterization is characterized by singularities. For instance, the Cartesian rotation vector leads to a singular parameterization when $\psi = \pm 2\pi$. However, at least three solutions are possible to overcome this difficulty. Firstly, one may use a local parameterization of rotations with a standard time integrator and reparameterize the motion when appropriate, see e.g. the incremental approach by Cardona and Géradin (1989). Another strategy is to use a global but redundant parameterization, e.g. the four Euler parameters or, as suggested by Betsch and Steinmann (2001), the nine components of the director vectors. In this case, the parameters are linked by additional constraints. Finally, one may use a geometric approach, which only refers implicitly to rotation parameters, see e.g. Simo and Vu-Quoc (1988). For reasons of space, those techniques will not be described in more details here.

2.2 Kinematics of the hinge

In Figure 3, we consider two bodies A and B connected by a hinge joint. The orientations of the two body-attached frames $\{\mathbf{e}_{A1}, \mathbf{e}_{A2}, \mathbf{e}_{A3}\}$ and $\{\mathbf{e}_{B1}, \mathbf{e}_{B2}, \mathbf{e}_{B3}\}$ are represented by the Cartesian rotation vectors $\boldsymbol{\psi}_A$ and $\boldsymbol{\psi}_B$. In the reference configuration $\boldsymbol{\psi}_A = \boldsymbol{\psi}_B = \mathbf{0}$, the two triads are aligned with the inertial axis $\{\mathbf{E}_1, \mathbf{E}_2, \mathbf{E}_3\}$ and the hinge axis is along $\mathbf{E}_3$. During the motion, we have

$$\mathbf{e}_{Ai} = \mathbf{R}(\boldsymbol{\psi}_A)\mathbf{E}_i, \qquad \mathbf{e}_{Bi} = \mathbf{R}(\boldsymbol{\psi}_B)\mathbf{E}_i, \qquad (i = 1, 2, 3). \tag{12}$$

The hinge joint imposes some constraints on the relative motion of the two bodies. Firstly, the translation of any point on the hinge axis is the

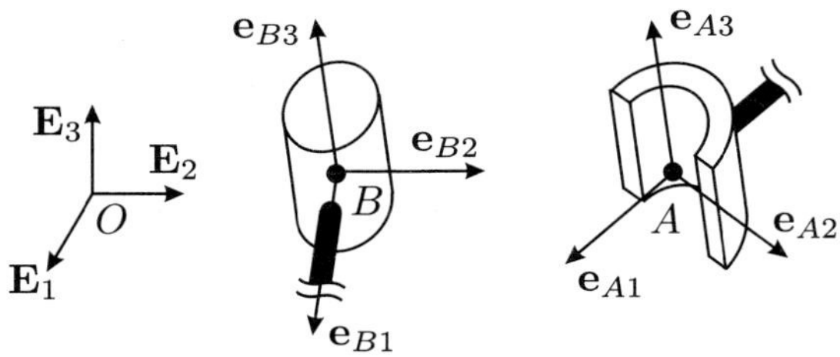

Figure 3. Kinematics of the hinge (exploded view).

same for both bodies, which is written as a constraint on the translations of nodes A and B

$$\mathbf{x}_A = \mathbf{x}_B. \tag{13}$$

Secondly, the axis $\mathbf{e}_{A1}$ and $\mathbf{e}_{A2}$ should remain orthogonal to the hinge axis $\mathbf{e}_{B3}$ during the motion

$$\mathbf{e}_{A1}^T \mathbf{e}_{B3} = 0, \qquad \mathbf{e}_{A2}^T \mathbf{e}_{B3} = 0, \tag{14}$$

which leads to two scalar kinematic constraints for the vectors $\boldsymbol{\psi}_A$ and $\boldsymbol{\psi}_B$

$$\mathbf{E}_1^T \mathbf{R}_A^T \mathbf{R}_B \mathbf{E}_3 = 0, \qquad \mathbf{E}_2^T \mathbf{R}_A^T \mathbf{R}_B \mathbf{E}_3 = 0 \tag{15}$$

with $\mathbf{R}_A = \mathbf{R}(\boldsymbol{\psi}_A)$ and $\mathbf{R}_B = \mathbf{R}(\boldsymbol{\psi}_B)$.

2.3 Kinematics of multibody systems : example

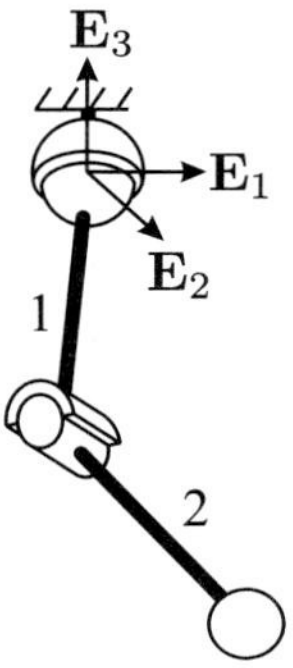

Figure 4. The spherical double pendulum.

Let us consider the spherical double pendulum in Figure 4. The system is composed of two rigid bars, a hinge joint and a spherical joint. The rigid bars 1 and 2 are characterized by their length l_1 and l_2, their mass m_1 and m_2, their inertia tensors $\mathbf{J}_1$ and $\mathbf{J}_2$ and their centers of gravity which are located at their tip. The system moves in the gravity field $\boldsymbol{\gamma} = -g\mathbf{E}_3$.

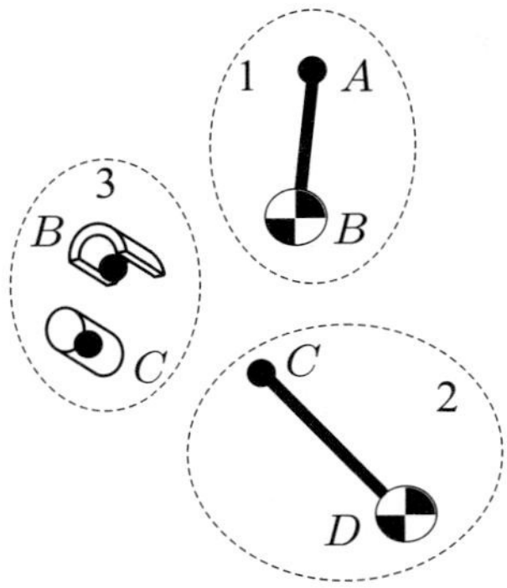

Figure 5. Finite element description of the spherical double pendulum.

The mesh is composed of three elements: two rigid bodies and one hinge, as illustrated in Figure 5. For each rigid body, one node is defined at the center of gravity and one additional node is defined at the other attachment point, which gives a total of four nodes (A to D). For node I ($I = A, ..., D$), the vector $\mathbf{x}_I$ denotes the position, $\{\mathbf{e}_{I1}, \mathbf{e}_{I2}, \mathbf{e}_{I3}\}$, the body-attached frame, and $\boldsymbol{\psi}_I$, the Cartesian rotation vector. In the reference configuration, we have

$$\mathbf{x}_B = \mathbf{x}_C = [l_1 \quad 0 \quad 0]^T, \tag{16}$$

$$\mathbf{x}_D = [l_1 + l_2 \quad 0 \quad 0]^T \tag{17}$$

and $\boldsymbol{\psi}_I = \mathbf{0}$ ($I = A, ..., D$).

Since 6 coordinates are defined for each node, one could expect 24 absolute nodal coordinates. However, all these coordinates do not need to be treated as independent variables in the model as it will be clear in the following.

Firstly, the spherical joint which connects the system to the ground is modelled as a boundary condition

$$\mathbf{x}_A = \mathbf{0}. \tag{18}$$

Secondly, for each rigid body, the center of gravity is considered as the master node and 6 rigid body constraints are defined for each additional node. For node A on bar 1, the translational constraint is

$$\mathbf{x}_A = \mathbf{x}_B + \mathbf{R}_B \mathbf{X}_A \tag{19}$$

where $\mathbf{X}_A = -l_1\mathbf{E}_1$ is the position of node A with respect to the center of gravity in the reference configuration. The rotational constraint is

$$\boldsymbol{\psi}_A = \boldsymbol{\psi}_B. \tag{20}$$

Likewise, for node C on bar 2, we have

$$\mathbf{x}_C = \mathbf{x}_D + \mathbf{R}_D\mathbf{X}_C \tag{21}$$

with $\mathbf{X}_C = -l_2\mathbf{E}_1$ and

$$\boldsymbol{\psi}_C = \boldsymbol{\psi}_D. \tag{22}$$

Thirdly, for the hinge joint, two rotational constraints are given by Equations (15) whereas the translational constraint is

$$\mathbf{x}_C = \mathbf{x}_B. \tag{23}$$

Considering the trivial relations in Equations (18), (20), (22) and (23) only $\mathbf{x}_B$, $\boldsymbol{\psi}_B$, $\mathbf{x}_D$ and $\boldsymbol{\psi}_D$ should appear in the vector of generalized coordinates, since the other variables can be directly identified from their values. The system is thus described using 12 generalized coordinates collected in a vector $\mathbf{q}$

$$\mathbf{q} = [\mathbf{x}_B^T \quad \boldsymbol{\psi}_B^T \quad \mathbf{x}_D^T \quad \boldsymbol{\psi}_D^T]^T. \tag{24}$$

A vector of element generalized coordinates can be defined for each element. For rigid bodies, this vector includes the translation and rotation of the center of gravity, as well as the translations of each additional node

$$\mathbf{q}_1 = [\mathbf{x}_B^T \quad \boldsymbol{\psi}_B^T \quad \mathbf{x}_A^T]^T, \tag{25}$$

$$\mathbf{q}_2 = [\mathbf{x}_D^T \quad \boldsymbol{\psi}_D^T \quad \mathbf{x}_C^T]^T. \tag{26}$$

For the hinge, the vector of generalized coordinates includes the rotations of the two nodes

$$\mathbf{q}_3 = [\boldsymbol{\psi}_B^T \quad \boldsymbol{\psi}_C^T]^T. \tag{27}$$

With those definitions, we will see that the contribution of each element to the equations of motion can be computed from its own element generalized coordinates.

Connectivity relations between element and global generalized coordinates play a key role in the finite element assembly procedure; they can be written as

$$\mathbf{q}_e = \mathbf{L}_e\mathbf{q}, \qquad (e = 1, 2, 3) \tag{28}$$

where $\mathbf{L}_e$ are Boolean localization matrices defined by

$$\mathbf{L}_1 = \begin{bmatrix} \mathbf{I} & \mathbf{0} & \mathbf{0} & \mathbf{0} \\ \mathbf{0} & \mathbf{I} & \mathbf{0} & \mathbf{0} \\ \mathbf{0} & \mathbf{0} & \mathbf{0} & \mathbf{0} \end{bmatrix}, \tag{29}$$

$$\mathbf{L}_2 = \begin{bmatrix} \mathbf{0} & \mathbf{0} & \mathbf{I} & \mathbf{0} \\ \mathbf{0} & \mathbf{0} & \mathbf{0} & \mathbf{I} \\ \mathbf{I} & \mathbf{0} & \mathbf{0} & \mathbf{0} \end{bmatrix}, \tag{30}$$

$$\mathbf{L}_3 = \begin{bmatrix} \mathbf{0} & \mathbf{I} & \mathbf{0} & \mathbf{0} \\ \mathbf{0} & \mathbf{0} & \mathbf{0} & \mathbf{I} \end{bmatrix}. \tag{31}$$

Note that the boundary conditions (18) as well as the identification relations in Equations (20), (22) and (23) are then implicitly taken into account in the connectivity relationships (28).

The kinematic constraints in Equations (19), (21) and in the hinge formulation (15) can be organized in elementary contributions according to

$$\mathbf{\Phi}_1(\mathbf{q}_1) = \mathbf{x}_A - \mathbf{x}_B - \mathbf{R}_B\mathbf{X}_A, \tag{32}$$

$$\mathbf{\Phi}_2(\mathbf{q}_2) = \mathbf{x}_C - \mathbf{x}_D - \mathbf{R}_D\mathbf{X}_C, \tag{33}$$

$$\mathbf{\Phi}_3(\mathbf{q}_3) = [\mathbf{E}_1^T\mathbf{R}_B^T\mathbf{R}_C\mathbf{E}_3 \quad \mathbf{E}_2^T\mathbf{R}_B^T\mathbf{R}_C\mathbf{E}_3]^T. \tag{34}$$

where $\mathbf{R}_I = \mathbf{R}(\psi_I)$, $I = A, ..., D$. The whole set of kinematic constraints is thus

$$\mathbf{\Phi}(\mathbf{q}) = \begin{bmatrix} \mathbf{\Phi}_1(\mathbf{L}_1\mathbf{q}) \\ \mathbf{\Phi}_2(\mathbf{L}_2\mathbf{q}) \\ \mathbf{\Phi}_3(\mathbf{L}_3\mathbf{q}) \end{bmatrix} = \mathbf{0} \tag{35}$$

which can be explicitly written as

$$\mathbf{\Phi}(\mathbf{q}) = \begin{bmatrix} -\mathbf{x}_B + l_1\mathbf{R}_B\mathbf{E}_1 \\ \mathbf{x}_B - \mathbf{x}_D + l_2\mathbf{R}_D\mathbf{E}_1 \\ \mathbf{E}_1^T\mathbf{R}_B^T\mathbf{R}_D\mathbf{E}_3 \\ \mathbf{E}_2^T\mathbf{R}_B^T\mathbf{R}_D\mathbf{E}_3 \end{bmatrix} = \mathbf{0}. \tag{36}$$

In summary, the kinematics of the spherical double pendulum is represented using 12 generalized coordinates and 8 scalar kinematic constraints (2 subsets of 3 constraints and 2 scalar constraints). The system has thus four degrees-of-freedom, which is in agreement with the Grübler formula.

3 Dynamics of multibody systems

3.1 General form of the equations of motion

The general form of the equations of motion of a flexible multibody system is now derived from classical principles of mechanics. According to the kinematic analysis in the previous section, a flexible multibody system can be represented using a set of n generalized coordinates $\mathbf{q}$, which have to satisfy a set of m holonomic constraints

$$\boldsymbol{\Phi}(\mathbf{q}) = \mathbf{0}. \tag{37}$$

The more general case of non-holonomic constraints of type $\boldsymbol{\Phi}(\mathbf{q}, \dot{\mathbf{q}}, t) = \mathbf{0}$ may be treated as described in Géradin and Cardona (2001).

Let us assume that we know the potential energy of external forces $\mathcal{V}(\mathbf{q})$ (e.g. the potential of gravity forces), the deformation energy of the elastic bodies $\mathcal{W}(\mathbf{q})$ and the kinetic energy

$$\mathcal{K}(\mathbf{q}, \dot{\mathbf{q}}) = \frac{1}{2}\dot{\mathbf{q}}^T \mathbf{M}(\mathbf{q})\dot{\mathbf{q}} \tag{38}$$

where the $n \times n$ symmetric mass matrix $\mathbf{M}$ may depend on the generalized coordinates. The Lagrangian function of the conservative system is defined by

$$\mathcal{L}(\mathbf{q}, \dot{\mathbf{q}}) = \mathcal{K}(\mathbf{q}, \dot{\mathbf{q}}) - \mathcal{V}(\mathbf{q}) - \mathcal{W}(\mathbf{q}). \tag{39}$$

From this expression, the equations of motion of the system can be derived using the Hamilton principle. In a first step, the unconstrained case is considered ($m = 0$). If the mechanical system is conservative and if the generalized coordinates are independent, the Hamilton principle states that, if the motion is specified at times t_1 and t_2, the integral action

$$A = \int_{t_1}^{t_2} \mathcal{L}(\mathbf{q}, \dot{\mathbf{q}})dt \tag{40}$$

is stationary for the real trajectory of the system between t_1 and t_2. If the generalized coordinates are not independent ($m > 0$), the trajectory of the system is then a solution of the constrained minimization problem

$$\min_{\mathbf{q}(t)} A \qquad \text{subject to} \qquad \boldsymbol{\Phi}(\mathbf{q}) = \mathbf{0}. \tag{41}$$

Let us introduce m Lagrange multipliers $\boldsymbol{\lambda}$, so that the constrained minimization problem is replaced by the equivalent unconstrained form

$$\min_{\mathbf{q}(t), \boldsymbol{\lambda}(t)} A^* \tag{42}$$

where A^* is the augmented functional

$$A^* = \int_{t_1}^{t_2} (\mathcal{L}(\mathbf{q}, \dot{\mathbf{q}}) - \boldsymbol{\lambda}^T \boldsymbol{\Phi})dt. \tag{43}$$

After integration by parts, the stationarity condition $\delta A^* = 0$ becomes, using the summation convention,

$$\left[\frac{\partial \mathcal{L}}{\partial \dot{q}_i}\delta q_i\right]_{t_1}^{t_2} - \int_{t_1}^{t_2}\left(\frac{d}{dt}\left(\frac{\partial \mathcal{L}}{\partial \dot{q}_i}\right) - \frac{\partial \mathcal{L}}{\partial q_i} + \frac{\partial \mathbf{\Phi}}{\partial q_i}^T \boldsymbol{\lambda}\right)\delta q_i dt - \int_{t_1}^{t_2} \delta \boldsymbol{\lambda}^T \mathbf{\Phi} dt = 0.$$

Since the motion is specified at instants t_1 and t_2, the first term vanishes and the motion needs to satisfy

$$\left(\frac{d}{dt}\left(\frac{\partial \mathcal{L}}{\partial \dot{q}_i}\right) - \frac{\partial \mathcal{L}}{\partial q_i} + \frac{\partial \mathbf{\Phi}}{\partial q_i}^T \boldsymbol{\lambda}\right)\delta q_i + \delta \boldsymbol{\lambda}^T \mathbf{\Phi} = 0.$$

Using Equation (39), this expression becomes

$$\left(\frac{d}{dt}\left(\frac{\partial \mathcal{K}}{\partial \dot{q}_i}\right) - \frac{\partial \mathcal{K}}{\partial q_i} + \frac{\partial \mathcal{W}}{\partial q_i} + \frac{\partial \mathcal{V}}{\partial q_i} + \frac{\partial \mathbf{\Phi}}{\partial q_i}^T \boldsymbol{\lambda}\right)\delta q_i + \delta \boldsymbol{\lambda}^T \mathbf{\Phi} = 0,$$

which is restated as

$$\delta \mathbf{q}^T(\mathbf{M}(\mathbf{q})\ddot{\mathbf{q}} + \mathbf{g}^{gyr}(\mathbf{q}, \dot{\mathbf{q}}) + \mathbf{g}^{int}(\mathbf{q}) - \mathbf{g}^{ext}(\mathbf{q}) + \mathbf{\Phi}_{\mathbf{q}}^T(\mathbf{q})\boldsymbol{\lambda}) + \delta \boldsymbol{\lambda}^T \mathbf{\Phi}(\mathbf{q}) = 0 \quad (44)$$

with the components of the internal elastic forces $[\mathbf{g}^{int}]_i = \partial \mathcal{W}/\partial q_i$ and of the external forces $[\mathbf{g}^{ext}]_i = -\partial \mathcal{V}/\partial q_i$. The inertia forces in Equation (44) are derived using the expression of the kinetic energy in Equation (38). After some algebraic manipulations, the vector of gyroscopic and centrifugal forces is given by

$$[\mathbf{g}^{gyr}]_i = \frac{1}{2}\left(\frac{\partial [\mathbf{M}]_{ij}}{\partial q_k} + \frac{\partial [\mathbf{M}]_{ik}}{\partial q_j} - \frac{\partial [\mathbf{M}]_{jk}}{\partial q_i}\right)\dot{q}_j \dot{q}_k. \quad (45)$$

Finally, the equations of motion take the general form of a differential-algebraic system (DAE)

$$\mathbf{M}(\mathbf{q})\ddot{\mathbf{q}} + \mathbf{g}^{gyr}(\mathbf{q}, \dot{\mathbf{q}}) + \mathbf{g}^{int}(\mathbf{q}) + \mathbf{\Phi}_{\mathbf{q}}^T(\mathbf{q})\boldsymbol{\lambda} = \mathbf{g}^{ext}(\mathbf{q}), \quad (46)$$

$$\mathbf{\Phi}(\mathbf{q}) = \mathbf{0}. \quad (47)$$

3.2 Finite element assembly procedure

The finite element concept is now exploited to derive in a systematic way the equations of motion obtained in the previous section. As in Section 2.3, the vectors $\mathbf{q}_e$ and $\mathbf{q}$ satisfy connectivity relations

$$\mathbf{q}_e = \mathbf{L}_e \mathbf{q}, \qquad e = 1, ..., n_e \quad (48)$$

where n_e is the number of elements in the finite element mesh. The Lagrangian of the system can be computed by summing up the contributions of the elements

$$\mathcal{L}(\mathbf{q}, \dot{\mathbf{q}}) = \sum_{e=1}^{n_e} \mathcal{L}_e(\mathbf{q}_e, \dot{\mathbf{q}}_e) \tag{49}$$

with the definition of $\mathcal{L}_e$ from the kinetic energy, the potential energy, and the internal energy of each element

$$\mathcal{L}_e(\mathbf{q}_e, \dot{\mathbf{q}}_e) = \mathcal{K}_e(\mathbf{q}_e, \dot{\mathbf{q}}_e) - \mathcal{V}_e(\mathbf{q}_e) - \mathcal{W}_e(\mathbf{q}_e). \tag{50}$$

The kinematic constraints can also be decomposed into elementary contributions, see e.g. Equation (35), so as the Lagrange multipliers

$$\boldsymbol{\Phi}(\mathbf{q}) = \begin{bmatrix} \boldsymbol{\Phi}_1(\mathbf{q}_1) \\ \vdots \\ \boldsymbol{\Phi}_{n_e}(\mathbf{q}_e) \end{bmatrix}, \qquad \boldsymbol{\lambda} = \begin{bmatrix} \boldsymbol{\lambda}_1 \\ \vdots \\ \boldsymbol{\lambda}_{n_e} \end{bmatrix}. \tag{51}$$

Hence, we have

$$\boldsymbol{\lambda}^T \boldsymbol{\Phi}(\mathbf{q}) = \sum_{e=1}^{n_e} \boldsymbol{\lambda}_e^T \boldsymbol{\Phi}_e(\mathbf{q}_e). \tag{52}$$

The augmented action integral in Equation (43) is then splitted in elementary contributions

$$A^* = \int_{t_1}^{t_2} \sum_{e=1}^{n_e} (\mathcal{L}_e(\mathbf{q}_e, \dot{\mathbf{q}}_e) - \boldsymbol{\lambda}_e^T \boldsymbol{\Phi}_e) dt \tag{53}$$

so that Equation (44) becomes

$$\sum_{e=1}^{n_e} \delta\mathbf{q}_e^T \left(\mathbf{M}_e \ddot{\mathbf{q}}_e + \mathbf{g}_e^{gyr} + \mathbf{g}_e^{int} - \mathbf{g}_e^{ext} + \left(\boldsymbol{\Phi}_{\mathbf{q}}^T \boldsymbol{\lambda}\right)_e \right) + \sum_{e=1}^{n_e} \delta\boldsymbol{\lambda}_e^T \boldsymbol{\Phi}_e = 0 \tag{54}$$

where $\mathbf{M}_e$, $\mathbf{g}_e^{gyr}$, $\mathbf{g}_e^{int}$ and $\mathbf{g}_e^{ext}$ are defined from $\mathcal{K}_e$, $\mathcal{W}_e$ and $\mathcal{V}_e$. In addition, $(\boldsymbol{\Phi}_{\mathbf{q}}^T \boldsymbol{\lambda})_e$ is defined by

$$(\boldsymbol{\Phi}_{\mathbf{q}}^T \boldsymbol{\lambda})_e = \left(\frac{\partial \boldsymbol{\Phi}_e}{\partial \mathbf{q}_e} \right)^T \boldsymbol{\lambda}_e. \tag{55}$$

Using Equations (48) and (54), the assembly relations

$$\mathbf{M} = \sum_{e=1}^{n_e} \mathbf{L}_e^T \mathbf{M}_e \mathbf{L}_e, \tag{56}$$

$$\mathbf{g}^{gyr} = \sum_{e=1}^{n_e} \mathbf{L}_e^T \mathbf{g}_e^{gyr}, \tag{57}$$

$$\mathbf{g}^{int} = \sum_{e=1}^{n_e} \mathbf{L}_e^T \mathbf{g}_e^{int}, \tag{58}$$

$$\mathbf{g}^{ext} = \sum_{e=1}^{n_e} \mathbf{L}_e^T \mathbf{g}_e^{ext}, \tag{59}$$

$$\mathbf{\Phi}_{\mathbf{q}}^T \boldsymbol{\lambda} = \sum_{e=1}^{n_e} \mathbf{L}_e^T (\mathbf{\Phi}_{\mathbf{q}}^T \boldsymbol{\lambda})_e \tag{60}$$

allow to compute numerically each term in the equations of motion from elementary formulations.

Using a similar assembly procedure, the linearized form of the nonlinear equations of motion can be efficiently evaluated from analytical expressions defined at the element level.

3.3 Example

In the spherical double pendulum example (Figures 4 and 5), the finite element mesh was composed of three elements: two rigid bodies and one hinge. Let us compute the various forces in the global dynamic equations from element force vectors. The Lagrangian of the system is decomposed into

$$\mathcal{L}(\mathbf{q}, \dot{\mathbf{q}}) = \mathcal{L}_1(\mathbf{q}_1, \dot{\mathbf{q}}_1) + \mathcal{L}_2(\mathbf{q}_2, \dot{\mathbf{q}}_2) + \mathcal{L}_3(\mathbf{q}_3, \dot{\mathbf{q}}_3) \tag{61}$$

where the vectors of generalized coordinates $\mathbf{q}$, $\mathbf{q}_1$, $\mathbf{q}_2$ and $\mathbf{q}_3$ were defined in Equations (24–27). Firstly, the Lagrangian of the rigid bodies is computed as

$$\mathcal{L}_1(\mathbf{q}_1, \dot{\mathbf{q}}_1) = \mathcal{K}_1(\mathbf{q}_1, \dot{\mathbf{q}}_1) - \mathcal{V}(\mathbf{q}_1), \tag{62}$$

$$\mathcal{L}_2(\mathbf{q}_2, \dot{\mathbf{q}}_2) = \mathcal{K}_2(\mathbf{q}_2, \dot{\mathbf{q}}_2) - \mathcal{V}(\mathbf{q}_2). \tag{63}$$

Indeed, $\mathcal{W}_1(\mathbf{q}_1) = \mathcal{W}_2(\mathbf{q}_2) = 0$ since there is no deformation energy in a rigid body. Secondly, we have

$$\mathcal{L}_3(\mathbf{q}_3, \dot{\mathbf{q}}_3) = 0 \tag{64}$$

since no kinetic, potential or deformation energy is associated with the hinge element. From those observations, one readily obtains

$$\mathbf{g}_e^{int}(\mathbf{q}_e) = \mathbf{0}, \qquad (e = 1, 2, 3), \tag{65}$$

$$\mathbf{M}_3(\mathbf{q}_3) = \mathbf{0}, \tag{66}$$

$$\mathbf{g}_3^{gyr}(\mathbf{q}_3, \dot{\mathbf{q}}_3) = \mathbf{0}, \tag{67}$$

$$\mathbf{g}_3^{ext}(\mathbf{q}_3) = \mathbf{0}. \tag{68}$$

Inertia forces The kinetic energy of each rigid body is computed as

$$\mathcal{K}_1(\mathbf{q}_1, \dot{\mathbf{q}}_1) = \frac{1}{2} m_1 \dot{\mathbf{x}}_B^T \dot{\mathbf{x}}_B + \frac{1}{2} \boldsymbol{\Omega}_B^T \mathbf{J}_1 \boldsymbol{\Omega}_B, \tag{69}$$

$$\mathcal{K}_2(\mathbf{q}_2, \dot{\mathbf{q}}_2) = \frac{1}{2} m_2 \dot{\mathbf{x}}_D^T \dot{\mathbf{x}}_D + \frac{1}{2} \boldsymbol{\Omega}_D^T \mathbf{J}_2 \boldsymbol{\Omega}_D, \tag{70}$$

where the angular velocity vectors $\boldsymbol{\Omega}_B$ and $\boldsymbol{\Omega}_D$ in body coordinates depend on the rotation parameters and on their time derivative (see Equation (10))

$$\boldsymbol{\Omega}_B = \mathbf{T}(\boldsymbol{\psi}_B)\dot{\boldsymbol{\psi}}_B, \qquad \boldsymbol{\Omega}_D = \mathbf{T}(\boldsymbol{\psi}_D)\dot{\boldsymbol{\psi}}_D. \tag{71}$$

The element kinetic energies can be written as in Equation (38) provided that the elementary mass matrices are defined as

$$\mathbf{M}_1 = \begin{bmatrix} m_1\mathbf{I} & \mathbf{0} & \mathbf{0} \\ \mathbf{0} & \mathbf{T}_B^T \mathbf{J}_1 \mathbf{T}_B & \mathbf{0} \\ \mathbf{0} & \mathbf{0} & \mathbf{0} \end{bmatrix}, \quad \mathbf{M}_2 = \begin{bmatrix} m_2\mathbf{I} & \mathbf{0} & \mathbf{0} \\ \mathbf{0} & \mathbf{T}_D^T \mathbf{J}_2 \mathbf{T}_D & \mathbf{0} \\ \mathbf{0} & \mathbf{0} & \mathbf{0} \end{bmatrix}. \tag{72}$$

Using Equation (56), the mass matrix of the system is

$$\mathbf{M}(\mathbf{q}) = \begin{bmatrix} m_1\mathbf{I} & \mathbf{0} & \mathbf{0} & \mathbf{0} \\ \mathbf{0} & \mathbf{T}_B^T \mathbf{J}_1 \mathbf{T}_B & \mathbf{0} & \mathbf{0} \\ \mathbf{0} & \mathbf{0} & m_2\mathbf{I} & \mathbf{0} \\ \mathbf{0} & \mathbf{0} & \mathbf{0} & \mathbf{T}_D^T \mathbf{J}_2 \mathbf{T}_D \end{bmatrix}. \tag{73}$$

The derivation of complementary inertia forces, which is described in detail by Géradin and Cardona (2001), leads to

$$\mathbf{g}_1^{gyr} = \begin{bmatrix} \mathbf{0} \\ \mathbf{T}_B^T(\boldsymbol{\Omega}_B \times \mathbf{J}_1 \boldsymbol{\Omega}_B + \mathbf{J}_1 \dot{\mathbf{T}}_B \dot{\boldsymbol{\psi}}_B) \\ \mathbf{0} \end{bmatrix}, \tag{74}$$

$$\mathbf{g}_2^{gyr} = \begin{bmatrix} \mathbf{0} \\ \mathbf{T}_D^T(\boldsymbol{\Omega}_D \times \mathbf{J}_2 \boldsymbol{\Omega}_D + \mathbf{J}_2 \dot{\mathbf{T}}_D \dot{\boldsymbol{\psi}}_D) \\ \mathbf{0} \end{bmatrix} \tag{75}$$

where the time derivative of the operator $\mathbf{T}_I$ is a function $\dot{\mathbf{T}}_I(\boldsymbol{\psi}_I, \dot{\boldsymbol{\psi}}_I)$ $(I = B, D)$. Using Equation (57), the gyroscopic forces of the system are

$$\mathbf{g}^{gyr}(\mathbf{q}, \dot{\mathbf{q}}) = \begin{bmatrix} \mathbf{0} \\ \mathbf{T}_B^T(\boldsymbol{\Omega}_B \times \mathbf{J}_1\boldsymbol{\Omega}_B + \mathbf{J}_1\dot{\mathbf{T}}_B\dot{\boldsymbol{\psi}}_B) \\ \mathbf{0} \\ \mathbf{T}_D^T(\boldsymbol{\Omega}_D \times \mathbf{J}_2\boldsymbol{\Omega}_D + \mathbf{J}_2\dot{\mathbf{T}}_D\dot{\boldsymbol{\psi}}_D) \end{bmatrix} \tag{76}$$

External forces The potential energy of each rigid body in the gravity field γ is given by

$$\mathcal{V}_1(\mathbf{q}_1) = -m_1\mathbf{x}_A^T\gamma, \tag{77}$$
$$\mathcal{V}_2(\mathbf{q}_2) = -m_2\mathbf{x}_B^T\gamma, \tag{78}$$

so that the elementary external forces are computed according to

$$\mathbf{g}_1^{ext}(\mathbf{q}_1) = \begin{bmatrix} m_1\gamma \\ \mathbf{0} \\ \mathbf{0} \end{bmatrix}, \qquad \mathbf{g}_2^{ext}(\mathbf{q}_2) = \begin{bmatrix} m_2\gamma \\ \mathbf{0} \\ \mathbf{0} \end{bmatrix}. \tag{79}$$

Using Equation (59), the global vector of external forces is

$$\mathbf{g}^{ext}(\mathbf{q}) = \begin{bmatrix} m_1\gamma \\ \mathbf{0} \\ m_2\gamma \\ \mathbf{0} \end{bmatrix}. \tag{80}$$

Reaction forces The element kinematic constraints were defined in Equations (32–34). The vector of Lagrange multipliers is splitted as

$$\boldsymbol{\lambda} = \begin{bmatrix} \boldsymbol{\lambda}_1 \\ \boldsymbol{\lambda}_2 \\ \boldsymbol{\lambda}_3 \end{bmatrix} \tag{81}$$

where the 3×1 vector $\boldsymbol{\lambda}_1$ is associated with the 3 rigidity constraints of bar 1, the 3×1 vector $\boldsymbol{\lambda}_2$ with the 3 rigidity constraints of bar 2, and the 2×1 vector $\boldsymbol{\lambda}_3$ with the 2 constraints of the hinge.

In the next subsection, technical details will be developed for the derivation of the elementary reaction forces. The final result is

$$(\mathbf{\Phi}_{\mathbf{q}}^T\boldsymbol{\lambda})_1 = \begin{bmatrix} -\boldsymbol{\lambda}_1 \\ -\mathbf{T}_B^T\mathbf{R}_B^T((\mathbf{R}_B\mathbf{X}_A)\times\boldsymbol{\lambda}_1) \\ \boldsymbol{\lambda}_1 \end{bmatrix}, \tag{82}$$

$$(\mathbf{\Phi}_{\mathbf{q}}^T\boldsymbol{\lambda})_2 = \begin{bmatrix} -\boldsymbol{\lambda}_2 \\ -\mathbf{T}_D^T\mathbf{R}_D^T((\mathbf{R}_D\mathbf{X}_C)\times\boldsymbol{\lambda}_2) \\ \boldsymbol{\lambda}_2 \end{bmatrix}, \tag{83}$$

$$(\mathbf{\Phi}_{\mathbf{q}}^T\boldsymbol{\lambda})_3 = \begin{bmatrix} \mathbf{T}_B^T\mathbf{R}_B^T(-(\mathbf{e}_{C3}\times\mathbf{e}_{B1})[\boldsymbol{\lambda}_3]_1 + (\mathbf{e}_{B2}\times\mathbf{e}_{C3})[\boldsymbol{\lambda}_3]_2) \\ \mathbf{T}_C^T\mathbf{R}_C^T((\mathbf{e}_{C3}\times\mathbf{e}_{B1})[\boldsymbol{\lambda}_3]_1 - (\mathbf{e}_{B2}\times\mathbf{e}_{C3})[\boldsymbol{\lambda}_3]_2) \end{bmatrix}. \tag{84}$$

At the equilibrium, the constraint equations can be used to simplify this last expression

$$(\mathbf{\Phi}_{\mathbf{q}}^T\boldsymbol{\lambda})_3 = \begin{bmatrix} \mathbf{T}_B^T\mathbf{R}_B^T(-\mathbf{e}_{B2}[\boldsymbol{\lambda}_3]_1 + \mathbf{e}_{B1}[\boldsymbol{\lambda}_3]_2) \\ \mathbf{T}_C^T\mathbf{R}_C^T(\mathbf{e}_{B2}[\boldsymbol{\lambda}_3]_1 - \mathbf{e}_{B1}[\boldsymbol{\lambda}_3]_2) \end{bmatrix} \tag{85}$$

In Equations (82) and (83), the multipliers $\boldsymbol{\lambda}_1$ and $\boldsymbol{\lambda}_2$ are thus interpreted as internal reaction forces in the rigid bodies 1 and 2, whereas $[\boldsymbol{\lambda}_3]_1$ and $[\boldsymbol{\lambda}_3]_2$ are interpreted as internal reaction torques in the hinge along axis $\mathbf{e}_{B2}$ and $\mathbf{e}_{B1}$.

After numerical assembly, we get at equilibrium:

$$\mathbf{\Phi}_{\mathbf{q}}^T\boldsymbol{\lambda} = \begin{bmatrix} \boldsymbol{\lambda}_2 - \boldsymbol{\lambda}_1 \\ \mathbf{T}_B^T\mathbf{R}_B^T\,(\mathbf{x}_B\times\boldsymbol{\lambda}_1 - \mathbf{e}_{B2}[\boldsymbol{\lambda}_3]_1 + \mathbf{e}_{B1}[\boldsymbol{\lambda}_3]_2) \\ -\boldsymbol{\lambda}_2 \\ \mathbf{T}_D^T\mathbf{R}_D^T\,((\mathbf{x}_D-\mathbf{x}_B)\times\boldsymbol{\lambda}_2 + \mathbf{e}_{B2}[\boldsymbol{\lambda}_3]_1 - \mathbf{e}_{B1}[\boldsymbol{\lambda}_3]_2) \end{bmatrix}. \tag{86}$$

Computation of element reaction forces This subsection demonstrates the result in Equation (82–84). Firstly, in order to evaluate the constraint gradient in the contribution $(\mathbf{\Phi}_{\mathbf{q}}^T\boldsymbol{\lambda})_e$, the expression $\partial(\mathbf{RX})/\partial\boldsymbol{\psi}$ should be evaluated. For that purpose, let us observe that

$$\frac{d(\mathbf{RX})}{dt} = \frac{\partial(\mathbf{RX})}{\partial\boldsymbol{\psi}}\dot{\boldsymbol{\psi}} \tag{87}$$

holds for any spherical motion $\boldsymbol{\psi}(t)$. On the other hand, using Equation (8), the anticommutativity of the cross product and Equation (10), we have

$$\frac{d(\mathbf{RX})}{dt} = \dot{\mathbf{R}}\mathbf{X} = \mathbf{R}\widetilde{\boldsymbol{\Omega}}\mathbf{X} = -\mathbf{R}\widetilde{\mathbf{X}}\boldsymbol{\Omega} = -\mathbf{R}\widetilde{\mathbf{X}}\mathbf{T}(\boldsymbol{\psi})\dot{\boldsymbol{\psi}}. \tag{88}$$

By identification between Equations (87) and (88), we deduce

$$\frac{\partial(\mathbf{RX})}{\partial\boldsymbol{\psi}} = -\mathbf{R}\widetilde{\mathbf{X}}\mathbf{T}(\boldsymbol{\psi}) = -\widetilde{\mathbf{RX}}\mathbf{RT}(\boldsymbol{\psi}) \tag{89}$$

where the frame transformation formula $\widetilde{\mathbf{RX}} = \mathbf{R}\widetilde{\mathbf{X}}\mathbf{R}^T$ has been used. This equation gives the following useful formulae

$$\frac{\partial(\mathbf{R}_I\mathbf{X})}{\partial\boldsymbol{\psi}_I} = -\widetilde{\mathbf{R}_I\mathbf{X}}\mathbf{R}_I\mathbf{T}_I \tag{90}$$

and

$$\frac{\partial(\mathbf{R}_I\mathbf{E}_j)}{\partial\boldsymbol{\psi}_I} = -\widetilde{\mathbf{R}_I\mathbf{E}_j}\mathbf{R}_I\mathbf{T}_I = -\widetilde{\mathbf{e}_{Ij}}\mathbf{R}_I\mathbf{T}_I, \tag{91}$$

for $I = A, B$ and $j = 1, 2, 3$.

Using those results, we obtain

$$\frac{\partial\boldsymbol{\Phi}_1}{\partial\mathbf{q}_1} = \begin{bmatrix} -\mathbf{I} & \widetilde{\mathbf{R}_B\mathbf{X}_A}\mathbf{R}_B\mathbf{T}_B & \mathbf{I} \end{bmatrix}, \tag{92}$$

$$\frac{\partial\boldsymbol{\Phi}_2}{\partial\mathbf{q}_2} = \begin{bmatrix} -\mathbf{I} & \widetilde{\mathbf{R}_D\mathbf{X}_C}\mathbf{R}_D\mathbf{T}_D & \mathbf{I} \end{bmatrix}, \tag{93}$$

$$\frac{\partial\boldsymbol{\Phi}_3}{\partial\mathbf{q}_3} = \begin{bmatrix} -\mathbf{e}_{C3}^T\widetilde{\mathbf{e}_{B1}}\mathbf{R}_B\mathbf{T}_B & -\mathbf{e}_{B1}^T\widetilde{\mathbf{e}_{C3}}\mathbf{R}_C\mathbf{T}_C \\ -\mathbf{e}_{C3}^T\widetilde{\mathbf{e}_{B2}}\mathbf{R}_B\mathbf{T}_B & -\mathbf{e}_{B2}^T\widetilde{\mathbf{e}_{C3}}\mathbf{R}_C\mathbf{T}_C \end{bmatrix}, \tag{94}$$

which yields the expression in Equations (82–84).

3.4 Summary

In this section, the dynamic equations of multibody systems have been formulated in a general setting. We have shown that the finite element approach allows a systematic implementation of the equations of motion. The finite element mesh may include rigid body elements, kinematic joints and flexible bodies, which allows to describe mechanisms of arbitrary topological complexity. The coupled equations of motion of the flexible multibody system are obtained using the finite element assembly procedure. In addition, the linearized equations can also be evaluated by numerical assembly from analytical expressions at the element level.

The next two sections describe in more details the formulation of flexible bodies. Two different techniques are considered, namely, the nonlinear finite element technique and the superelement technique.

4 Flexible body dynamics: nonlinear finite element formulation

According to the nonlinear finite element approach, absolute coordinates are used to represent both the motions and the deformations of the flexible body. Section 4.1 addresses the definition of strains in a deformed configuration. The spatial discretization using absolute nodal coordinates is discussed in Section 4.2 and the discretized form of the dynamic equilibrium is developed in Section 4.3. Two special elements are then considered: the planar elastic bar (Section 4.4) and the elastic beam (Section 4.5).

4.1 Strain analysis

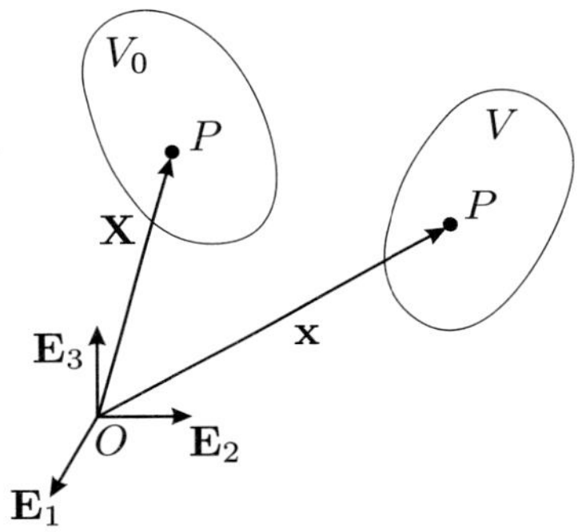

Figure 6. Kinematics of the flexible body.

Let us consider the spatial motion of a flexible body illustrated in Figure 6. In the reference undeformed configuration V_0, the position of the material point P is represented by the material position vector $\mathbf{X}$, whereas in the current configuration (after motion and deformation), the position of the same point P is represented by the spatial position vector $\mathbf{x}$. The configuration mapping φ defines the relationship between $\mathbf{X}$ and $\mathbf{x}$ for any point of the flexible body, i.e.

$$\mathbf{x} = \varphi(\mathbf{X}, t), \tag{95}$$

This section recalls classical definitions of the 3×3 strain tensor which are based on the 3×3 Jacobian matrix $\mathbf{F}$ of the configuration mapping

$$[\mathbf{F}]_{ij} = \frac{\partial [\mathbf{x}]_i}{\partial [\mathbf{X}]_j}. \tag{96}$$

The operator $\mathbf{F}$ maps an infinitesimal material vector from the reference to the current configuration according to $d\mathbf{x} = \mathbf{F}d\mathbf{X}$.

For small displacements problems, the *Cauchy strain tensor* is defined by

$$\widehat{\boldsymbol{\epsilon}}^C = \frac{1}{2}(\mathbf{F} + \mathbf{F}^T) - \mathbf{I}. \tag{97}$$

For large displacements problems, a more general definition should be used, such as the *Green strain tensor* defined by

$$\widehat{\boldsymbol{\epsilon}}^G = \frac{1}{2}\left(\mathbf{F}^T\mathbf{F} - \mathbf{I}\right). \tag{98}$$

Let us note that Cauchy strains are linear with respect to $\mathbf{F}$ whereas Green strains are nonlinear. As a consequence, even though a linear formulation of elastic forces is possible for small displacements problems, a geometrically nonlinear formulation is mandatory for large displacements problems. These geometric nonlinearities have a direct influence on the formulation of elastic forces in flexible multibody dynamics.

Both strain tensors are symmetric and thus characterized by 6 independent components. For convenience, for any 3×3 strain tensor $\widehat{\boldsymbol{\epsilon}}$, the 6×1 strain vector $\boldsymbol{\epsilon}$ is defined by

$$\boldsymbol{\epsilon} = \left[[\widehat{\boldsymbol{\epsilon}}]_{11} \quad [\widehat{\boldsymbol{\epsilon}}]_{22} \quad [\widehat{\boldsymbol{\epsilon}}]_{33} \quad 2[\widehat{\boldsymbol{\epsilon}}]_{12} \quad 2[\widehat{\boldsymbol{\epsilon}}]_{13} \quad 2[\widehat{\boldsymbol{\epsilon}}]_{23}\right]^T. \tag{99}$$

4.2 Finite element discretization

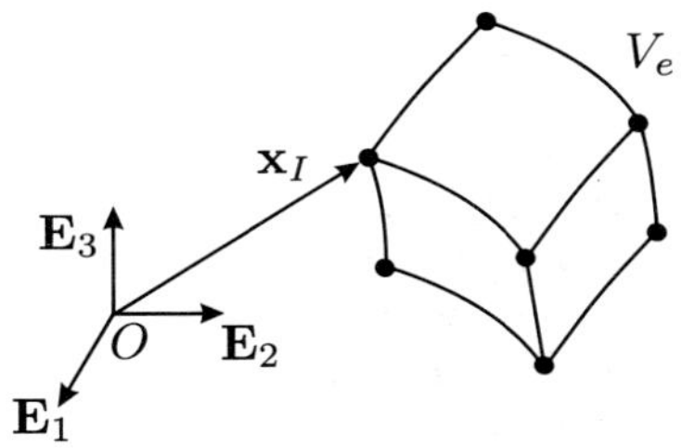

Figure 7. Description of a finite element.

According to the finite element approach, the volume of the flexible body is discretized in space. In each finite element e, the configuration mapping is expressed in terms of shape functions N_I and nodal coordinates $\mathbf{x}_I$

$$\mathbf{x} = \varphi_e(\mathbf{X}, t) = \sum_{I=1}^{n_n} N_I(\mathbf{X})\mathbf{x}_I(t) \tag{100}$$

where n_n is the number of nodes. The vector of generalized coordinates of the element is defined as

$$\mathbf{q}_e = [\mathbf{x}_1^T \; \dots \; \mathbf{x}_{n_n}^T]^T. \tag{101}$$

The relation between the position and the generalized coordinates is a linear one

$$\mathbf{x} = \varphi_e(\mathbf{X}, t) = \mathbf{Q}_e(\mathbf{X})\mathbf{q}_e(t) \tag{102}$$

where $\mathbf{Q}_e = [N_1\mathbf{I} \; \dots \; N_{n_n}\mathbf{I}]$ is the matrix of shape functions of the element.

4.3 Discretized form of the dynamic equilibrium

In order to derive the equations of motion, the discretization in Equation (102) is introduced into the expression of the kinetic energy, of the potential energy and of the internal elastic energy.

The kinetic energy becomes

$$\mathcal{K}_e = \frac{1}{2}\int_{V_e} \rho \dot{\mathbf{x}}^T \dot{\mathbf{x}} dV = \frac{1}{2}\dot{\mathbf{q}}_e^T \left(\int_{V_e} \rho \mathbf{Q}_e^T \mathbf{Q}_e dV \right) \dot{\mathbf{q}}_e. \tag{103}$$

The mass matrix is thus constant in time and given by

$$\mathbf{M}_e = \int_{V_e} \rho \mathbf{Q}_e^T \mathbf{Q}_e dV. \tag{104}$$

The variation of the potential energy of external forces is

$$\delta \mathcal{V}_e = \int_{V_e} \delta \mathbf{x}^T \mathbf{b} dV + \int_{S_e} \delta \mathbf{x}^T \mathbf{t} dS = \delta \mathbf{q}_e^T \left(\int_{V_e} \mathbf{Q}_e^T \mathbf{b} dV + \int_{S_e} \mathbf{Q}_e^T \mathbf{t} dS \right) \tag{105}$$

where $\mathbf{b}$ is the vector of body forces, $\mathbf{t}$ is the vector of surface loads and S_e is the external surface of the element. The vector of external forces is thus given by

$$\mathbf{g}_e^{ext}(\mathbf{q}) = -\int_{V_e} \mathbf{Q}_e^T \mathbf{b} dV - \int_{S_e} \mathbf{Q}_e^T \mathbf{t} dS \tag{106}$$

The virtual work of internal elastic forces is expressed as

$$\delta \mathcal{W}_e = \int_{V_e} \delta \boldsymbol{\epsilon}^T \boldsymbol{\sigma} dV \tag{107}$$

where $\boldsymbol{\epsilon}$ is the 6×1 vector of strain components and $\boldsymbol{\sigma}$ is the 6×1 vector of stress components. In general, $\delta\boldsymbol{\epsilon}$ is related to $\delta\mathbf{q}_e$ by a relation

$$\delta \boldsymbol{\epsilon} = \mathbf{B}(\mathbf{q}_e)\delta \mathbf{q}_e \tag{108}$$

where $\mathbf{B}(\mathbf{q}_e)$ is the so-called strain matrix. Hence, Equation (107) becomes

$$\delta\mathcal{W}_e = \delta\mathbf{q}_e^T \int_{V_e} \mathbf{B}^T \boldsymbol{\sigma} dV \tag{109}$$

which leads to the internal elastic forces

$$\mathbf{g}_e^{int}(\mathbf{q}_e) = \int_{V_e} \mathbf{B}^T \boldsymbol{\sigma} dV. \tag{110}$$

For a linear elastic material, $\boldsymbol{\epsilon}$ and $\boldsymbol{\sigma}$ are related by the Hooke law

$$\boldsymbol{\sigma} = \mathbf{C}\boldsymbol{\epsilon} \tag{111}$$

with $\mathbf{C}$, the matrix of Hooke material coefficients.

The internal elastic forces in Equation (110) depend on the generalized coordinates in a nonlinear way. It is instructive to develop the tangential force-displacement relationship

$$\frac{\partial \mathbf{g}_e^{int}(\mathbf{q}_e)}{\partial \mathbf{q}_e} = \mathbf{K}_e^m + \mathbf{K}_e^g \tag{112}$$

where the material stiffness matrix $\mathbf{K}_m$ is defined by

$$\mathbf{K}_e^m = \int_{V_e} \mathbf{B}^T \mathbf{C} \mathbf{B} dV \tag{113}$$

and the geometric stiffness matrix $\mathbf{K}_g$ is defined by its components

$$[\mathbf{K}_e^g]_{ij} = \int_{V_e} \sum_{k=1}^{6} \frac{\partial [\mathbf{B}]_{ki}}{\partial [\mathbf{q}_e]_j} [\boldsymbol{\sigma}]_k dV. \tag{114}$$

Let us remark that a linear formulation of elastic forces would be obtained under the assumption of infinitesimal rotations of the body. In this case, the strain matrix $\mathbf{B}$ is constant, the geometric stiffness matrix vanishes and the internal forces simply reduce to

$$\mathbf{g}_e^{int}(\mathbf{q}_e) = \mathbf{K}_m(\mathbf{q}_e - \mathbf{q}_{e0}) \tag{115}$$

where $\mathbf{q}_{e0}$ is the vector of generalized coordinates in a reference configuration for which the system is undeformed. This linear formulation is, however, not valid for the analysis of flexible multibody systems with large rotations.

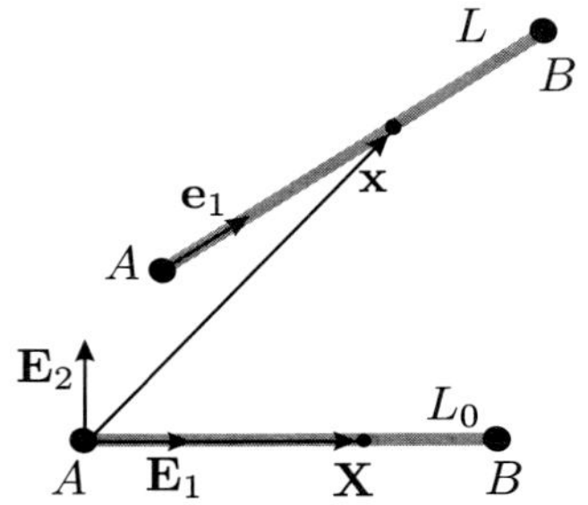

Figure 8. Elastic bar.

4.4 Formulation of a planar elastic bar element

Let us consider the planar formulation of an elastic bar element in Figure 8. The element is characterized by two nodes and the 4×1 vector of generalized coordinates is

$$\mathbf{q} = [\mathbf{x}_A^T \quad \mathbf{x}_B^T]^T \tag{116}$$

where $\mathbf{x}_A$ and $\mathbf{x}_B$ are the 2×1 vectors of nodal positions. We remark that $\mathbf{q}$ now denotes the vector of generalized coordinates of the element and not of the whole system. L_0 is the length in the reference undeformed configuration, whereas the current length is computed from the nodal coordinates

$$L = \|\mathbf{x}_B - \mathbf{x}_A\|. \tag{117}$$

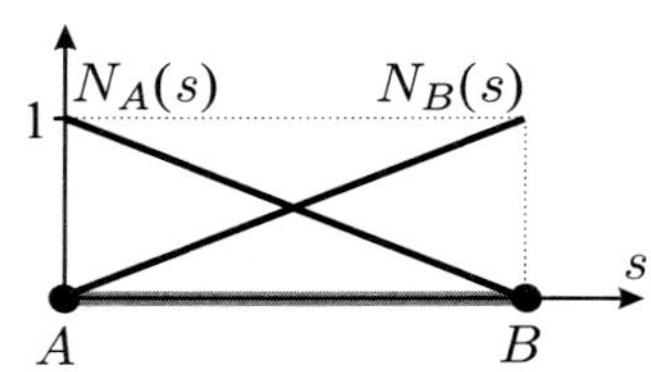

Figure 9. Shape functions for the elastic bar.

In the reference configuration, the bar is aligned along $\mathbf{E}_1$ and the position of an arbitrary material point is represented by the abscissa s

$$\mathbf{X} = s\mathbf{E}_1. \tag{118}$$

After motion, the position of the same point is described using the shape functions of the element

$$\mathbf{x} = N_A(s)\mathbf{x}_A + N_B(s)\mathbf{x}_B \tag{119}$$

where the linear shape functions, illustrated in Figure 9, are defined by

$$N_A(s) = 1 - s/L_0, \tag{120}$$
$$N_B(s) = s/L_0. \tag{121}$$

The expression (119) is rewritten as

$$\mathbf{x} = \mathbf{Q}(s)\mathbf{q}(t) \tag{122}$$

with the 4×2 matrix of shape functions

$$\mathbf{Q}(s) = \begin{bmatrix} N_A(s) & 0 & N_B(s) & 0 \\ 0 & N_A(s) & 0 & N_B(s) \end{bmatrix}. \tag{123}$$

From Equation (104), the mass matrix of the element is computed as

$$\mathbf{M} = \int_0^{L_0} \rho A \mathbf{Q}^T \mathbf{Q} ds = \frac{\rho A L_0}{6} \begin{bmatrix} 2 & 0 & 1 & 0 \\ 0 & 2 & 0 & 1 \\ 1 & 0 & 2 & 0 \\ 0 & 1 & 0 & 2 \end{bmatrix} \tag{124}$$

where A is the area of the cross section and where the assumption of small deformations was used to evaluate the integral in the reference configuration.

The internal energy of deformation is computed from the axial deformation ϵ according to

$$\delta \mathcal{W} = \int_0^{L_0} \delta\epsilon \, \sigma A \, ds = \int_0^{L_0} \delta\epsilon \, N \, ds. \tag{125}$$

In this equation, the resultant of axial stresses $N = \sigma A$ may be computed from the linear constitutive relation

$$N = EA\epsilon \tag{126}$$

where E is the Young modulus of the material. In the following, the expression of internal elastic forces is derived using the Cauchy and Green strain tensors. The Cauchy strain tensor leads to a linear elastic model but we will show that it cannot be used in case of large rotations of the bar. In contrast, the Green strain tensor leads to a geometrically nonlinear model which is valid for arbitrary motions.

Linear formulation The Cauchy strain tensor is restricted to the analysis of small displacements problems. Starting from Equation (97), the component $\epsilon^C = [\widehat{\boldsymbol{\epsilon}}^C]_{11}$ is computed as

$$\epsilon^C = \frac{\partial[\mathbf{x}]_1}{\partial[\mathbf{X}]_1} - 1 = [\mathbf{Q}'\mathbf{q}]_1 - 1 = \mathbf{B}^C\mathbf{q} - 1 \tag{127}$$

where $\mathbf{Q}'$ is the derivative of the shape function matrix with respect to s and $\mathbf{B}^C$ is the constant strain matrix

$$\mathbf{B}^C = [-1/L_0 \quad 0 \quad 1/L_0 \quad 0]. \tag{128}$$

The internal energy in Equation (125) becomes

$$\delta \mathcal{W} = \delta \mathbf{q}^T \int_0^{L_0} (\mathbf{B}^C)^T N \, ds = \delta \mathbf{q}^T (\mathbf{B}^C)^T N L_0 \tag{129}$$

and the vector of internal forces is obtained as

$$\mathbf{g}^{int} = (\mathbf{B}^C)^T N L_0, \tag{130}$$

which gives $\mathbf{g}^{int} = N[-\mathbf{E}_1^T \quad \mathbf{E}_1^T]^T$. As shown in Figure 10(a), this result only makes sense if the rotation angle α is infinitesimal. Using Equations (126) and (127), we get the linear model for the elastic forces

$$\mathbf{g}^{int} = \mathbf{K}(\mathbf{q} - \mathbf{q}_0) \tag{131}$$

with the stiffness matrix $\mathbf{K} = EAL_0(\mathbf{B}^C)^T\mathbf{B}^C$ and the reference configuration $\mathbf{q}_0 = [0 \ \ 0 \ \ L_0 \ \ 0]^T$.

Let us show that the Cauchy strain does not satisfy the objectivity requirement if the bar undergoes a pure rotation around node A. In this case, we have $\mathbf{q} = [0 \ \ 0 \ \ L_0 \cos\alpha \ \ L_0 \sin\alpha]^T$ and the Cauchy strain

$$\epsilon^C = \cos\alpha - 1 \tag{132}$$

is different from zero even though the motion is purely rigid. For this reason, the linear formulation of elastic forces based on the Cauchy strain tensor is not valid for the analysis of flexible multibody systems with large rotations.

Nonlinear formulation Unlike the Cauchy strain tensor, the Green strain tensor can be used even for large displacements problems. Starting from Equation (98), the component $\epsilon^G = [\widehat{\boldsymbol{\epsilon}}^G]_{11}$ is computed as

$$\epsilon^G = \frac{1}{2}\left(\left(\frac{\partial \mathbf{x}}{\partial s}\right)^T \frac{\partial \mathbf{x}}{\partial s} - 1\right) = \frac{1}{2}\left(\mathbf{q}^T(\mathbf{Q}')^T\mathbf{Q}'\mathbf{q} - 1\right). \tag{133}$$

After development, we get

$$\epsilon^G = \frac{1}{2}\left(\left(\frac{\mathbf{x}_B - \mathbf{x}_A}{L_0}\right)^T \frac{\mathbf{x}_B - \mathbf{x}_A}{L_0} - 1\right) = \frac{1}{2}\left((L/L_0)^2 - 1\right). \tag{134}$$

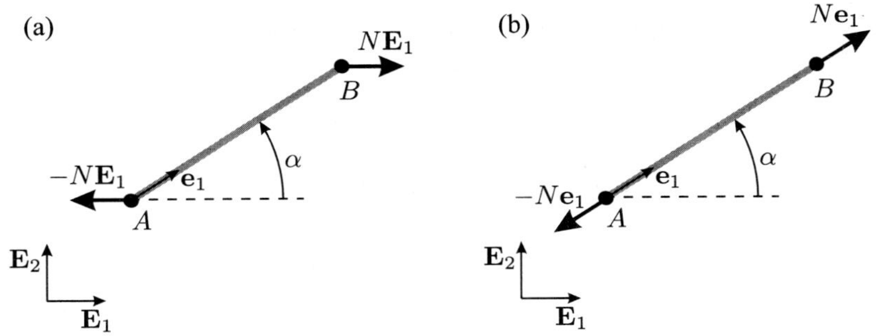

Figure 10. Internal elastic forces in the bar: (a) linear formulation, (b) nonlinear formulation.

By derivation of Equation (133) with respect to $\mathbf{q}$, the strain matrix is

$$\mathbf{B}^G = \mathbf{q}^T(\mathbf{Q}')^T\mathbf{Q}'. \tag{135}$$

The internal elastic forces are computed as

$$\mathbf{g}^{int} = (\mathbf{B}^G)^T N L_0. \tag{136}$$

One can verify that $\mathbf{g}^{int} = N[-\mathbf{e}_1^T \quad \mathbf{e}_1^T]^T$, which makes sense even for large rotations, see Figure 10(b). The relation between $\mathbf{g}^{int}$ and $\mathbf{q}$ is nonlinear and the tangent stiffness matrix is

$$\frac{\partial \mathbf{g}^{int}}{\partial \mathbf{q}} = (\mathbf{B}^G)^T L_0 \frac{\partial N}{\partial \mathbf{q}} + \frac{\partial (\mathbf{B}^G)^T}{\partial \mathbf{q}} N L_0 \tag{137}$$

$$= EAL_0(\mathbf{B}^G)^T\mathbf{B}^G + NL_0(\mathbf{Q}')^T\mathbf{Q}' \tag{138}$$

where the first term is the material stiffness matrix and the second term is the geometric stiffness matrix, which is proportional to the resultant of axial stresses N.

From Equation (134), we observe that $\epsilon^G = 0$ for any large amplitude rigid motion. Hence, the nonlinear formulation of elastic forces based on the Green strain tensor allows a consistent analysis of large displacements problems in flexible multibody dynamics.

4.5 Formulation of an elastic beam element

According to the geometrically exact formulation by Simo (1985), the motion of the beam is directly referred to the inertial frame. Independent translation and rotation fields along the beam axis are used to represent both the motion and the deformation in agreement with Timoshenko theory. Compared to an Euler-Bernoulli model, the Timoshenko model has two main advantages: firstly, shear deformations can be represented and, secondly, the same interpolation formulae are used in any spatial direction. Hence, the beam orientation does not influence the finite element interpolation procedure, which allows to simplify considerably the computations.

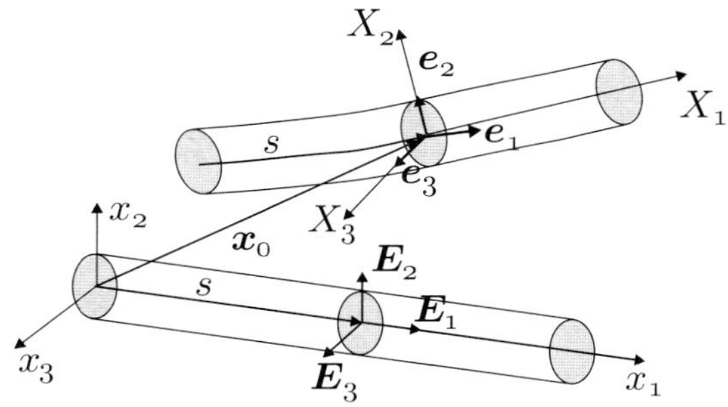

Figure 11. Description of the beam kinematics.

Let us assume that the beam is initially straight and that the beam cross sections remain plane and do not deform during elastic deformation (see Figure 11). The kinematics of the beam is described by the position of the point of the neutral axis noted $\mathbf{x}_0(s)$ and by a material frame $\{\mathbf{e}_1(s), \mathbf{e}_2(s), \mathbf{e}_3(s)\}$ attached to the cross section. We have

$$\mathbf{e}_i(s) = \mathbf{R}(s)\mathbf{E}_i, \qquad (i = 1, 2, 3) \tag{139}$$

where $\mathbf{R}(s)$ is the rotation operator which represents the orientation of the cross-section.

Hence, the location of a material point in the reference configuration is given by

$$\mathbf{X} = s\mathbf{E}_1 + X_2\mathbf{E}_2 + X_3\mathbf{E}_3 = \mathbf{X}_0 + \mathbf{Y}, \tag{140}$$

where $\mathbf{X}_0 = s\mathbf{E}_1$ and $\mathbf{Y} = X_2\mathbf{E}_2 + X_3\mathbf{E}_3$. After motion, this point moves to the new position

$$\mathbf{x} = \mathbf{x}_0(s) + \mathbf{R}(s)\mathbf{Y}. \tag{141}$$

At any point s, the rotation field $\mathbf{R}(s)$ can be parameterized using the Cartesian rotation vector $\boldsymbol{\psi}(s)$. As a result, the configuration of the beam is represented by a one-dimensional translation field $\mathbf{x}_0(s)$ and a one-dimensional rotation field $\boldsymbol{\psi}(s)$. Since both fields are treated as independent quantities, the cross-section is not necessarily perpendicular to the neutral axis and shear deformation is allowed.

The beam element is thus composed of two nodes and the set of nodal coordinates is gathered in the vector

$$\mathbf{q} = \begin{bmatrix} \mathbf{x}_A^T & \boldsymbol{\psi}_A^T & \mathbf{x}_B^T & \boldsymbol{\psi}_B^T \end{bmatrix}^T . \tag{142}$$

The finite element discretization of the translational and rotational fields leads to

$$\mathbf{x}_0(s) = N_A(s)\mathbf{x}_A + N_B(s)\mathbf{x}_B, \tag{143}$$

$$\boldsymbol{\psi}(s) = N_A(s)\boldsymbol{\psi}_A + N_B(s)\boldsymbol{\psi}_B, \tag{144}$$

where linear shape functions can be used (see Figure 9). Let us note that the linear interpolation of rotational parameters in Equation (144) has been criticized by several authors, since the resulting measure of deformation is not truly invariant under a rigid motion. However, this problem of objectivity can be eliminated using a simple refinement of the interpolation procedure. For example, a fully consistent treatment is obtained if the interpolation is based on incremental rotations with respect to a close reference configuration, see Cardona and Géradin (1988), or if it is based on the relative rotation between nodes A and B, see Crisfield and Jelenic (1999) or Lens and Cardona (2008).

The position of an arbitrary material point of the beam is computed from the nodal coordinates using

$$\mathbf{x} = N_A\mathbf{x}_A + N_B\mathbf{x}_B + \mathbf{R}(N_A\boldsymbol{\psi}_A + N_B\boldsymbol{\psi}_B)\mathbf{Y}. \tag{145}$$

Without proof, we shall simply give the expression of the kinetic and of the strain energy of the beam. The kinetic energy is given by

$$\mathcal{K} = \frac{1}{2}\int_0^{L_0} (\mu\dot{\mathbf{x}}_0^T\dot{\mathbf{x}}_0 + \boldsymbol{\Omega}^T\mathbf{J}\boldsymbol{\Omega})ds \tag{146}$$

where $\mu = \rho A$ is the mass per unit length, $\mathbf{J}$ is the tensor of inertia of the cross section and $\boldsymbol{\Omega}(s)$ is the angular velocity of the cross section, which can be computed from $\boldsymbol{\psi}(s)$ and $\dot{\boldsymbol{\psi}}(s)$ using Equation (10). Let us observe the similarity between the kinetic energy of the beam element in Equation (146) and the kinetic energy of a rigid body: as for a rigid body, the inertia forces of the beam will involve gyroscopic and centrifugal terms.

The strain energy is given by

$$\delta\mathcal{W} = \int_0^{L_0} (\delta\boldsymbol{\Gamma}^T\mathbf{N} + \delta\mathbf{K}^T\mathbf{M})ds. \tag{147}$$

This expression involves the following terms which are all evaluated in the material frame $\{\mathbf{e}_1(s), \mathbf{e}_2(s), \mathbf{e}_3(s)\}$

- $\mathbf{N}$: the 3×1 vector of axial and shear forces,
- $\mathbf{M}$: the 3×1 vector of torsion and bending moments,
- $\mathbf{\Gamma}$: the 3×1 vector of axial and shear deformations,
- $\mathbf{K}$: the 3×1 vector of torsion and bending deformations.

The constitutive relation is written as

$$\begin{bmatrix} \mathbf{N} \\ \mathbf{M} \end{bmatrix} = \mathbf{C} \begin{bmatrix} \mathbf{\Gamma} \\ \mathbf{K} \end{bmatrix}. \tag{148}$$

For an orthotropic material, the 6×6 matrix of elastic coefficients $\mathbf{C}$ is defined by

$$\mathbf{C} = \text{diag}(EA \quad GA_2 \quad GA_3 \quad GJ \quad EI_2 \quad EI_3) \tag{149}$$

where EA is the axial stiffness, GA_2 and GA_3 are the shear stiffnesses along the tranverse axes, GJ is the torsional stiffness, and EI_2 and EI_3 are the bending stiffnesses.

From those expressions, the inertia and elastic forces can be derived, as explained in Géradin and Cardona (2001). This reference also discusses the shear locking phenomenon and the reduced integration technique that need to be implemented to eliminate the problem.

4.6 Summary

In this section, the nonlinear finite element approach was used for the modelling of elastic components. The formulation is based on absolute nodal coordinates and the strains are directly extracted from the configuration mapping. A nonlinear measure of deformation is therefore needed, e.g. using the Green strain tensor, and geometrical nonlinearities appear in the elastic force model. The inertia forces are easily computed using a constant mass matrix, at least if the formulation of the element does not involve any rotational field.

In the elastic beam formulation, the position of the neutral axis is represented using a translational field, whereas the orientation of cross sections is represented using an independent rotational field. Both fields are interpolated linearly from the absolute translation and rotation parameters of the two nodes of the element. A consistent interpolation of rotations can be implemented provided special care. The rotational inertia of cross sections leads to nonlinear gyroscopic and centrifugal contributions to the inertia forces.

5 Flexible body dynamics: superelement approach

The finite element approach requires a geometrically nonlinear formulation of elastic forces and it involves a rather large number of nodal coordinates. As a result, the equations of motion take the form of a sparse but large set of differential-algebraic equations. In this section, the superelement technique is proposed in order to reduce the complexity of the model.

Geometric nonlinearities, which appear in case of large amplitude motions, are present even though the deformations are small. However, in this particular case, the small displacements assumption can be recovered in a floating or corotational frame that follows the gross motion of the flexible body. In this moving frame, a much simpler linear elastic model can be used to compute the elastic forces. A further simplification is then possible if a component-mode synthesis is exploited to reduce the order of the linear elastic model.

Among classical model reduction techniques in linear structural dynamics, let us mention the methods of Craig and Bampton (1968), Mac Neal (1971) and Rubin (1975). Alternative techniques coming from system theory, such as balanced truncation methods by Moore (1981) and Glover (1984) or Krylov subspace techniques by Grimme (1997), have also been adapted for structural dynamics problems, see e.g. Su and Craig (1991); Meyer and Srinivasan (1996); Bai and Su (2005).

The superelement technique described in the following relies on a nonlinear corotational formulation of a reduced-order model and it is fully compatible with the finite element approach. The resulting superelement, which is used to model a whole flexible body with large motions, has its own interface nodes so that it can be assembled with other components of the system (e.g. joints, force elements or other bodies) as any other element of the mesh.

The reduction methods of Craig and Bampton, Mac Neal and Rubin appear especially well-suited for the formulation of a superelement. Indeed, unlike balanced truncation and Krylov subspace techniques, those methods allow the definition of interface or boundary coordinates which are preserved in the reduced-order model. In addition to the interface nodal coordinates, internal modal coordinates are used in the superelement in order to represent the internal dynamics of the body with a sufficient level of accuracy.

In the following, linear reduction techniques are first reviewed. Then, the nonlinear corotational formulation of the superelement is followed by discussions on the computation of elastic and inertia forces.

5.1 Linear reduction method

Let us consider the *small amplitude motion* of a dynamic structure characterized by its kinetic energy $\mathcal{K}$, its internal energy of deformation $\mathcal{W}$ and the virtual work of applied forces $\delta\mathcal{V}$

$$\mathcal{K} = \frac{1}{2}\dot{\mathbf{y}}^T\mathbf{M}\dot{\mathbf{y}}, \qquad \mathcal{W} = \frac{1}{2}\mathbf{y}^T\mathbf{K}\mathbf{y}, \qquad \delta\mathcal{V} = \delta\mathbf{y}^T\mathbf{g} \tag{150}$$

with $\mathbf{y}$, the $n \times 1$ vector of generalized coordinates, $\mathbf{M}$, the mass matrix, $\mathbf{K}$, the stiffness matrix, and $\mathbf{g}$, the vector of applied forces. From the Hamilton principle, the equations of motion are derived in the form

$$\mathbf{M}\ddot{\mathbf{y}} + \mathbf{K}\mathbf{y} = \mathbf{g}. \tag{151}$$

In order to reduce the order of this initial model, we assume that the vector $\mathbf{y}$ can be expanded in terms of $\overline{n}$ modal coordinates $\boldsymbol{\eta}$ ($\overline{n} < n$) according to

$$\mathbf{y} = \boldsymbol{\Psi}\boldsymbol{\eta} \tag{152}$$

where $\boldsymbol{\Psi}$ is the $n \times \overline{n}$ matrix of component-modes. The variations of $\mathbf{y}$ are thereby restricted to the subspace spanned by the component-modes and the accuracy of the reduced model depends on the ability of the component-modes to describe the actual motion of the system.

In order to formulate the reduced equations of motion, the coordinate transformation is introduced into the expressions of $\mathcal{K}$, $\mathcal{W}$ and $\delta\mathcal{V}$ in Equation (150)

$$\mathcal{K} = \frac{1}{2}\dot{\boldsymbol{\eta}}^T\overline{\mathbf{M}}\dot{\boldsymbol{\eta}}, \qquad \mathcal{W} = \frac{1}{2}\boldsymbol{\eta}^T\overline{\mathbf{K}}\boldsymbol{\eta}, \qquad \delta\mathcal{V} = \delta\boldsymbol{\eta}^T\overline{\mathbf{g}} \tag{153}$$

with the reduced mass matrix, stiffness matrix and equivalent force vector

$$\overline{\mathbf{M}} = \boldsymbol{\Psi}^T\mathbf{M}\boldsymbol{\Psi}, \qquad \overline{\mathbf{K}} = \boldsymbol{\Psi}^T\mathbf{K}\boldsymbol{\Psi}, \qquad \overline{\mathbf{g}} = \boldsymbol{\Psi}^T\mathbf{g}. \tag{154}$$

The equations of motion of the reduced system follow as

$$\overline{\mathbf{M}}\ddot{\boldsymbol{\eta}} + \overline{\mathbf{K}}\boldsymbol{\eta} = \overline{\mathbf{g}}. \tag{155}$$

Before the application of the reduction method, one should identify the boundary or interface generalized coordinates $\mathbf{y}_B$, which are needed to assembly the superelement with the other components of the system. Therefore, the initial vector of generalized coordinates is first partitioned

$$\mathbf{y} = \begin{bmatrix}\mathbf{y}_B^T & \mathbf{y}_I^T\end{bmatrix}^T \tag{156}$$

where $\mathbf{y}_I$ is the vector of internal coordinates which disappear in the procedure. For each boundary generalized coordinate in $\mathbf{y}_B$, a static boundary mode is defined and the modal transformation in Equation (152) is splitted as

$$\mathbf{y} = \mathbf{\Psi}_B \mathbf{y}_B + \mathbf{\Psi}_I \boldsymbol{\eta}_I \tag{157}$$

where $\mathbf{\Psi}_B$ is the matrix of static boundary modes and $\mathbf{\Psi}_I$ is the matrix of internal modes. We have $\mathbf{\Psi} = [\mathbf{\Psi}_B \ \mathbf{\Psi}_I]$ and $\boldsymbol{\eta} = [\mathbf{y}_B^T \ \boldsymbol{\eta}_I^T]^T$, i.e. the generalized coordinates of the boundary explicitly appear in the reduced model.

5.2 Kinematic description of a superelement

Now, let us consider that the superelement undergoes *large amplitude motions*. The vector of generalized coordinates of the element is defined as

$$\mathbf{q} = \begin{bmatrix} \mathbf{q}_B^T & \boldsymbol{\eta}_I^T \end{bmatrix}^T \tag{158}$$

where $\mathbf{q}_B$ represents the absolute coordinates of the interface nodes and $\boldsymbol{\eta}_I$, the coordinates of the internal modes. The vector $\mathbf{q}_B$ is composed of nodal positions $\mathbf{x}_i$ and nodal rotations $\boldsymbol{\psi}_i$

$$\mathbf{q}_B = \begin{bmatrix} \mathbf{x}_1^T & \boldsymbol{\psi}_1^T & \dots & \mathbf{x}_{n_n}^T & \boldsymbol{\psi}_{n_n}^T \end{bmatrix}^T. \tag{159}$$

According to the corotational approach, the elastic forces are computed in a corotational frame $\{\mathbf{e}_1, \mathbf{e}_2, \mathbf{e}_3\}$ that follows the gross motion of the body (Figure 12). This frame is represented by a translation vector $\mathbf{x}_0$ and a rotation matrix $\mathbf{R}_0$, which is parameterized using the rotation vector $\boldsymbol{\psi}_0$. The frame transformation between the inertial and the corotational frame can be analysed. In particular, the relationship between the absolute position and orientation of any boundary node i ($i = 1, ..., n_n$) and its local displacement $\mathbf{u}_i$ and local rotation $\boldsymbol{\phi}_i$ with respect to the corotational frame is given by

$$\begin{aligned} \mathbf{x}_i &= \mathbf{x}_0 + \mathbf{R}_0(\mathbf{X}_i + \mathbf{u}_i), & (160)\\ \mathbf{R}_i &= \mathbf{R}_0 \mathbf{R}(\boldsymbol{\phi}_i). & (161) \end{aligned}$$

Definition of the corotational frame For given values of the absolute positions and orientations of the boundary nodes, the corotational frame is

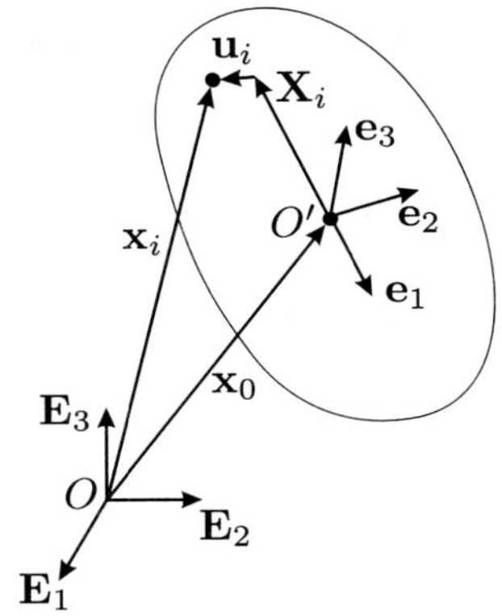

Figure 12. Kinematic description of a superelement ($\mathbf{X}_i$ and $\mathbf{u}_i$ are referred to the local frame).

defined by linear combinations

$$\mathbf{x}_0 = \sum_{i=1}^{n_n} \alpha_i \mathbf{x}_i, \tag{162}$$

$$\boldsymbol{\psi}_{0,inc} = \sum_{i=1}^{n_n} \alpha_i \boldsymbol{\psi}_{i,inc} \tag{163}$$

where α_i are nodal weight coefficients and $\boldsymbol{\psi}_{i,inc}$ denotes small amplitude incremental rotation with respect to a close reference configuration $\boldsymbol{\psi}_{ref}$. For instance, this reference can be defined in a previously computed configuration. The incremental rotation is defined such that

$$\mathbf{R}(\boldsymbol{\psi}_i) = \mathbf{R}(\boldsymbol{\psi}_{ref})\mathbf{R}(\boldsymbol{\psi}_{i,inc}). \tag{164}$$

This last equation is written in compact form

$$\boldsymbol{\psi}_i = \boldsymbol{\psi}_{ref} \circ \boldsymbol{\psi}_{i,inc} \tag{165}$$

where the operator $\circ$ denotes the composition of rotations. Note that the use of incremental rotations in Equation (163) is mandatory because a linear combination of large rotations would be inconsistent.

5.3 Corotational computation of the elastic forces

In the corotational frame, the internal deformation energy is computed according to

$$\mathcal{W} = \frac{1}{2}\boldsymbol{\eta}^T\overline{\mathbf{K}}\boldsymbol{\eta} \tag{166}$$

where $\overline{\mathbf{K}}$ is the reduced-order stiffness matrix of the superelement and $\boldsymbol{\eta}$ is the vector of local coordinates

$$\boldsymbol{\eta} = \begin{bmatrix}\mathbf{u}_1^T & \boldsymbol{\phi}_1^T & \dots & \mathbf{u}_{n_n}^T & \boldsymbol{\phi}_{n_n}^T & \boldsymbol{\eta}_I^T\end{bmatrix}^T. \tag{167}$$

We have

$$\delta\mathcal{W} = \delta\boldsymbol{\eta}^T\overline{\mathbf{K}}\boldsymbol{\eta}. \tag{168}$$

Given a tangential relationship between local coordinates $\boldsymbol{\eta}$ and absolute coordinates $\mathbf{q}$ in the form

$$\delta\boldsymbol{\eta} = \mathbf{P}(\mathbf{q})\delta\mathbf{q}, \tag{169}$$

the internal energy becomes

$$\delta\mathcal{W} = \delta\mathbf{q}^T\mathbf{P}^T\overline{\mathbf{K}}\boldsymbol{\eta}, \tag{170}$$

which leads to the expression of the internal forces

$$\mathbf{g}^{int} = \mathbf{P}^T\overline{\mathbf{K}}\boldsymbol{\eta}. \tag{171}$$

The operator $\mathbf{P}$ can be evaluated from Equations (158–161) and (167). The full derivation, which is described in Géradin and Cardona (2001), is not reproduced here for reasons of space.

In summary, the evaluation of the elastic forces $\mathbf{g}^{int}$ for a given configuration $\mathbf{q}$ involves the successive computation of

1. the corotational frame (Equations (162–163)),
2. the local coordinate vector $\boldsymbol{\eta}$ (in Equation (167), local displacements and rotations are extracted from Equations (160–161)),
3. the tangential operator $\mathbf{P}(\mathbf{q})$ (not given here),
4. the internal forces $\mathbf{g}^{int}$ (Equation (171)).

5.4 Corotational computation of the inertia forces

In the previous section, we have shown that the elastic forces can be computed using the stiffness matrix $\overline{\mathbf{K}}$. Likewise, a procedure is described to compute the inertia forces using the mass matrix $\overline{\mathbf{M}}$ of the linear reduced-order model. The time derivative of the generalized coordinates $\dot{\mathbf{q}}$ and the vector of corotated generalized velocities $\mathbf{v}$ are then introduced

$$\dot{\mathbf{q}} = \begin{bmatrix} \dot{\mathbf{x}}_1 \\ \dot{\boldsymbol{\psi}}_1 \\ \vdots \\ \dot{\mathbf{x}}_{n_n} \\ \dot{\boldsymbol{\psi}}_{n_n} \\ \dot{\boldsymbol{\eta}}_I \end{bmatrix}, \qquad \mathbf{v} = \begin{bmatrix} \mathbf{R}_0^T \dot{\mathbf{x}}_1 \\ \boldsymbol{\Omega}_1 \\ \vdots \\ \mathbf{R}_0^T \dot{\mathbf{x}}_{n_n} \\ \boldsymbol{\Omega}_{n_n} \\ \dot{\boldsymbol{\eta}}_I \end{bmatrix}. \tag{172}$$

Each contribution $\mathbf{R}_0^T \dot{\mathbf{x}}_i$ has the meaning of an *absolute* velocity expressed in the corotational frame. $\boldsymbol{\Omega}_i$ is the angular velocity in the material frame which, in first approximation, also corresponds to the corotational frame. Let us observe that the corotated velocities are non-integrable quantities. The relation between $\mathbf{v}$ and $\dot{\mathbf{q}}$ can be put in the form

$$\mathbf{v} = \mathbf{S}(\mathbf{q})\dot{\mathbf{q}} \tag{173}$$

with the block diagonal tangent operator

$$\mathbf{S}(\mathbf{q}) = \operatorname{diag}\left(\mathbf{R}_0^T\,,\ \mathbf{T}(\boldsymbol{\psi}_1)\,,\ \ldots\,,\ \mathbf{R}_0^T\,,\ \mathbf{T}(\boldsymbol{\psi}_{n_n})\,,\ \mathbf{I}\right). \tag{174}$$

The kinetic energy is then evaluated in the corotational frame. Starting from Equation (103) but performing the finite element interpolation of velocities in the corotational frame yields

$$\mathcal{K} = \frac{1}{2}\mathbf{v}^T\overline{\mathbf{M}}\mathbf{v} = \frac{1}{2}\dot{\mathbf{q}}^T\mathbf{S}^T\overline{\mathbf{M}}\mathbf{S}\dot{\mathbf{q}}. \tag{175}$$

Hence, the interpolation of velocities is not fully consistent with the interpolation of displacements, which is performed in the inertial frame. From Equation (175), the non-constant mass matrix is given by

$$\mathbf{M}(\mathbf{q}) = \mathbf{S}^T(\mathbf{q})\,\overline{\mathbf{M}}\,\mathbf{S}(\mathbf{q}) \tag{176}$$

and the gyroscopic forces can be computed from Equation (45). Again, we refer to Géradin and Cardona (2001) for a more comprehensive treatment.

5.5 Summary

In case of large displacements but small deformations, the superelement technique allows to simplify the computation of elastic forces and to reduce the number of generalized coordinates. The corotational computation of elastic and inertia forces only requires the reduced stiffness and mass matrices which are obtained by a linear analysis of the flexible body. Model

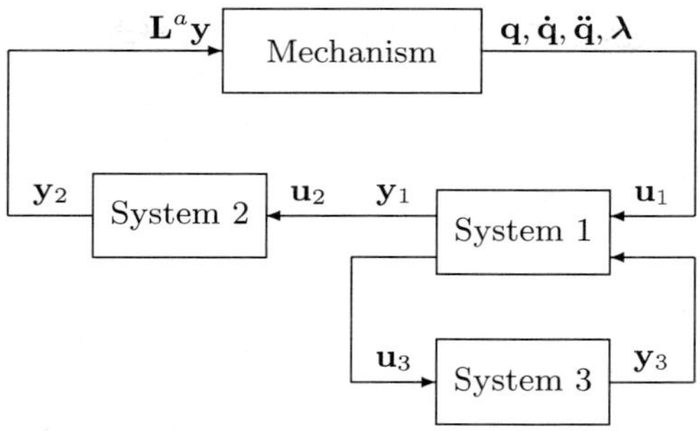

Figure 13. Modular approach for the description of the control system.

reduction techniques, such as the Craig and Bampton method, lead to compact formulations of those structural matrices. The interface nodal coordinates of the superelement allow the finite element assembly with the rest of the mechanical system.

6 Dynamics of mechatronic systems

In this section, the finite element formulation is extended to the integrated dynamic analysis of mechatronic systems. The analysis should account for various technological components: mechanical elements, control units, actuators, sensors, etc.

We have seen that the finite element method allows the modelling of complex flexible mechanisms. In order to describe mechatronic systems, we propose to integrate the *block diagram language* within the finite element framework. Relying on a nonlinear state-space description, this approach presents the advantages of modularity and generality with a language very familiar to control engineers. Each block of the block diagram model is treated as a special type of finite element so that the strongly coupled mechanical and state equations can be obtained by numerical assembly.

6.1 Formulation of a block element

As illustrated in Figure 13, the block diagram language allows a modular description of a control system. A block element e can be considered as a special type of finite element, whose state space dynamics is expressed in terms of input variables $\mathbf{u}_e$, internal state variables $\mathbf{x}_e$ and output variables $\mathbf{y}_e$

$$\dot{\mathbf{x}}_e = \mathbf{f}_e(\mathbf{u}_e, \mathbf{x}_e, t), \quad (177)$$

$$\mathbf{y}_e = \mathbf{h}_e(\mathbf{u}_e, \mathbf{x}_e, t). \quad (178)$$

The state and output variables are considered as new independent coordinates in the dynamic model, whereas the vector of inputs is constructed by localization relationships

$$\mathbf{u}_e = \mathbf{L}_e^{\mathbf{uq}}\mathbf{q} + \mathbf{L}_e^{\mathbf{u}\dot{\mathbf{q}}}\dot{\mathbf{q}} + \mathbf{L}_e^{\mathbf{u}\ddot{\mathbf{q}}}\ddot{\mathbf{q}} + \mathbf{L}_e^{\mathbf{u}\boldsymbol{\lambda}}\boldsymbol{\lambda} + \mathbf{L}_e^{\mathbf{uy}}\mathbf{y}. \quad (179)$$

where $\mathbf{L}_e^{\star}$ are Boolean matrices. This equation means that any input variable is either a generalized displacement, a velocity, an acceleration, an internal force (Lagrange multiplier) or the output of another block of the control system.

6.2 Integrated modelling of the mechatronic system

If n_b denotes the number of blocks, the global state and output vectors are defined by

$$\mathbf{x} = \begin{bmatrix} \mathbf{x}_1 \\ \vdots \\ \mathbf{x}_{n_b} \end{bmatrix}, \quad \mathbf{y} = \begin{bmatrix} \mathbf{y}_1 \\ \vdots \\ \mathbf{y}_{n_b} \end{bmatrix} \quad (180)$$

and the global state space equation is written as

$$\dot{\mathbf{x}} = \mathbf{f}(\mathbf{q}, \dot{\mathbf{q}}, \ddot{\mathbf{q}}, \boldsymbol{\lambda}, \mathbf{x}, \mathbf{y}, t), \quad (181)$$

$$\mathbf{y} = \mathbf{h}(\mathbf{q}, \dot{\mathbf{q}}, \ddot{\mathbf{q}}, \boldsymbol{\lambda}, \mathbf{x}, \mathbf{y}, t) \quad (182)$$

with the functions $\mathbf{f}$ and $\mathbf{h}$

$$\mathbf{f}(\mathbf{q}, \dot{\mathbf{q}}, \ddot{\mathbf{q}}, \boldsymbol{\lambda}, \mathbf{x}, \mathbf{y}, t) = \begin{bmatrix} \mathbf{f}_1(\mathbf{u}_1, \mathbf{x}_1, t) \\ \vdots \\ \mathbf{f}_{n_b}(\mathbf{u}_{n_b}, \mathbf{x}_{n_b}, t) \end{bmatrix}, \quad (183)$$

$$\mathbf{h}(\mathbf{q}, \dot{\mathbf{q}}, \ddot{\mathbf{q}}, \boldsymbol{\lambda}, \mathbf{x}, \mathbf{y}, t) = \begin{bmatrix} \mathbf{h}_1(\mathbf{u}_1, \mathbf{x}_1, t) \\ \vdots \\ \mathbf{h}_{n_b}(\mathbf{u}_{n_b}, \mathbf{x}_{n_b}, t) \end{bmatrix}. \quad (184)$$

In these equations, the input variables $\mathbf{u}_i$ have been expressed in terms of $\mathbf{q}$, $\dot{\mathbf{q}}$, $\ddot{\mathbf{q}}$, $\boldsymbol{\lambda}$ and $\mathbf{y}$ according to Equation (179). We remark that the output equation is thus implicit with respect to the output variable $\mathbf{y}$.

The control system is influenced by the dynamics of the mechanical system, and conversely, the mechanical system is influenced by control forces. We consider that the generalized control forces appear explicitly in the output vector

$$\mathbf{g}^a = \mathbf{L}^a \mathbf{y} \tag{185}$$

where $\mathbf{L}^a$ is the actuator localization matrix. Those control forces produce the mechanical virtual work $\delta\mathcal{W}^a = \delta\mathbf{q}^T\mathbf{g}^a$ and the coupled dynamic equations take the general form

$$\begin{aligned}
\mathbf{M}(\mathbf{q})\ddot{\mathbf{q}} + \mathbf{g}^{gyr}(\mathbf{q},\dot{\mathbf{q}}) + \mathbf{g}^{int}(\mathbf{q}) + \mathbf{\Phi}_{\mathbf{q}}^T\boldsymbol{\lambda} &= \mathbf{g}^{ext}(\mathbf{q}) + \mathbf{L}^a\mathbf{y}, && (186)\\
\mathbf{\Phi}(\mathbf{q}) &= \mathbf{0}, && (187)\\
\dot{\mathbf{x}} &= \mathbf{f}(\mathbf{q},\dot{\mathbf{q}},\ddot{\mathbf{q}},\boldsymbol{\lambda},\mathbf{x},\mathbf{y},t), && (188)\\
\mathbf{y} &= \mathbf{h}(\mathbf{q},\dot{\mathbf{q}},\ddot{\mathbf{q}},\boldsymbol{\lambda},\mathbf{x},\mathbf{y},t). && (189)
\end{aligned}$$

6.3 Example : PID control of the spherical double pendulum

In the spherical double pendulum example, let us now consider that the hinge is motorized according to a Proportional-Integral-Derivative (PID) law

$$\begin{aligned}
\dot{x} &= \theta - \theta^{ref}, && (190)\\
\tau &= -P\,(\theta - \theta^{ref}) - I\,x - D\,\dot{\theta}. && (191)
\end{aligned}$$

θ^{ref} is the reference value for the joint angle, θ is the joint angle, x is the integral of the error, and τ is the output torque exerted by the motor. The virtual work exerted by the motor is $\delta\mathcal{W}^a = \delta\theta\,\tau$.

Definition of relative coordinates in the finite element formulation
The joint angle θ is not explicitly available in the absolute coordinate formulation described before. The formulation of the hinge joint element can, however, be slightly adapted for that purpose. Indeed, the joint angle satisfies

$$\begin{aligned}
\sin\theta &= \mathbf{e}_{B2}^T\mathbf{e}_{C1} = \mathbf{E}_2^T\mathbf{R}_B^T\mathbf{R}_C\mathbf{E}_1, && (192)\\
\cos\theta &= \mathbf{e}_{B1}^T\mathbf{e}_{C1} = \mathbf{E}_1^T\mathbf{R}_B^T\mathbf{R}_C\mathbf{E}_1. && (193)
\end{aligned}$$

Let us introduce an additional generalized coordinate α which should satisfy the constraint $\alpha = \theta$. This equation can be conveniently rewritten as

$$\sin(\theta - \alpha) = 0, \tag{194}$$

which becomes

$$\mathbf{E}_2^T \mathbf{R}_B^T \mathbf{R}_C \mathbf{E}_1 \cos\alpha - \mathbf{E}_1^T \mathbf{R}_B^T \mathbf{R}_C \mathbf{E}_1 \sin\alpha = 0. \tag{195}$$

Hence, the constraint vector of the hinge element is redefined as

$$\mathbf{\Phi}_3(\mathbf{q}_3) = \begin{bmatrix} \mathbf{E}_1^T \mathbf{R}_B^T \mathbf{R}_C \mathbf{E}_3 \\ \mathbf{E}_2^T \mathbf{R}_B^T \mathbf{R}_C \mathbf{E}_3 \\ \mathbf{E}_2^T \mathbf{R}_B^T \mathbf{R}_C \mathbf{E}_1 \cos\alpha - \mathbf{E}_1^T \mathbf{R}_B^T \mathbf{R}_C \mathbf{E}_1 \sin\alpha \end{bmatrix} \tag{196}$$

with

$$\mathbf{q}_3 = [\boldsymbol{\psi}_B^T \quad \boldsymbol{\psi}_C^T \quad \alpha]^T. \tag{197}$$

Note that the global vector of generalized coordinates $\mathbf{q}$ has now 13 components

$$\mathbf{q} = [\mathbf{x}_B^T \quad \boldsymbol{\psi}_B^T \quad \mathbf{x}_D^T \quad \boldsymbol{\psi}_D^T \quad \alpha]^T, \tag{198}$$

whereas the vector of Lagrange multipliers has 9 components ($\boldsymbol{\lambda}_3$ has a third component). At the equilibrium, the vector of reaction forces becomes, see Géradin and Cardona (2001),

$$\mathbf{\Phi}_{\mathbf{q}}^T \boldsymbol{\lambda} = \begin{bmatrix} \boldsymbol{\lambda}_2 - \boldsymbol{\lambda}_1 \\ \mathbf{T}_B^T \mathbf{R}_B^T \left(\mathbf{x}_B \times \boldsymbol{\lambda}_1 - \mathbf{e}_{B2}[\boldsymbol{\lambda}_3]_1 + \mathbf{e}_{B1}[\boldsymbol{\lambda}_3]_2 - \mathbf{e}_{B3}[\boldsymbol{\lambda}_3]_3\right) \\ -\boldsymbol{\lambda}_2 \\ \mathbf{T}_D^T \mathbf{R}_D^T \left((\mathbf{x}_D - \mathbf{x}_B) \times \boldsymbol{\lambda}_2 + \mathbf{e}_{B2}[\boldsymbol{\lambda}_3]_1 - \mathbf{e}_{B1}[\boldsymbol{\lambda}_3]_2 + \mathbf{e}_{B3}[\boldsymbol{\lambda}_3]_3\right) \\ -[\boldsymbol{\lambda}_3]_3 \end{bmatrix}. \tag{199}$$

Clearly, $[\boldsymbol{\lambda}_3]_3$ has the meaning of an internal torque in the direction of the hinge axis.

Formulation of the PID block Firstly, Equations (190–191) can be recasted as Equations (177–178) with the block inputs and outputs

$$\mathbf{u}_1 = [\alpha \quad \dot{\alpha}]^T, \quad \mathbf{y}_1 = [\tau], \tag{200}$$

and the input localization matrices

$$\mathbf{L}_1^{\mathbf{uq}} = \begin{bmatrix} 0 & \dots & 1 \\ 0 & \dots & 0 \end{bmatrix}, \qquad \mathbf{L}_1^{\mathbf{u}\dot{\mathbf{q}}} = \begin{bmatrix} 0 & \dots & 0 \\ 0 & \dots & 1 \end{bmatrix}, \tag{201}$$

$$\mathbf{L}_1^{\mathbf{u}\ddot{\mathbf{q}}} = \mathbf{0}, \qquad \mathbf{L}_1^{\mathbf{u}\boldsymbol{\lambda}} = \mathbf{0}, \qquad \mathbf{L}_1^{\mathbf{uy}} = \mathbf{0}. \tag{202}$$

The actuator localization matrix is defined by

$$\mathbf{L}^a = \begin{bmatrix} 0 & \dots & 1 \end{bmatrix}^T \tag{203}$$

and the thirteenth dynamic equation associated with the generalized coordinate α yields

$$-[\boldsymbol{\lambda}_3]_3 = \tau \tag{204}$$

which means that the internal torque in the direction of the hinge axis equilibrates the motor torque.

7 Time integration of coupled DAEs

In this section, numerical methods to solve the equations of motion of the system are discussed. Time integration methods can be classified in three families: algorithms for first-order ODEs (multistep and Runge-Kutta methods), algorithms for second-order ODEs (methods of the Newmark family) and structure-preserving methods (energy preserving schemes and variational integrators). In this chapter, we focus on algorithms of the Newmark family, see Newmark (1959), the HHT method by Hilber et al. (1977) and the generalized-α method by Chung and Hulbert (1993).

These methods have several well-known advantages for the simulation of flexible multibody systems represented by stiff differential-algebraic equations. In particular, the generalized-α method combines the advantages of one-step implementation, second-order accuracy, unconditional stability (A-stability) and adjustable high-frequency numerical damping. Numerical damping is especially important to eliminate spurious high-frequency effects caused by the higher-order modes of a finite element model. Erlicher et al. (2002) have studied the generalized-α scheme in the nonlinear regime, Cardona and Géradin (1989) have proposed to use this algorithm for constrained problems in flexible multibody dynamics and they have shown the importance of numerical damping in this context, and Arnold and Brüls (2007) have achieved a full convergence analysis in the DAE case. An extension of the generalized-α method to mechatronic problems was analyzed in Brüls and Golinval (2008) and in Brüls and Arnold (2008).

Let us describe the algorithm proposed in this last reference. In this formulation, the dynamic equilibrium represented by Equations (186–189) is enforced exactly at each time step. Hence, the algorithm does not exploit a particular structure of those equations and, unlike the classical generalized-α method, it can be used even though the mass matrix is not constant and the control system includes nonlinear acceleration feedbacks.

At time step $n+1$, the numerical variables $\mathbf{q}_{n+1}$, $\dot{\mathbf{q}}_{n+1}$, $\ddot{\mathbf{q}}_{n+1}$, $\boldsymbol{\lambda}_{n+1}$, $\mathbf{x}_{n+1}$, $\dot{\mathbf{x}}_{n+1}$ and $\mathbf{y}_{n+1}$ satisfy the coupled Equations (186–189), whereas the vector $\mathbf{a}$ of acceleration-like variables is defined by the recurrence relation

$$(1-\alpha_m)\mathbf{a}_{n+1} + \alpha_m \mathbf{a}_n = (1-\alpha_f)\ddot{\mathbf{q}}_{n+1} + \alpha_f \ddot{\mathbf{q}}_n, \qquad \mathbf{a}_0 = \ddot{\mathbf{q}}_0. \tag{205}$$

We emphasize that $\mathbf{a}$ is an auxiliary variable, which is not equal to the true accelerations $\ddot{\mathbf{q}}$. Since $\mathbf{M}$ depends on $\mathbf{q}$, this equation cannot be restated as a weighted form of the mechanical dynamic equilibrium in Equation (186). The treatment of the state variables $\mathbf{x}$ and their derivative $\dot{\mathbf{x}}$ is similar to the treatment of the velocities $\dot{\mathbf{q}}$ and their derivative $\ddot{\mathbf{q}}$. Therefore, the auxiliary variable $\mathbf{w}$ is defined by

$$(1-\delta_m)\mathbf{w}_{n+1} + \delta_m \mathbf{w}_n = (1-\delta_f)\dot{\mathbf{x}}_{n+1} + \delta_f \dot{\mathbf{x}}_n, \qquad \mathbf{w}_0 = \dot{\mathbf{x}}_0. \tag{206}$$

The generalized-α scheme is obtained using $\mathbf{a}$ in the Newmark integration formulae

$$\mathbf{q}_{n+1} = \mathbf{q}_n + h\dot{\mathbf{q}}_n + h^2(0.5-\beta)\mathbf{a}_n + h^2\beta\mathbf{a}_{n+1}, \tag{207}$$

$$\dot{\mathbf{q}}_{n+1} = \dot{\mathbf{q}}_n + h(1-\gamma)\mathbf{a}_n + h\gamma\mathbf{a}_{n+1} \tag{208}$$

and $\mathbf{w}$ in the state integration formula

$$\mathbf{x}_{n+1} = \mathbf{x}_n + h(1-\theta)\mathbf{w}_n + h\theta\mathbf{w}_{n+1}. \tag{209}$$

Seven algorithmic parameters α_m, α_f, β, γ, δ_m, δ_f and θ should thus be selected in order to obtain suitable accuracy and numerical stability properties. One option is to define α_m, α_f, β and γ according to the classical generalized-α method and to impose $\delta_m = \alpha_m$, $\delta_f = \alpha_f$ and $\gamma = \theta$. However, we will show that different integration formulae for the state variables are required to achieve an optimal compromise between accuracy and stability. Therefore, in order to allow $\delta_f \neq \alpha_f$, $\delta_m \neq \alpha_m$ and $\theta \neq \gamma$, it is interesting to treat all algorithmic parameters as independent variables.

An algorithm to solve Equations (205–209) together with the dynamic equilibrium at each time step t_{n+1} is described in Figure 14. In the Newton iterations, the correction step involves the parameters

$$\beta' = \frac{1-\alpha_m}{h^2\beta(1-\alpha_f)}, \qquad \gamma' = \frac{\gamma}{h\beta}, \qquad \theta' = \frac{1-\delta_m}{h\theta(1-\delta_f)}$$

which satisfy the properties

$$\frac{\partial \ddot{\mathbf{q}}_{n+1}}{\partial \mathbf{q}_{n+1}} = \mathbf{I}\beta', \qquad \frac{\partial \dot{\mathbf{q}}_{n+1}}{\partial \mathbf{q}_{n+1}} = \mathbf{I}\gamma', \qquad \frac{\partial \dot{\mathbf{x}}_{n+1}}{\partial \mathbf{x}_{n+1}} = \mathbf{I}\theta'.$$

The iteration matrix is given by

$$\mathbf{S}_t = \left[\begin{array}{cc|cc} (\mathbf{M}\beta' + \mathbf{g}^{gyr}_{\dot{\mathbf{q}}}\gamma' + \mathbf{K}_t) & \mathbf{\Phi}^T_{\mathbf{q}} & \mathbf{0} & -\mathbf{L}^a \\ \mathbf{\Phi}_{\mathbf{q}} & \mathbf{0} & \mathbf{0} & \mathbf{0} \\ \hline (-\mathbf{f}_{\ddot{\mathbf{q}}}\beta' - \mathbf{f}_{\dot{\mathbf{q}}}\gamma' - \mathbf{f}_{\mathbf{q}}) & -\mathbf{f}_{\boldsymbol{\lambda}} & (\mathbf{I}\theta' - \mathbf{f}_{\mathbf{x}}) & -\mathbf{f}_{\mathbf{y}} \\ (-\mathbf{h}_{\ddot{\mathbf{q}}}\beta' - \mathbf{h}_{\dot{\mathbf{q}}}\gamma' - \mathbf{h}_{\mathbf{q}}) & -\mathbf{h}_{\boldsymbol{\lambda}} & -\mathbf{h}_{\mathbf{x}} & (\mathbf{I} - \mathbf{h}_{\mathbf{y}}) \end{array}\right]$$

Initial values $\mathbf{q}_0$, $\dot{\mathbf{q}}_0$, $\ddot{\mathbf{q}}_0$, $\mathbf{x}_0$ and $\dot{\mathbf{x}}_0$
$\mathbf{a}_0 := \ddot{\mathbf{q}}_0$
$\mathbf{w}_0 := \dot{\mathbf{x}}_0$
for $n = 0$ to $n_{final} - 1$ **do**
  $\mathbf{a}_{n+1} := 1/(1-\alpha_m)(\alpha_f \ddot{\mathbf{q}}_n - \alpha_m \mathbf{a}_n)$
  $\mathbf{q}_{n+1} := \mathbf{q}_n + h\dot{\mathbf{q}}_n + h^2(0.5-\beta)\mathbf{a}_n + h^2\beta\mathbf{a}_{n+1}$
  $\dot{\mathbf{q}}_{n+1} := \dot{\mathbf{q}}_n + h(1-\gamma)\mathbf{a}_n + h\gamma\mathbf{a}_{n+1}$
  $\ddot{\mathbf{q}}_{n+1} := \mathbf{0}$
  $\boldsymbol{\lambda}_{n+1} := \mathbf{0}$
  $\mathbf{w}_{n+1} := 1/(1-\delta_m)(\delta_f \dot{\mathbf{x}}_n - \delta_m \mathbf{w}_n)$
  $\mathbf{x}_{n+1} := \mathbf{x}_n + h(1-\theta)\mathbf{w}_n + h\theta\mathbf{w}_{n+1}$
  $\dot{\mathbf{x}}_{n+1} := \mathbf{0}$
  $\mathbf{y}_{n+1} := \mathbf{0}$
  for $i = 1$ to i_{max} **do**
    Compute the residuals $\mathbf{r}^{\mathbf{q}}$, $\mathbf{r}^{\boldsymbol{\lambda}}$, $\mathbf{r}^{\mathbf{x}}$, $\mathbf{r}^{\mathbf{y}}$ using Equations (186-189)
    if $\sqrt{\|\mathbf{r}^{\mathbf{q}}\|^2 + \|\mathbf{r}^{\boldsymbol{\lambda}}\|^2 + \|\mathbf{r}^{\mathbf{x}}\|^2 + \|\mathbf{r}^{\mathbf{y}}\|^2} < tol$ **then**
      break
    end if

$$\left[\begin{array}{c} \Delta\mathbf{q} \\ \Delta\boldsymbol{\lambda} \\ \hline \Delta\mathbf{x} \\ \Delta\mathbf{y} \end{array}\right] := -\mathbf{S}_t^{-1} \left[\begin{array}{c} \mathbf{r}^{\mathbf{q}} \\ \mathbf{r}^{\boldsymbol{\lambda}} \\ \hline \mathbf{r}^{\mathbf{x}} \\ \mathbf{r}^{\mathbf{y}} \end{array}\right]$$

    $\mathbf{q}_{n+1} := \mathbf{q}_{n+1} + \Delta\mathbf{q}$
    $\dot{\mathbf{q}}_{n+1} := \dot{\mathbf{q}}_{n+1} + \gamma'\Delta\mathbf{q}$
    $\ddot{\mathbf{q}}_{n+1} := \ddot{\mathbf{q}}_{n+1} + \beta'\Delta\mathbf{q}$
    $\boldsymbol{\lambda}_{n+1} := \boldsymbol{\lambda}_{n+1} + \Delta\boldsymbol{\lambda}$
    $\mathbf{x}_{n+1} := \mathbf{x}_{n+1} + \Delta\mathbf{x}$
    $\dot{\mathbf{x}}_{n+1} := \dot{\mathbf{x}}_{n+1} + \theta'\Delta\mathbf{x}$
    $\mathbf{y}_{n+1} := \mathbf{y}_{n+1} + \Delta\mathbf{y}$
  end for
  $\mathbf{a}_{n+1} := \mathbf{a}_{n+1} + (1-\alpha_f)/(1-\alpha_m)\ddot{\mathbf{q}}_{n+1}$
  $\mathbf{w}_{n+1} := \mathbf{w}_{n+1} + (1-\delta_f)/(1-\delta_m)\dot{\mathbf{x}}_{n+1}$
end for

Figure 14. Generalized-α time integration scheme for mechatronic systems.

with the tangent stiffness matrix $\mathbf{K}_t = \partial(\mathbf{M}\ddot{\mathbf{q}}+\mathbf{g}^{gyr}+\mathbf{g}^{int}+\mathbf{\Phi}_\mathbf{q}^T\boldsymbol{\lambda}-\mathbf{g}^{ext})/\partial\mathbf{q}$.

Let us briefly discuss the choice of the algorithmic parameters. For a mechanical system represented by second-order ODEs, the generalized-α algorithm is second-order accurate provided that (Chung and Hulbert (1993))

$$\gamma = 0.5 + \alpha_f - \alpha_m. \tag{210}$$

Likewise, for a control system represented by first-order ODEs, this condition becomes (Jansen et al. (2000))

$$\theta = 0.5 + \delta_f - \delta_m. \tag{211}$$

For stiff problems, the solution should be computed accurately only in the low-frequency range, whereas the high-frequency solution should rather be damped out by the algorithm. The high-frequency numerical damping is represented by the spectral radius of the algorithm at infinity ρ_∞: an undamped scheme is characterized by $\rho_\infty = 1$, whereas $\rho_\infty = 0$ means asymptotic annihilation of the high-frequency response.

In the case of uncoupled mechanical or state-space systems, optimal parameters can be defined from a frequency domain analysis. For a user-defined value of $\rho_\infty \in [0, 1]$, Chung and Hulbert (1993) have proposed optimal algorithmic parameters for second-order mechanical systems

$$\alpha_m = \frac{2\rho_\infty - 1}{\rho_\infty + 1}, \quad \alpha_f = \frac{\rho_\infty}{\rho_\infty + 1}, \quad \beta = 0.25\,(\gamma + 0.5)^2. \tag{212}$$

For first-order oscillatory state-space models, Jansen et al. (2000) obtained

$$\delta_m = \frac{1}{2}\left(\frac{3\rho_\infty - 1}{\rho_\infty + 1}\right), \quad \delta_f = \frac{\rho_\infty}{\rho_\infty + 1}. \tag{213}$$

The algorithms defined by Equations (210) and (212) or by Equations (211) and (213) are stable when $h \to 0$ (zero-stability) and when $h \to \infty$ (stability). A frequency domain analysis of the algorithm for fully coupled mechatronic problems is described in Brüls and Golinval (2008).

8 Applications

8.1 Landing of an aircraft

The landing gear of a modern airplane (split into main and nose landing gears) is a highly critical component of modern civil airplanes. It has four main functions: take-off (producing the largest rolling speed), landing (including impact on the runway), taxiing and breaking, and it undergoes dynamic loads during each phase. It must be retractable, an operation controlled by a powerful hydraulic system; soft landing implies the action of appropriate dampers; tyres have particular and nonlinear mechanical properties; the strong influence of structural flexibility exists during landing and take-off phases, etc. The deformation of the wings of the aircraft can be taken into account using superelements.

Figure 15. Nose landing gear of a large civil airplane at touch down.

Figure 15 displays in a few frames the behaviour of the nose landing gear of a large civil airplane during landing, immediately before and after impact. It mainly exhibits the strong deformation of tyres and the absorbing role of the articulated structure (work accomplished under the Brite-Euram project ELGAR).

8.2 Semi-active vehicle suspension

The second application is a car equipped with a semi-active suspension. This benchmark has been analyzed in the framework of the Belgian Inter-University Attraction Pole on Advanced Mechatronic Systems (AMS-IAP5/06)[2].

Using a semi-active actuator in parallel with a passive spring, a control action is obtained by varying the restriction in current controlled valves. The control law exploits information from accelerometers mounted on the four corners of the chassis (Figure 16), and from linear displacement sensors measuring the extension of the four shock-absorbers. The model of the mechatronic system is composed of a mechanical model, a model of the

[2]Website of the AMS-IAP5/06: http://www.mech.kuleuven.ac.be/pma/project/ams

Figure 16. Audi A6 - corner accelerometers.

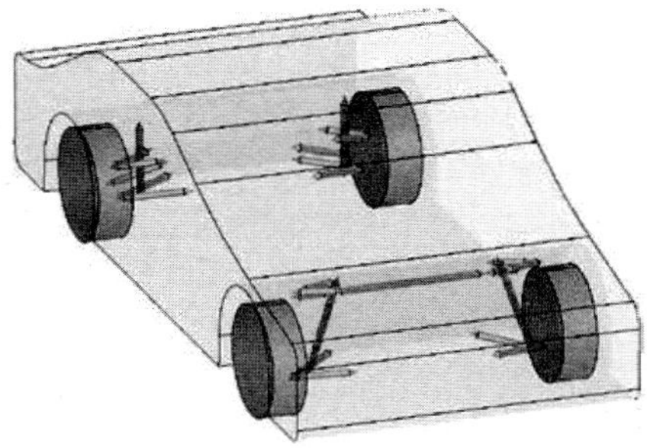

Figure 17. Mechanical model of the car.

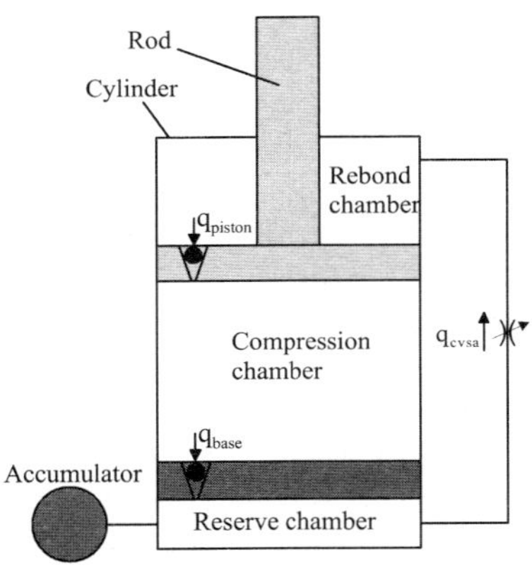

Figure 18. Semi-active damper.

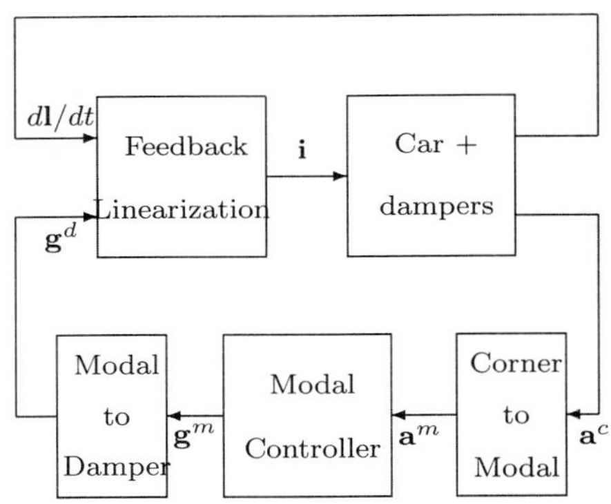

Figure 19. Semi-active control strategy.

actuators, and a model of the controller. A detailed description of the actuators and of the controller is presented by Lauwerys et al. (2004).

The rigid-body model of the car is illustrated in Figure 17, and it includes the car-body, the suspension mechanisms, the slider-crank direction mechanism for the front wheels, and a model for the wheel/ground contact (lateral force, vertical force and yaw torque). The mechanical model involves about 600 degrees-of-freedom, and it could be later extended to include the stiffness of the suspension bushings, the flexibility of the chassis, and a longitudinal model for the wheels.

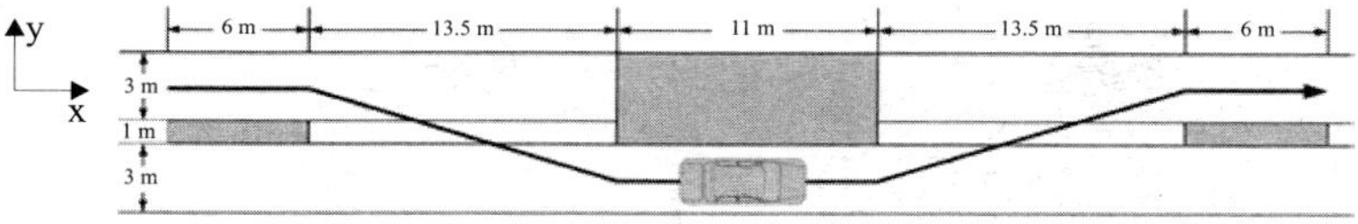

Figure 20. Lane change maneuver (standard qualification test).

Figure 18 illustrates the principle of a semi-active actuator. The actuator model is available in nonlinear state-space format:

$$\dot{\mathbf{x}}^{(damp)} = \mathbf{f}^{(damp)}(\mathbf{u}^{(damp)}, \mathbf{x}^{(damp)}),$$
$$\mathbf{y}^{(damp)} = \mathbf{h}^{(damp)}(\mathbf{u}^{(damp)}, \mathbf{x}^{(damp)}),$$

with the inputs, states and output:

$$\mathbf{u}^{(damp)} = [l \quad dl/dt \quad i]^T, \quad \mathbf{x}^{(damp)} = [p^{reb} \quad p^{comp}]^T, \quad \mathbf{y}^{(damp)} = [g^a].$$

l is the actuator extension; i, the electrical current in the valve; p^{reb} and p^{comp}, the pressures in the rebound and compression chambers; and g^a, the force exerted by the damper. The functions $\mathbf{f}^{(damp)}$ and $\mathbf{h}^{(damp)}$ are given by the manufacturer of the shock-absorber as C-functions.

The control law, available as a block diagram model, consists of three stages (Figure 19): a feedback linearization (inverse actuator model), a transformation of the actuator forces and of the acceleration measurements into modal space (heave, roll, and pitch), and a linear integral control. In the figure, the vector $\mathbf{a}^c$ denotes the accelerations measured at the four corners of the car-body; $\mathbf{a}^m$, the modal accelerations; $\mathbf{g}^m$, the desired modal forces; $\mathbf{g}^d$, the desired damper forces; $d\mathbf{l}/dt$, the extension rates of the dampers; and $\mathbf{i}$, the electrical currents.

A lane change maneuver has been simulated (Figure 20). The car has a 10 m/s initial velocity, and a driver applies an open-loop steering command, without any real-time correction (blind driver assumption). The motor and the brakes do not produce any torque on the wheels. The time-step of the simulation is 0.01 s and the algorithmic parameters are defined from the desired spectral radius $\rho_\infty = 0.9$. The simulation was carried out within a reasonable computational time (a few minutes on a desk computer).

Figure 21 illustrates the horizontal trajectory of the car. Due to the lateral sliding of the wheels and the open-loop nature of the driving command, a lateral drift is observed in the trajectory. Figure 22 presents the pitch and roll angles of the car-body. The dynamic behavior of the semi-active shock absorbers is analyzed in Figures 23 and 24. Saturation phenomena

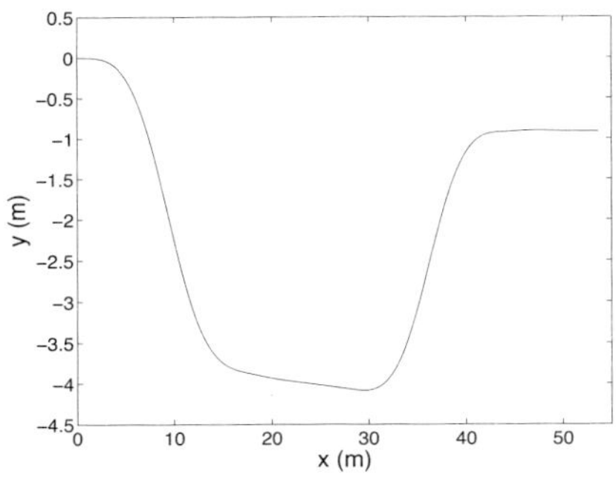

Figure 21. Horizontal trajectory of the car-body.

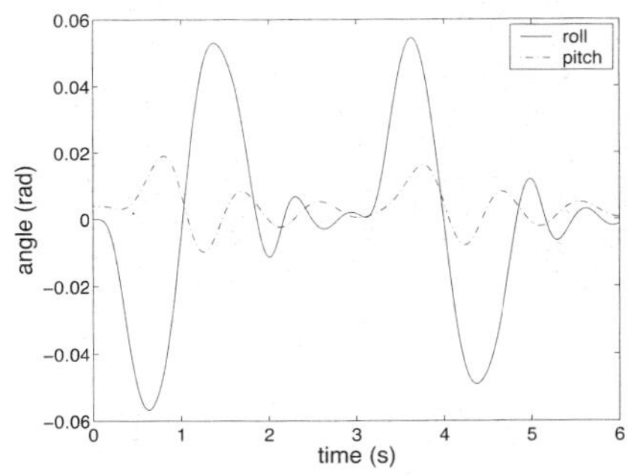

Figure 22. Angles of the car-body.

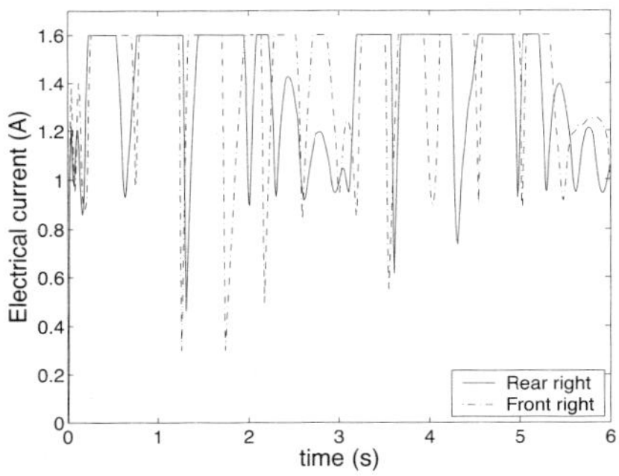

Figure 23. Electrical currents in the valves.

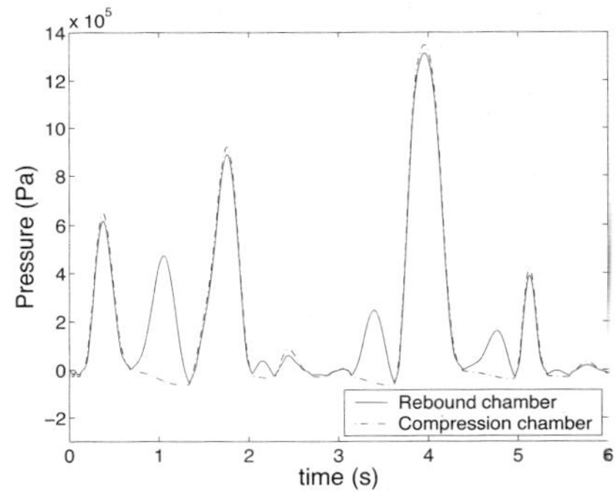

Figure 24. Hydraulic pressures in the rear right damper.

dominate the variations of the electrical currents. The pressure in the compression chamber is very low in the extension phase, but it is close to the pressure in the rebound chamber in the compression phase. All those results are physically consistent and can be of great use in a pre-prototyping design phase or to improve the control law.

8.3 Control of an experimental flexible manipulator

The second example concerns the motion and vibration control of an experimental robotic arm. The long-reach manipulator Ralf, shown in Figure 25, has been developed at the Georgia Institute of Technology, see Wil-

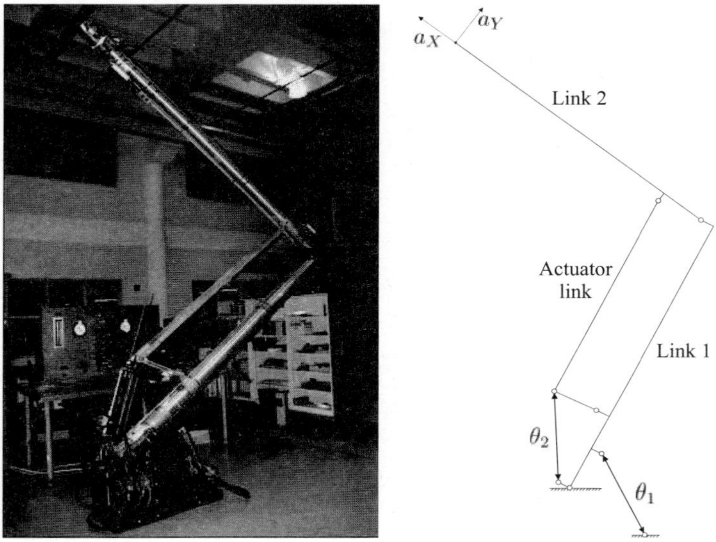

Figure 25. Ralf. θ_1 and θ_2 are the actuated degrees-of-freedom, whereas a_X and a_Y denote the measured accelerations.

son (1986); Huggins (1988). It has a high payload to weight ratio, and it is stiff enough to achieve real-world applications. However, flexible effects in the links affect the positioning accuracy. Ralf has two kinematic degrees-of-freedom in a vertical plane, and it is actuated by two hydraulic cylinders. Two linear position sensing transducers, fixed to the hydraulic cylinder, measure the cylinder extension. Moreover, in order to detect the vibrations of the mechanism, two accelerometers are placed at the tip, in orthogonal directions.

As shown in Figure 26, the control law is based on a composite two-time-scale strategy, which involves (i) a slow collocated controller, responsible for the trajectory tracking, (ii) a fast controller, which relies on both collocated measurements and tip acceleration measurements, for the active damping of the flexible modes. This vibration controller exploits a real-time model of the mechanism, whose formulation is explained in Brüls et al. (2007).

The simulation model of the mechatronic system includes a model of the flexible mechanisms (link 1, link 2 and the actuator link are modeled as flexible beams), a linear dynamic model of the actuators obtained by experimental identification, and a block diagram model for the control law (equivalent to the model implemented in the real-time controller).

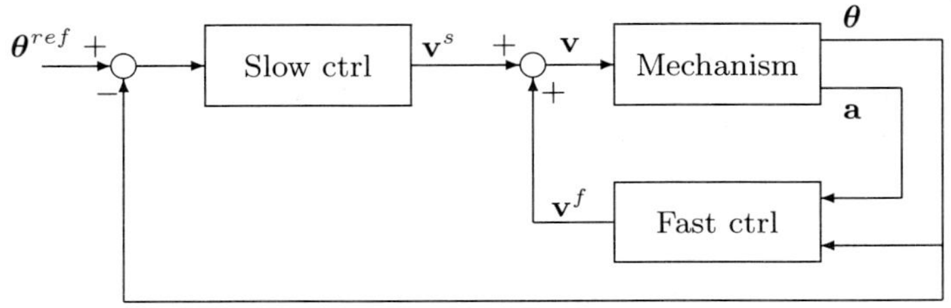

Figure 26. Two-time-scale control: **v** denotes the voltages applied to the hydraulic actuators.

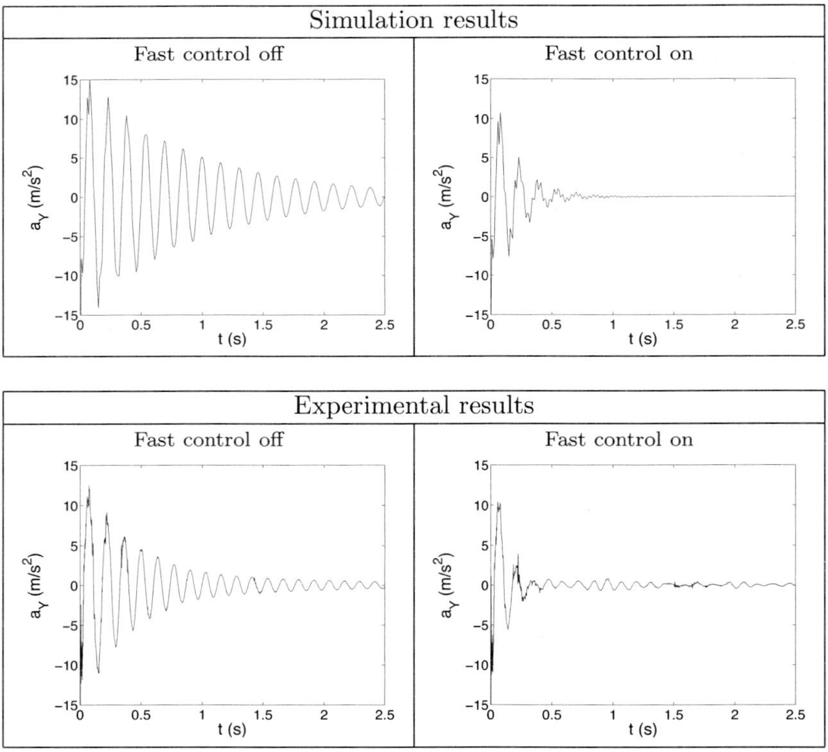

Figure 27. Tip response to a tip disturbance.

The step response to a tip vertical force is analyzed in Figure 27. The time-step of the simulation is 0.01 s, and the algorithmic parameters are obtained for the desired spectral radius $\rho_\infty = 0.7$. From these results, we conclude that the simulation tool predicts accurately the dynamic behaviour of the actual mechatronic system.

9 Summary and conclusion

In this chapter, three approaches are first reviewed for the dynamic analysis of flexible multibody system, i.e. the inertial frame, the corotational frame and the floating frame of reference approaches. Each of them has its own merits and drawbacks; we believe that a suitable simulation environment should allow a combination of those methods.

A general modelling framework based on the finite element concept is then described in detail. We have shown that absolute nodal coordinates allow a systematic assembly of various types of elements: rigid bodies, kinematic joints, bars, beams, superelements, etc. It is also possible to account for components of a control system using a block diagram formulation.

The finite element model may involve a rather large set of nodal coordinates. However, the resulting equations of motion are usually sparse and can be solved efficiently using a sparse solver. The superelement technique also allows to improve the efficiency by reducing the size of the model. This technique, which is based on a corotational formulation of a reduced-order model, is applicable for structural components with large displacements but small deformations.

The nonlinear equations of motion are derived using the numerical finite element assembly procedure. The couplings between motion, deformation and controller dynamics are thus represented in a fully consistent way. A monolithic generalized-α time integration scheme is described for the strongly coupled simulation of flexible multibody systems and mechatronic systems.

Finally, three multibody and mechatronic systems are analyzed: a landing gear, a semi-active car suspension and a flexible manipulator. Those examples demonstrate the ability of the proposed method to deal with industrial problems.

Acknowledgements

Olivier Brüls is supported by the Belgian National Fund for Scientific Research (FNRS) which is gratefully acknowledged. The authors thank Prof. J.C. Golinval and Prof. P. Duysinx from the University of Liège for their support in the preparation of this work. The research groups of Prof. P. Fisette (Université Catholique de Louvain, Belgium) and Prof. J. Swevers (Katholieke Universiteit Leuven, Belgium) have participated in the semi-active car suspension benchmark; they are gratefully acknowledged. We also thank Prof. W.J. Book and his team (Georgia Institute of Technology, U.S.) for making the Ralf setup available and for contributing to the design and implementation of its vibration controller.

Bibliography

M. Arnold and O. Brüls. Convergence of the generalized-α scheme for constrained mechanical systems. *Multibody System Dynamics*, 18(2):185–202, 2007.

Z. Bai and Y. Su. Dimension reduction of large-scale second-order dynamical systems via a second-order Arnoldi method. *SIAM Journal on Scientific Computing*, 26(5):1692–1709, 2005.

O. Bauchau and L. Trainelli. The vectorial parameterization of rotation. *Nonlinear Dynamics*, 32:71–92, 2003.

T. Belytschko and J. Hsieh. Non-linear transient finite element analysis with convected co-ordinates. *International Journal for Numerical Methods in Engineering*, 7:255–271, 1973.

P. Betsch and P. Steinmann. Constrained integration of rigid body dynamics. *Computer Methods in Applied Mechanics and Engineering*, 191: 467–488, 2001.

O. Brüls and M. Arnold. The generalized-α scheme as a linear multistep integrator: Towards a general mechatronic simulator. *ASME Journal of Computational and Nonlinear Dynamics*, 13 pages, accepted, 2008.

O. Brüls and J.-C. Golinval. On the numerical damping of time integrators for coupled mechatronic systems. *Computer Methods in Applied Mechanics and Engineering*, 197(6-8):577–588, 2008.

O. Brüls, P. Duysinx, and J.-C. Golinval. The global modal parameterization for nonlinear model-order reduction in flexible multibody dynamics. *International Journal for Numerical Methods in Engineering*, 69(5):948–977, 2007.

A. Cardona and M. Géradin. A beam finite element non-linear theory with finite rotations. *International Journal for Numerical Methods in Engineering*, 26:2403–2438, 1988.

A. Cardona and M. Géradin. Time integration of the equations of motion in mechanism analysis. *Computers and Structures*, 33:801–820, 1989.

A. Cardona and M. Géradin. A superelement formulation for mechanism analysis. *International Journal for Numerical Methods in Engineering*, 32(8):1565–1594, 1991.

J. Chung and G.M. Hulbert. A time integration algorithm for structural dynamics with improved numerical dissipation: The generalized-α method. *ASME Journal of Applied Mechanics*, 60:371–375, 1993.

R. Craig and M. Bampton. Coupling of substructures for dynamic analysis. *AIAA Journal*, 6(7):1313–1319, 1968.

M.A. Crisfield and G. Jelenic. Objectivity of strain measures in the geometrically exact three-dimensional beam theory and its finite-element implementation. *Proceedings of the Royal Society of London A*, 455: 1125–1147, 1999.

M.A. Crisfield, U. Galvanetto, and G. Jelenic. Dynamics of 3-D corotational beams. *Computational Mechanics*, 20:507–519, 1997.

S. Erlicher, L. Bonaventura, and O.S. Bursi. The analysis of the generalized-α method for non-linear dynamic problems. *Computational Mechanics*, 28:83–104, 2002.

B. Fraeijs de Veubeke. The dynamics of flexible bodies. *International Journal of Engineering Science*, 14:895–913, 1976.

M. Géradin and A. Cardona. *Flexible Multibody Dynamics: A Finite Element Approach*. John Wiley & Sons, New York, 2001.

K.D. Glover. All optimal Hankel-norm approximation of linear multivariable systems and their L^∞-error bounds. *International Journal of Control*, 39(6):1115-1193, 1984.

E.J. Grimme. *Krylov projection methods for model reduction*. PhD thesis, ECE Deparment, Univ. Of Illinois, Urbana-Champaign, 1997.

H. Hilber, T. Hughes, and R. Taylor. Improved numerical dissipation for time integration algorithms in structural dynamics. *Earthquake Engineering and Structural Dynamics*, 5:283–292, 1977.

J.D. Huggins. Experimental verification of a model of a two-link flexible, lightweight manipulator. Master's thesis, Georgia Institute of Technology, 1988.

K.E. Jansen, C.H. Whiting, and G.M. Hulbert. A generalized-α method for integrating the filtered Navier-Stokes equations with a stabilized finite element method. *Computer Methods in Applied Mechanics and Engineering*, 190:305–319, 2000.

H.-J. Lai, E.J. Haug, S.S. Kim, and D.-S. Bae. A decoupled flexible-relative co-ordinate recursive approach for flexible multibody dynamics. *International Journal for Numerical Methods in Engineering*, 32:1669–1690, 1991.

C. Lauwerys, J. Swevers, and P. Sas. Model free design for a semi-active suspension of a passenger car. In *Proc. of ISMA 2004*, Leuven, Belgium, September 2004.

E.V. Lens and A. Cardona. A nonlinear beam element formulation in the framework of an energy preserving time integration scheme for constrained multibody systems dynamics. *Computers and Structures*, 86: 47–63, 2008.

R. Mac Neal. A hybrid method of component mode synthesis. *Computers and Structures*, 1:581–601, 1971.

L. Meirovitch and M.-K. Kwak. Rayleigh-Ritz based substructure synthesis for flexible multibody dynamics. *AIAA Journal*, 28:1709–1719, 1991.

D.G. Meyer and S. Srinivasan. Balancing and model reduction for second-order form linear systems. *IEEE Transactions on Automatic Control*, 41 (11):1632–1644, 1996.

B.C. Moore. Principal component analysis in linear systems: Controllability, observability and model reduction. *IEEE Transactions on Automatic Control*, 26:17–32, 1981.

N. Newmark. A method of computation for structural dynamics. *ASCE Journal of the Engineering Mechanics Division*, 85:67–94, 1959.

S. Rubin. Improved component mode representation for structural dynamics analysis. *AIAA Journal*, 13(8):995–1006, 1975.

A.A. Shabana and R.A. Wehage. A coordinate reduction technique for dynamic analysis of spatial substructures with large angular rotations. *Journal of Structural Mechanics*, 11:401–431, 1983.

J.C. Simo. A finite strain beam formulation. The three-dimensional dynamic problem. Part I. *Computer Methods in Applied Mechanics and Engineering*, 49:55–70, 1985.

J.C. Simo and L. Vu-Quoc. On the dynamics in space of rods undergoing large motions - a geometrically exact approach. *Computer Methods in Applied Mechanics and Engineering*, 66:125–161, 1988.

T.J. Su and R.J. Craig. Model reduction and control of flexible structures using Krylov vectors. *AIAA Journal of Guidance, Control, and Dynamics*, 14(2):260–267, 1991.

O. Wallrapp and R. Schwertassek. Representation of geometric stiffening in multibody system simulation. *International Journal for Numerical Methods in Engineering*, 32(8):1833-1850, 1991.

T.M. Wasfy and A.K. Noor. Computational strategies for flexible multibody systems. *Applied Mechanics Review*, 56(6):553–613, 2003.

T.R. Wilson. The design and construction of flexible manipulators. Master's thesis, Georgia Institute of Technology, 1986.

Modeling, simulation and control of mechatronical systems

Michael Valasek

Department of Mechanics, Biomechanics and Mechatronics,
Faculty of Mechanical Engineering,
Czech Technical University in Prague, Prague, Czech Republic

Abstract The mechatronical systems become the basis of new products. Their design requires the development of multiphysical simulation models and using them the model based control design. Both these areas are in rapid development. This contribution presents the following material: the concept of mechatronical system, the methods for development of simulation models of multiphysical systems, the methods of model based control design and examples of mechatronical applications in vehicles.

1 Introduction

The forecast (Tomizuka, 2000) for the 21st century argues that mechatronics will be one of the 5 most influencing engineering topics. Mechatronics can be treated as the future of at least mechanical engineering. However, the design of mechatronical systems is not an easy task. It is heavily based on modeling and simulation. The capability to develop valid and efficient simulation models in reasonable development time is essential for successful development of mechatronical products. Certainly the control represents the kernel of the mechatronical system. Recently two different approaches towards control of mechatronical systems occur. One approach can be called model based control and the other modelless control. The first approach is developing the control using the explicit knowledge of the model of the controlled system while the second one avoids any necessity of knowledge of the model of the controlled system. The pure version of the latter one is a topic of on-going research while the first one is ready for practical applications.

Vehicles are currently the most mechatronical products. They represent the ideal example for demonstration of the application of described approaches. Moreover the described approaches have been developed specifically based on the inspiration and demands of automotive engineering.

Therefore this contribution covers three main topics. The first topic is the efficient development of simulation models of mechatronical systems. The second topic is the model based control design of mechatronical systems. The third topic is a demonstration of application of the described approaches towards the development of integrated chassis control and the control of combustion engines of vehicles.

2 Mechatronical system

Mechatronics is one of the largest changes (revolutions) that have taken place in engineering at the end of the last century. It has undergone the evolution (Hewit, 1992) from integration of the mechanisms with electronics and the robot and NC machine-tools control, through more general joining of the mechanics (machines) and the electronics, as well as the control of many new high-tech products (faxes, video cameras, cars, aircrafts, etc.), up to the very general concept of synergistic combination of various technologies in a single product or a new design methodology (Figure 1). In the 20th century, the mechatronics has been qualified as the synergistic integration of the mechanics (machines) with the electronics and the intelligent computer control with the emphasis on the design of intelligent machines. In the 21st century, the mechatronics as the synergistic integration of the physical systems with the information technologies and complex decision processes is considered (Tomizuka, 2000) to have one of the greatest influences on the development on the engineering technology, i.e. machines and constructions. The mechatronics is expected to contribute deeply to the creation of new products and technical systems, which will outdo existing limitations of properties achievable by today's traditional products, and thus will be the motor of the wealth and the economy increase, of the serious technical problems' solution of the world but also just of better satisfying the human needs. The intelligent structures (Guran and Tzou, 1998; Janocha, 1999), intelligent materials (Kelly and Zweben, 2000; Wallace, 2003), mechatronic components (Neugebauer et al., 2007) are the natural parts of mechatronics. They not only possess the fundamental characteristics of mechatronics connecting the matter, sensors, and decision-making but above all they introduce the essential features of intelligence and smartness in the mechatronics. There are numbers of other designations like adaptronics (Janocha, 1999), structronics (Guran and Tzou, 1998), and so on.

Anybody who becomes familiar with the mechatronics acquires an impression that now anything is possible in technology. The sensors and drives (actuators) can be placed anywhere and they can be used to control things so that anything that is possible in the virtual world of human imagination

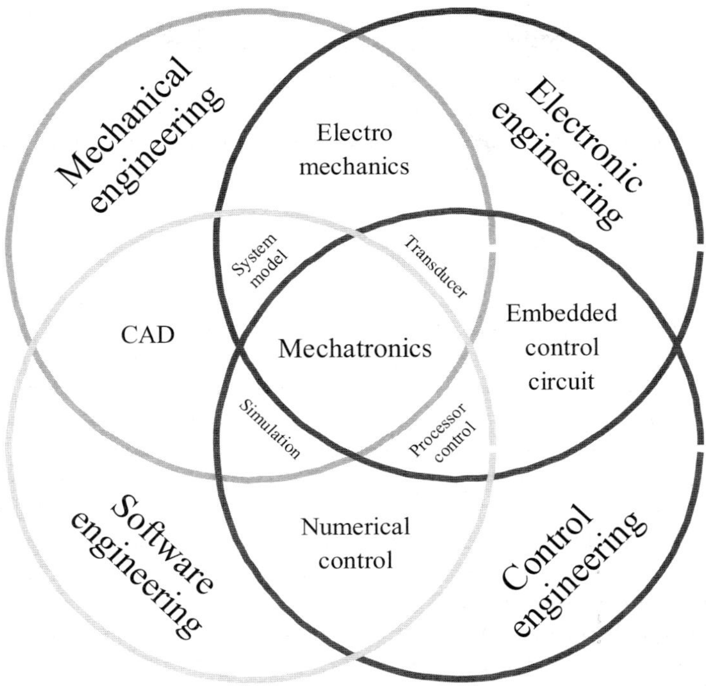

Figure 1. Mechatronics as synergistic combination of technologies.

and in the computer simulation models after it is possible to implement in the real material world. It is judged of the physical limitations fall down when the mechatronics is taken into account; that the products are limited only by the price and the market demand.

However, after a deeper examination it comes out that it's not true. In individual cases it is found out that the sensors are not suitable, that the actuators are missing or not suitable, that there is no support (fixed part of the frame) for the actuator to put forth a force according to the Newton's law of action and reaction, that the mechanical system under consideration is uncontrollable, etc. If the structure of the mechatronical system is described (see Figure 2), it is possible to formulate such problem in a way that the information or energy or force flow cannot be closed to realize the desired function. Analyzing the successful mechatronic applications we find out they have mostly been accomplished by a proper modification of the traditional physical objects, machines, structures to realize and close the mechatronic system structure. Therefore the basis of the origination of the smart structures, intelligent materials, mechatronic components, mechatronic applications, artefacts with intelligent behaviour generally is the inventive modification of the existing object, physical system so that the mechatronic structure is possible to be created. Thus the mechatronics has become a new source of inventions and design. A large amount of the solutions, new patents, new machine and constructions innovations is based on the creation of an inventive modification of the existing physical system to make the intelligent behaviour possible. This is the reason for

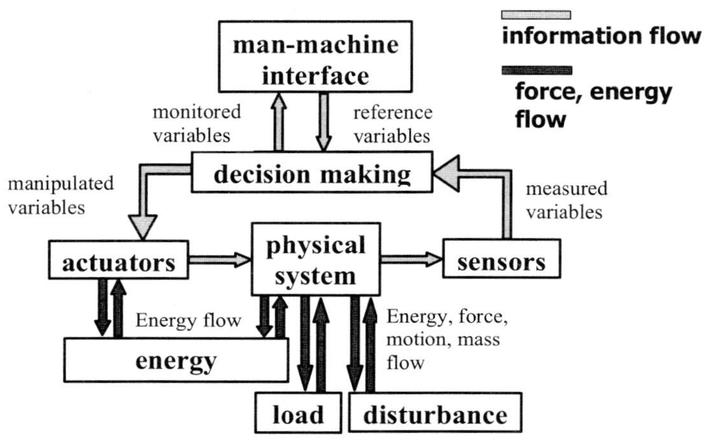

Figure 2. Structure of a mechatronic system.

the investigation of the principles of the smart structures, intelligent materials, intelligent behaviour, and mechatronic components as an invention for the modifications of the physical systems of machines, constructions, and general systems to allow the mechatronics application being the core subject of the proposed research project. The design of the new mechatronic products using these new principles requires the utilization of new advanced design procedures. Finally, the formulation and the aiming of the necessary principles research cannot be done without an outlook and an inspiration of the application of these new mechatronic solutions principles and integrated (concurrent) design procedures for the new generations of the machine concepts, constructions for the implementation of a function on higher qualitative level of parameters and for wider application range and the feedback for fundamental research of the new principles following from it.

According to the Figure 2 the mechatronical system is a multiphysical system as it consists of subsystems of different physical nature, e.g. mechanical, hydraulic, pneumatic, aerodynamic, electrical, magnetic, chemical, electronical, control etc. Because of feedbacks the mechatronical system must be designed using the virtual prototyping. The other reason for great importance of modelling and simulation is that mechatronical systems are not designed based on long-term previous experience. This means that the system is modelled as multiphysical system, analysed, modified, optimised and only after that physically designed and manufactured. The capability to develop simulation models of mechatronical systems is the basis of design of mechatronical systems and it is further investigated in details.

There is still no complete design methodology for mechatronic systems, only the experience from successful design solutions (e.g. (Kortüm et al., 1998)) and initial generalized mechatronical design methodology (Valasek, 1998), although a standard has been already formulated (VDI 2206, 2003). The key part of all existing approaches towards design of mechatronic systems is the modelling and simulation. Therefore again the essential problem is the capability to assemble the simulation model of the complete mechatronic system.

Vehicle industry is one of the industries where the application of mechatronic solution concepts has become the present design approach. The vehicles of today, both road and railway, include combination of components of mechanical, electrical, fluid-power, control and digital computing components. The advantages of such solutions as ABS, electronic fuel injection and ignition, electronically controlled adaptive suspension or tilting train system are the examples of fulfillment of the mechatronics definition. The present vehicles clearly demostrate the combination of components, solutions and technologies from the intersecting circles of mechanical, electrical, control and software engineering.

3 Modeling of multiphysical systems

In previous section it has been concluded that modelling is a key part of existing approaches of design methodology of mechatronic systems. The main problem of this section is to describe the ways how to assemble the simulation model of mechatronical system as a multiphysical system. It is supposed that the simulation model of particular subsystems of one physical nature can be assembled by the techniques of that particular physical branch. The problem is how to interconnect such submodels into one resulting multiphysical (multidisciplinary) model. Several approaches towards assembly of model of a (multiphysical, multidisciplinary) mechatronical system exist (Karnopp et al., 1990; Cellier, 1991; Veitl et al., 1999; Valasek et al., 1999; Valasek, 1994; Arnold et al., 2004). Two different classes of approaches can be distinguished (Figure 3) (Valasek et al., 1999). The first one is the co-simulation and the second one can be called uniform modeling. The co-simulation is based on interfacing the resulted models in different physical domains and the uniform modeling is based on the modeling of the whole multidisciplinary system within one uniform language.

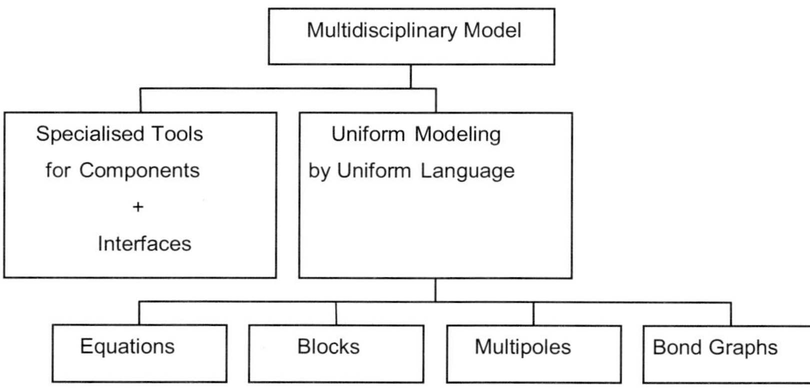

Figure 3. Approaches to Multidisciplinary Modeling.

3.1 Co-simulation

The first approach is the usage of tools for specialized (component) modelling in one particular physical domains and then interfacing the resulting models (Veitl et al., 1999). The output variables from one model are the input variables of another model and vice versa. This exchange of variables imitates the real interaction between two models of subsystems (Figure 4).

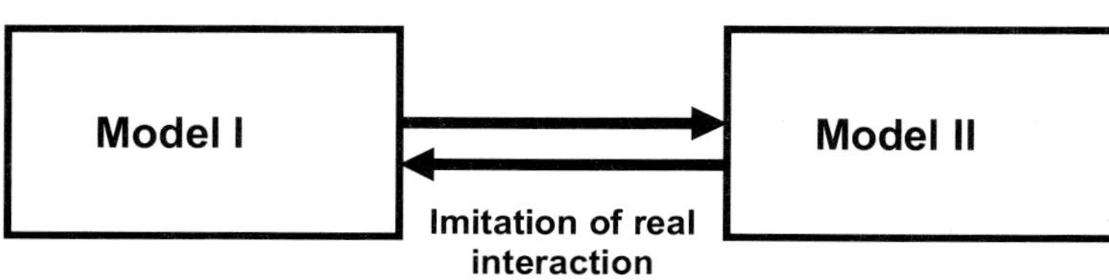

Figure 4. Principle of co-simulation.

Let us describe a simple example of dynamic interaction of two masses from Figure 5 (Valasek et al., 2006a). The interaction of two masses m_1 and m_2 by the stiffness k_1, k_2 and damping elements b_1, b_2 is described by simple simultaneous equations

$$\begin{aligned} m_2\ddot{z}_2 &= k_2(z_1 - z_2) + b_2(\dot{z}_1 - \dot{z}_2) \\ m_1\ddot{z}_1 &= k_2(z_2 - z_1) + b_2(\dot{z}_2 - \dot{z}_1) - k_1 z_1 - b_1\dot{z}_1 \end{aligned}$$

These equations can be split into two groups naturally concentrated around the masses m_1 and m_2

$$\begin{aligned} m_2\ddot{z}_2 &= -F_2 \\ F_2 &= k_2(z_2 - z_1) + b_2(\dot{z}_2 - \dot{z}_1) \\ m_1\ddot{z}_1 &= F_2 - k_1 z_1 - b_1\dot{z}_1 \end{aligned}$$

The model around the mass m_2 receives the values of variables z_1, $\dot{z}_1$ from the model around the mass m_1, computes the force F_2 and the right-hand side of the equation of motion of mass m_2 and transmits the value of the force F_2 to the model around the mass m_1. The model around the mass m_1 receives the value of the force F_2, computes the right-hand side of the equation of motion of mass m_1 and transmits the values of variables z_1, $\dot{z}_1$ to the model around the mass m_2. The equations of motion can be integrated by one common integrator (solver of differential equations) or separately by two integrators associated with the model around the mass m_1 and around the mass m_2. This example was used as a testing benchmark of co-simulation environment between Matlab/Simulink and in-house developed FEA tool in C-code (Figure 5). This example opens a lot of possibilities how to arrange the co-simulation.

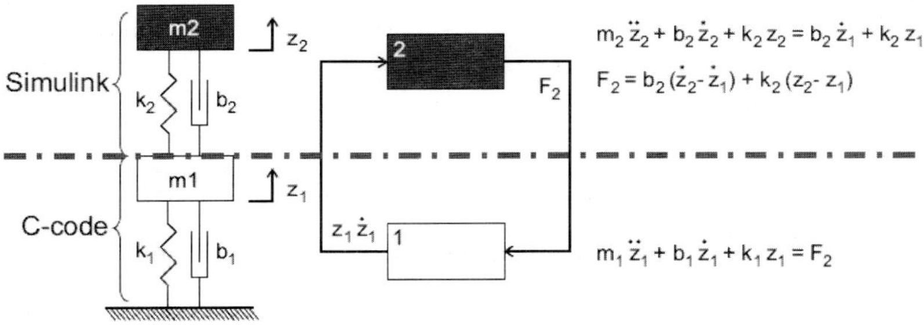

Figure 5. Simple example of co-simulation of two models.

This is a very promising approach for multiphysical (multidisciplinary) modeling and simulation because the methodologies and corresponding software tools for modeling and simulation the systems from one physical domain are very well established, immediately available and very well tested. The advantages are that the models are available in specialized modeling tools with specialized numerical solvers developed using specialist/expert knowledge often into large model complexity. The disadvantage of this approach is that many details of splitting the model and the solver are pragmatic without elaborated theoretical knowledge. The disadvantages are that development of interfaces is necessary leading to complex simulation environment and coping with different modelling techniques. Another disadvantage is that it is to great extent limited to the design by simulation. Recently great progress has been achieved in methodologies for interfacing of particular software tools. There are many possible methods for interfacing of software tools for mechatronic system simulation.

The coupling of models can be realized on different levels. Their classification uses the general consideration of coupling CAE software tools (Figure 6) (Veitl et al., 1999). The first distinction in this classification structure is between design tools and design models. Design model describes the behaviour functionality of the system being designed. The design tools can only include the computation of design parameters without modelling the whole system. The second distinction is between the descriptive and operational models. The descriptive model just describes the structure and content of the plant model. The operational model can be interpreted and generates new data about the plant (e.g. time sequence of the plant behaviour, the value of the performance index). Examples of descriptive models are CAD models, symbolic equations of motion, programme source. Examples of operational models are running simulation models, running computer codes. The next distinction is between simulation model coupling as the coupling of time sequence of values of system behaviour from different simulation models and the single value coupling as the coupling of single values produced by different models (e.g. exchange of values of performance index within optimisation). The simulation models can be coupled on the level of descriptive models (e.g. coupling of symbolic equations or symbolic source code) or on the level of operational models when the values generated during the simulation are exchanged. The next distinction is the number of independent integrators which are communicating. One integrator is the 'tight coupling' method (function-call) of the systems and two/more integrators result in the 'weak coupling' method (co-simulation) (Veitl et al., 1999). All coupling variants can be implemented as one or more processes. An overview of variants of co-simulation implementation is in Figure 7.

In particular two specific points of view are essential for co-simulation technology. It is the software/hardware and implementation issue and the

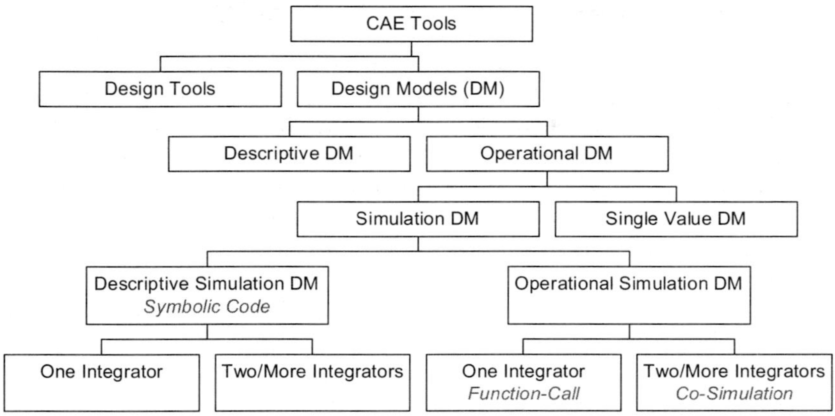

Figure 6. Interfacing/Coupling of CAE Tools.

numerical solution issue (Figure 7). The first issue is the question of how many processors are used and the software technology of programming (coupling of codes, coupling of libraries, single-multiple processes) and communication (file interface, shared memory, TCP/IP). The second issue is the most critical problem of co-simulation. The main problems of co-simulation are the stability, accuracy and efficiency of the integration (numerical solving) process. Especially critical one is the stability of numerical solving process. From this point of view two main co-simulation approaches can be distinguished (Figure 8). It is tight-coupling scheme where the integration is provided just once for all subsystems and the other subsystems are treated by function-call. The other is weak-coupling scheme where the integration is provided in each co-simulated subsystem and the subsystems are exchanging the information in equal way.

Type of models	Numerical solution				Software technology	
	One integrator	Multiple integrators				
Descriptive = source code	Symbolic code	Symbolic code			Coupling codes	
Operational = executable code	Function call	Weak co-simulation			Coupling libraries	Inter-process communication
		Direct	Alternated	Iterated		

Figure 7. Variants of co-simulation implementation.

The stability problem specifically arise from co-simulation concerns just the weak-coupling. Three different schemes of co-simulation integration process are currently used (Figure 9) (Valasek et al., 2006a; Arnold, 2004; Kübler and Schiehlen, 2000). These schemes describe one co-simulation time step when the subsystems exchange the information about their states, but between these co-simulation time steps (so called macro steps) both subsystems are numerically integrated in many micro-steps. This enables to increase the co-simulation efficiency by the different integration rates for different subsystems.

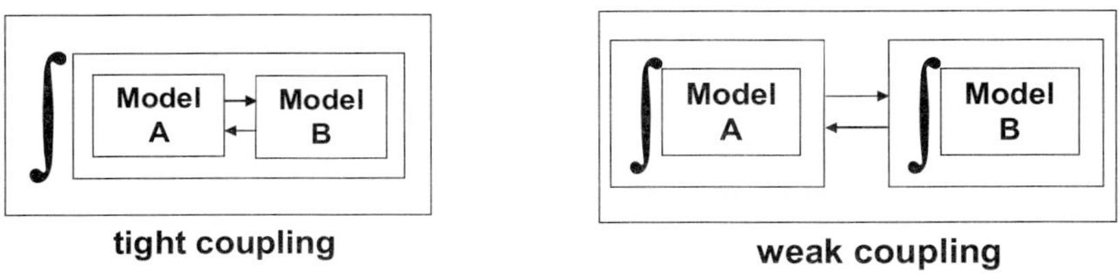

Figure 8. Principal co-simulation schemes.

In the direct scheme both subsystems exchange the information about their states just once and then they are independently simultaneously integrated until the next macro-step. In the alternated scheme both subsystems exchange the information about their states and the subsystems are integrated consecutively. One subsystem is integrated first within the macro-step based just on the information about the state of the other subsystem in the initial time instance of the macro step and possible extrapolation of previous time behaviour and then the other subsystem is integrated using the whole information of the integration of the first one. In the iterated scheme both subsystems exchange the information about their states in the initial time instance of the macro step, then they are independently simultaneously integrated until the next macro-step using possible extrapolation of previous time behaviour and finally the results are mutually iterated until required equality of integrated and extrapolated results are achieved.

The most often used scheme is the first one, the direct co-simulation (Valasek et al., 2006a). For the mechatronical systems where one subsystem is the physical plant and the other subsystem is the control system it is the natural simulation of the control of mechatronical system. The macro-step in such co-simulation process corresponds to the sampling period in the real system. The stability, accuracy and efficiency of the co-simulation of such system is equal to the stability, accuracy and efficiency of the control of the

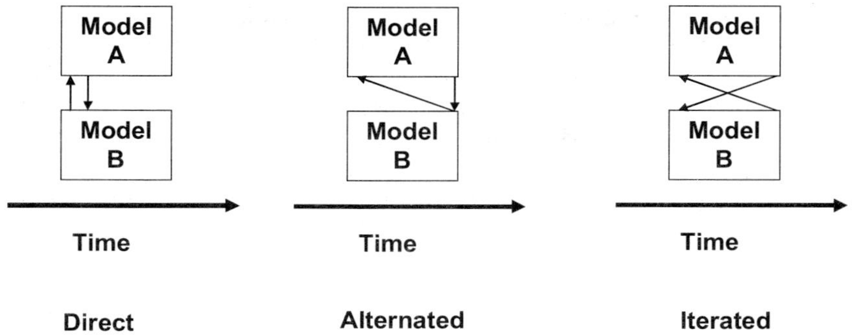

Figure 9. Three schemes of co-simulation integration process.

mechatronical system. From the theory of automatic control, in particular from the design of digital control the rules for the stability can be concluded. Direct co-simulation scheme corresponds to the design-by-emulation of digital control (Franklin et al., 1993). The rule is that the sampling frequency (period) should be 20 times higher than the bandwith of the system. This rule can be followed for the setting of direct co-simulation integration time step for the co-simulation of the control of the mechatronical system as well as in other cases. The experience from the digital control design has shown that the violation of this rule leads to severe inaccuracies of the simulation.

The other two schemes are used experimentally despite that several theorems about their stability have been proven (Arnold, 2004; Kübler and Schiehlen, 2000) and applied in practice (Arnold, 2004).

Example of variety of co-simulation interfaces can be demonstrated on the example of the co-simulation interface between Simpack (modeling of multibody systems) and Matlab/Simulink (modeling of dynamic block diagrams) in Figure 10. Both simulation packages can exchange the description of linear models as descriptive models, the simulation model in each of these packages can be symbolically generated into descriptive simulation model (symbolic code) and compiled as additional procedure into the other one. On the level of operational models the values of right-hand side of state space model in each of these packages can be used as procedure (function call) for evaluation of the right-hand sides of additional differential equations. This however does not enable to solve multibody systems described by differential-algebraic equations (DAE) by function call as the simple interface of right-hand sides of differential equations does not transfer required information of DAE and Simulink does not include a solver for DAE index 3 for multibody systems. Finally the weak co-simulation interface with simultaneous integration in both simulation packages is available.

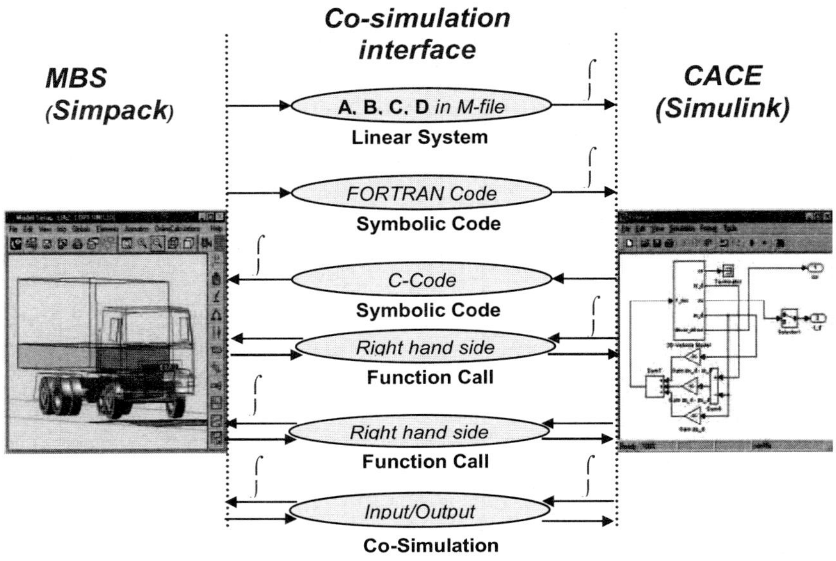

Figure 10. Example of variety of co-simulation interfaces.

The advantage of co-simulation is that it offers large support of library elements developed for particular physical domains as well as intuitive modelling and methodology for the given subproblem where the designer creates the models within the particular domain language. The disadvantages are connected with the problems of stability, accuracy and efficiency of the co-simulation.

3.2 Uniform modeling

The second approach tries to describe the complete multi-disciplinary system by uniform language. Traditionally the mathematical models and corresponding simulation models are being developed for systems from one physical domain. The basis of uniform modeling is that the models for all particular subsystems are modeled by the same description, in the same modeling language. The submodels are described by the same concepts, using the same description. Such models can be naturally interconnected. As the uniform language the four different formalisms can be used: equations (differential and algebraic), dynamic blocks, multipoles and bond graphs.

The fundamental uniform modeling is the language of mathematics. All models are described by the differential and algebraic equations and the interconnection of equations is always the final way of connecting the submodels. The procedure is relatively simple. The equations for all submodels are written, the relationships between the input-output variables in these submodels are added and the resulting set of equations is processed into a form suitable for numerical solution like state space description. This approach is supported for example by the simulation software ACSL (ACSL, web). Because the equations are very natural common language for all mathematical and thus simulation models, recently the equations have become the basis of the powerfull modelling language within VHDL-AMS language and standard (VHDL-AMS, web). The disadvantage of this approach is that either a symbolic equation generator based on another uniform modelling approach or the specific symbolic equation generators for each physical domain with prepared multidomain interfaces are needed. But the general problem is that this approach is tedious, more open to errors.

The similarity of physical principles, laws and resulting equations are well known and described, see, e.g. (Karnopp et al., 1990), and have lead to the development of higher concepts and model assembly procedures that enable to assemble the model from larger parts. The first group of such models of larger parts creates blocks of dynamic block diagrams. A block represents either a linear dynamic system described by linear ordinary differential equation or an algebraic (static) function that can be linear like gain or nonlinear like square root. Because any state space description can be treated as time integration of right-hand side consisting of nonlinear algebraic (static) function then any dynamic system can be described in dynamic block diagrams. The important fact is that blocks represent causal input-output relationship. Therefore the block diagram for a certain system can be created only after complete causal analysis of the model of the investigated system. The dynamic blocks as the uniform language have been used since the first simulation languages. Their disadvantage is that first the symbolic equations of the multidisciplinary model must be generated and second these equations must be sorted and modified into a causal sequence, both outside the formalism of dynamic blocks. The approach of dynamic blocks is supported in almost all simulation packages. The block diagrams were the technique of programming the analogue computers. It is for example the basis of simulation software Matlab/Simulink (MATLAB, web).

The second group of such models of larger parts creates multipoles. They belong together with bond graphs to the network approaches. Both these network approaches are based on the recognition that any physical interaction in or between any physical domain requires an exchange of energy. This means that parts of a physical system can be thought to interact via ports that allow energy to be exchanged, the so-called power-port. Secondly, it is recognized that any dynamic interaction can be described by two dynamically conjugated variables ('bilateral exchange of information'). Usually the product of these variables expresses the power or energy flow.

The multipole modelling approach is a very general approach based on the uniform treatment of any physical system as the interaction between elements (subsystems) described by two kind of variables: through variables t and across variables a. These variables are dual and in couples are associated with each interaction connection (constraint). These variables fulfill the analogy of Kirchhoff's laws the application of which suffices for the complete treatment of equations describing the connected elements (Figure 11)

$$\begin{aligned} t_1 + t_2 &= 0 \\ a_1 &= a_2 \end{aligned} \tag{1}$$

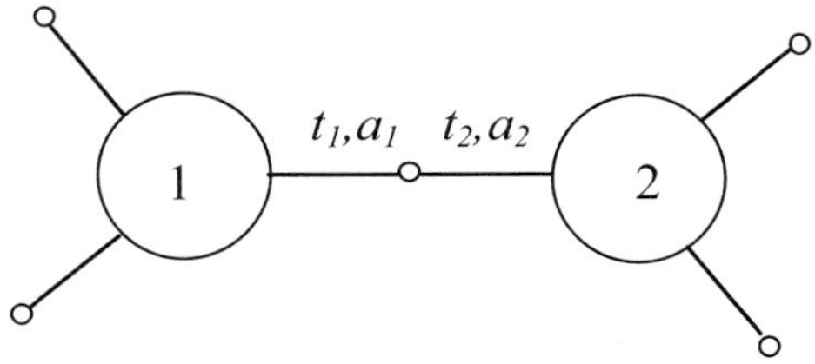

Figure 11. Multipole modelling approach.

It is possible to derive both analytically (Valasek, 1994) and empirically that the through and across variables can be introduced and have direct interpretation in each physical domain (Figure 14) and thus naturally enable uniformly to describe the multidomain physical systems. If the element 1 in Figure 10 is described by the internal variables $s_i^{(1)}$ and the element 2 in Figure 10 is described by the internal variables $s_j^{(2)}$ and both elements can be described by the variational principle of least action of the (Lagrange) functions $L^{(1)}(\dot{s}_i^{(1)}, s_i^{(1)})$, $L^{(2)}(\dot{s}_j^{(2)}, s_j^{(2)})$, and if the interaction of both elements is described by the constraint $f(s_i^{(1)}, s_j^{(2)})$ that can be derived as the equality of some boundary conditions $b^{(1)}(s_i^{(1)}), b^{(2)}(s_j^{(2)})$

$$f(s_i^{(1)}, s_j^{(2)}) = b^{(1)}(s_i^{(1)}) - b^{(2)}(s_j^{(2)}) = 0 \tag{2}$$

then the complete combined system can be described by the variational principle of least action of the (Lagrange) function $L^{(1+2)}(\dot{s}_i^{(1)}, s_i^{(1)}, \dot{s}_j^{(2)}, s_j^{(2)})$

$$L^{(1+2)}(\dot{s}_i^{(1)}, s_i^{(1)}, \dot{s}_j^{(2)}, s_j^{(2)}) = L^{(1)}(\dot{s}_i^{(1)}, s_i^{(1)}) + L^{(2)}(\dot{s}_j^{(2)}, s_j^{(2)}) + \lambda f(s_i^{(1)}, s_j^{(2)}) \tag{3}$$

where λ is Lagrange multiplier corresponding to the constraint $f(s_i^{(1)}, s_j^{(2)})$. The last expression can be modified

$$\lambda f(s_i^{(1)}, s_j^{(2)}) = \lambda(b^{(1)}(s_i^{(1)}) - b^{(2)}(s_j^{(2)})) = \lambda^{(1)} b^{(1)}(s_i^{(1)}) + \lambda^{(2)} b^{(2)}(s_j^{(2)}) = 0 \tag{4}$$

where new Lagrange multipliers $\lambda^{(1)}$, $\lambda^{(2)}$ are introduced

$$\lambda = \lambda^{(1)} = -\lambda^{(2)} \tag{5}$$

Then the dynamic equations of the complete combined system derived from the (Lagrange) function (3) are composed from the equations for each subsystem separately supplemented by the constraint relationship (2) and (5). Each subsystem 1 and 2 is described by the variational principle of least action of the (Lagrange) function $L^{(1)}(\dot{s}_i^{(1)}, s_i^{(1)}) + \lambda^{(1)} b^{(1)}(s_i^{(1)})$ and $L^{(2)}(\dot{s}_j^{(2)}, s_j^{(2)}) + \lambda^{(2)} b^{(2)}(s_j^{(2)})$. The constraint relationship (2) and (5) can be described as through and across variables (1)

$$\begin{aligned} &t_1 = \lambda^{(1)}, t_2 = \lambda^{(2)}, t_1 + t_2 = \lambda^{(1)} + \lambda^{(2)} = 0 \\ &a_1 = b^{(1)}(s_i^{(1)}), a_2 = b^{(2)}(s_j^{(2)}), a_1 = a_2 = b^{(1)}(s_i^{(1)}) = b^{(2)}(s_j^{(2)}) \end{aligned} \tag{6}$$

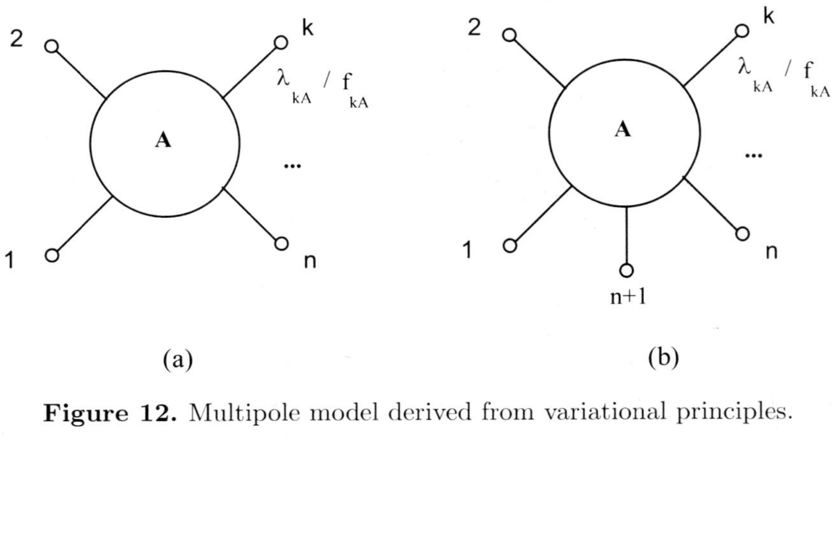

Figure 12. Multipole model derived from variational principles.

In case that the interaction cannot be described as the equality (2) of some boundary property of interacting elements, i.e. the external variable $s_j^{(2)}$ to the element 1 cannot be split in the constraint $f(s_i^{(1)}, s_j^{(2)})$ into an external additive term, then it is possible to introduce a new auxiliary variable(s) $s_{n+j}^{(1)}$ to the variables $s_i^{(1)}$, $i = 1, \ldots, n$ of the element 1 and instead of the indecomposable constraint an external constraint of interaction between the elements 1 and 2 is introduced

$$s_{n+j}^{(1)} - s_j^{(2)} = 0 \tag{7}$$

and an internal constraint inside the element 1 is introduced

$$f(s_i^{(1)}, s_{n+j}^{(1)}) = 0 \tag{8}$$

The intermediate result is that a multipole with n poles which have associated the through and across variables with local property (1) is introduced (Figure 12a). However, such multipoles do not satisfy the generalized Kirchhoff's laws for through and across variables of the whole multipole. Such satisfaction is achieved by the extension of the poles by the $(n+1)$-th auxiliary pole with the definition of the through and across variables (Figure 12b)

$$\lambda_{n+1}^{(1)} = -\sum_{i=1}^{n} \lambda_i^{(1)}, \quad f_{n+1}^{(1)}(s_i^{(1)}) = 0 \tag{9}$$

Then such multipole satisfies the generalized Kirchhoff's laws

$$\begin{aligned} &\sum_{i=1}^{n+1} \lambda_i^{(1)} = 0 \\ &u_{ij}^{(1)} = u_{ik}^{(1)} + u_{kj}^{(1)}, \quad u_{ij}^{(1)} = (f_i^{(1)} - f_{n+1}^{(1)}) - (f_j^{(1)} - f_{n+1}^{(1)}) \end{aligned} \tag{10}$$

Then the system being modelled is represented by the corresponding network of interacting elements. The elements are described by known internal equations and their interactions are described by the generalized Kirchhoff's laws for through and across variables. Such modeling approach enables to build analogy of any physical system and an electric network. An example of such analogy for one multipole is in Figure 13. The internal dynamic equations are described as

$$G_i^{(1)}(\dot{s}_i^{(1)}, \ldots, s_i^{(1)}, \ldots) = 0 \tag{11}$$

The longtime investigated analogies between different physical domains are in Figure 14.

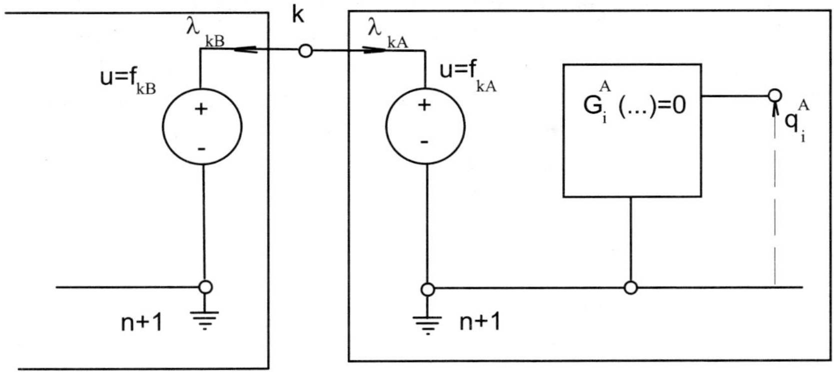

Figure 13. Analogy of a multipole and an electric network.

Physical Domain	**Through Variable**	**Across Variable**
Mechanics	Force	Displacement/Velocity
Electronics	Current	Voltage
Fluid-Power	Flow	Pressure

Figure 14. Through and across variables for different physical domains.

The advantage of this approach is the immediate capability for multidomain modelling and the natural modularity of the models. Another great advantage is that it is not necessary to carry out the causal analysis at all. The causal analysis is a natural part of the formulation of the dynamic equations describing the multipole model. The disadvantage is that this approach naturally leads to the usage of variables which describe the state of each element in the network independently from the others instead of usage of relative variables across the connections. The consequence of that for mechanical multibody systems is that the qualitatively less efficient equations of motion are assembled because the efficient recursive equations of motion are based on relative coordinates. This problem has been recently partially addressed by symbolic postprocessing of the dynamic equations. This approach for multiphysical modeling is currently being evolved by Modelica language (MODELICA, web) and Dymola software (DYMOLA, web).

The bond graph approach is another network approach (Karnopp et al., 1990). It simultaneously represents physical and computational structure of a system independent of the physical domain and it is purely topological and does *not* graphically represent the geometry of the modelled system. A bond graph is a directed graph with labeled nodes. The edges are bonds and represent a bilateral signal flow of the power-conjugated variables called effort e and flow f. Again these conjugated variables fulfill in the nodes the analogy of Kirchoff's laws. The half arrow represents just the positive orientation, i.e. not the direction of the flow. The edges only represent power interaction between the ports and carry themselves no information about the network structure. A crucial property of a bond graph is that it represents different types of structural constraints by different types of nodes, in the same manner as it represents different types of elements (storage and dissipative elements, sources, etc.) by different types of nodes. However, in the bond graphs a detailed distinction of connection elements is used. There are two fundamental cases called 0 junction and 1 junction where the kind of Kirchoff's laws for effort and flow are interchanged (Figure 15) in contrast to multipoles where in the nodes the across variables are always equal and the through variables are always summed to zero. The bond graph notation can be generalized to multiports and multibonds, thus increasing the expressing capabilities of bond graphs which is necessary for complex system models, like multibody systems. The interpretation of effort and flow variables in bond graph approach (Figure 16) is more complicated than in multipole approach as it follows from description of physical phenomena by storing quantity and keeping equilibrium (Karnopp et al., 1990). This approach for multiphysical modeling is currently being evolved by 20-Sim software (20-SIM, web), see more in the contribution of P. Breedveld to this book.

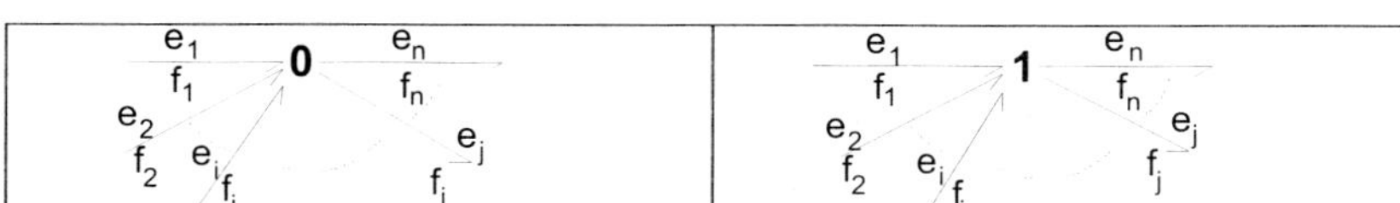

0 junction	1 junction
$\sum_{i=1}^{n} \varepsilon_i f_i = 0$, where $\varepsilon_i=+1$ for orientation towards junction and $\varepsilon_i=-1$ for orientation from junction $e_i = e_j$	$\sum_{i=1}^{n} \varepsilon_i e_i = 0$, where $\varepsilon_i=+1$ for orientation towards junction and $\varepsilon_i=-1$ for orientation from junction $f_i = f_j$

Figure 15. Bond graph modelling approach.

The general advantage of network approaches is the natural multidomain capability of modelling which has been for long time almost the only possibility for multidisciplinary modelling. The general disadvantage is that there have been already devoted too much effort and accumulated too much experience for modelling systems of particular specific physical domain and that the users in these specific domains use their specific language and schemes which are for them better and more naturally understandable than rather neutral and abstract network description. These disadvantages have lead to the development of interface methodologies for component modelling software tools which now possesses the full capability of multidomain modelling. The other disadvantage of network approaches is that specific solving algorithms for particular physical domain are sometimes difficult to be expressed in general network approach, e.g. recursive formalisms for multibody systems.

Domain	**extensive energy state (q)**	**rate of change flow**	**equilibrium determining variable, effort (intensive)**
Kinetic	momentum	Force	velocity
Elastic/potential	displacement	Velocity	force
Electrical	charge	Current	voltage
Magnetic	flux linkage	Voltage	current
Thermal	entropy	Entropy flow	temperature

Figure 16. Variable classification in a generalized bond graph approach.

4 Development of simulation model

The development process of simulation model is often underestimated. The result of the development seems to be just the code of the simulation program. But during the development process a lot of specific knowledge is derived that is used during the simulation model life cycle (Figure 17) and during knowledge reuse in the development of next simulation model. The development of simulation model can be considered as an analogy with a software program development (Valasek, 2006). This process has been analysed in details within the Clockwork project (Zdrahal et al., 2007; Steinbauer et al., 2003).

The development of a simulation model typically goes through a number of tasks. These tasks are illustrated in a modelling and simulation knowledge framework in Figure 18. The light arrows represent mappings between the different elements and the black arrows represent transformation processes between the tasks. The development of a simulation, as shown in Figure 18, involves five steps.

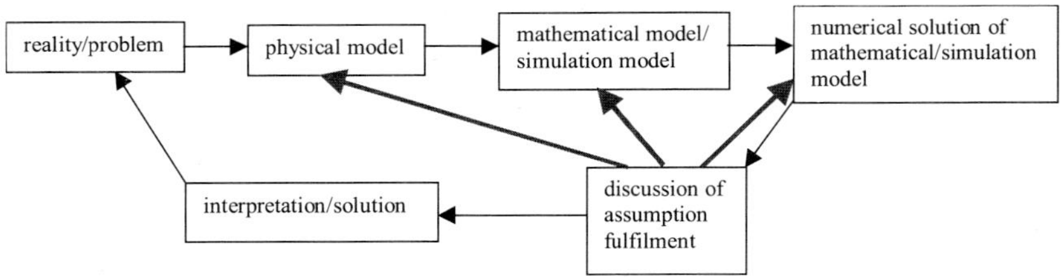

Figure 17. Traditional simulation model life cycle.

The first step is the analysis of the real world object (either actual or hypothesised) that is motivating the simulation and modelling task. This real world object is investigated within a certain experimental frame concentrated on the aspects that need to be investigated in order to answer some question. Answering this question is the objective of the investigation. The real world system is the model created in order to help answer the modelling question, such as how a real-world system would respond to particular stimuli. The real world object therefore has three elements:

- The real world system (either actual or hypothesised), of which we want to create a model to use to answer a modelling question, such as how a real-world system would respond to particular stimuli.
- The question (modelling question) is a question that is based within the real world and is to be answered using simulation.
- The solution is the interpretation of the interpreted output of the simulation task, this will be left blank until there is a set of results.

The second step is the conceptual task, where the conceptual model is developed. The transformation from real world into conceptual world consists of creating a hierarchical break-down of the real world system being investigated (e.g. hierarchical description of real product components and its environment). This scheme should be accompanied by the description of function (physical interaction) as it is the basis of causal and functional explanation. As soon as this description is completed the conceptualisation is completed. Different assumptions are raised about the granularity of the model being developed. A crucial design decision is determination of what factors influence system behaviour and what system behaviours are to be incorporated into this task. The result of the conceptualisation is the conceptual model investigated within the modelling environment in order to answer the modelling objective:

- The conceptual model is an abstracted view of the real world system.

- The modelling objective is the conceptualisation of the question and forms the objectives for the simulation task (i.e. what behaviour we are interested in? and how we will know that we have the correct answer from the results?)
- The modelling environment is the conceptualisation of the inputs of the real world systems, providing a simulation environment to work within in the simulation task, typically the inputs are specified here.

Validation is the process of determining whether the conceptual model is an adequately accurate representation of the real world task.

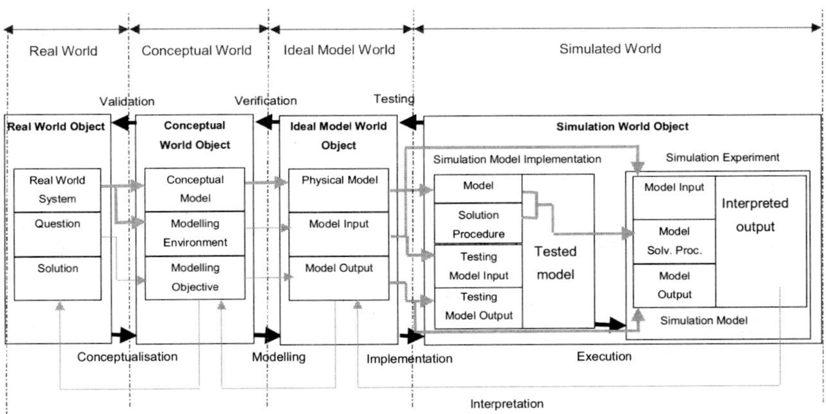

Figure 18. Simulation and modelling knowledge framework.

The next task is physical modelling. The third step is the transformation from conceptual world into ideal model world. It consists of replacement of each element on the conceptual level by corresponding elements on the physical level. The elements of the physical model are ideal modelling elements (e.g. rigid body, ideally flexible body, ideal gas, ideal incompressible fluid, ideal capacitor etc.). The engineering sciences deal precisely with the ideal objects. Mathematically described exact prediction of the behaviour is available only for ideal objects. The art of engineering modelling and simulation is the capability to translate the behaviour of real world objects into the behaviour of a system consisting of ideal objects.

Again as soon as this replacement is completed the physical modelling is completed. In this stage, the majority of modelling assumptions are raised and many modelling modifications are being done. It is the way of "modelling it" and representing the real world system in a physical modelling manner. For example two typical modifications are provided: *Either* an element is decomposed into more elements (a small subsystem) (e.g. a body cannot be treated as rigid body and it is modelled as flexible body being modelled as subsystem) *or* several elements are replaced by one element as the detailed treatment of interactions is neglected (e.g. the influence of a car suspension mechanism can be neglected and the model consists only of chassis and wheels with flexibility of tyre and suspension spring). The modelling task results into the physical model, its input and the investigated output:

- The physical model is an accurate physical view of the conceptual model.
- The model input is the modelling conceptualisation of the inputs of the real world systems, providing a physical modelling of the real world stimuli to the investigated system, i.e. detailed physical description of modelling environment from conceptual task.
- The model output is the modelling conceptualisation of the objectives for the simulation task (i.e. how the interested behaviour is measured and evaluated ?)

The process of determining whether the physical model is an adequately accurate representation of the conceptual model is the verification.

The results of the modelling task are implemented in the form of a computer-executable set of instructions known as the simulation model. It is the transformation from physical world into simulation world. It consists of the replacement of each element on the physical modelling level by corresponding elements on the simulation level. Usually this replacement can be straightforward as simulation packages are trying to directly contain the ideal objects of physical models. However, always some modifications are necessary. The simulation task consists of two stages. The first stage deals with the implementation of the physical model using a particular simulation environment (simulation language and simulation software). The second stage deals with the simulation experiment, i.e. proper usage of the developed simulation model for the solution of the real world problem. Testing is the process of determining whether the implemented version of the model is an accurate representation of the physical model.

The linear design process is made up of numerous transitions across world objects of different levels of the framework, producing different iterations at each level. Each transition from one world to the next involves the adoption of specific assumptions pertaining to how the model should be re-represented in the next world. An example is shown in Figure 19. The designer develops the real world model, and then constructs an associated conceptual model

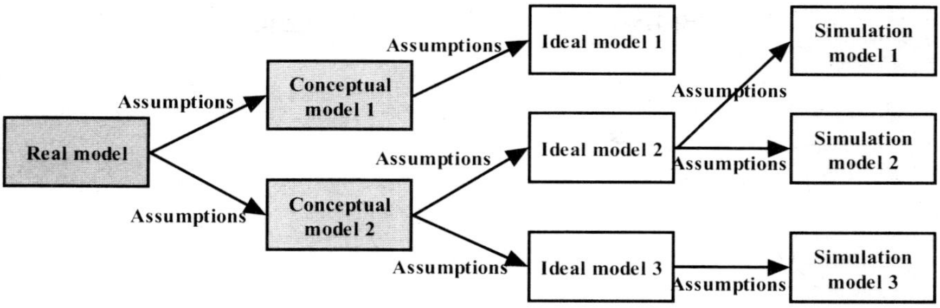

Figure 19. Iterative process of simulation model development.

which is transformed into an ideal model. A failure at this point leads the designer to construct an alternative model on the conceptual world model, meaning the problems identified in construction of the first ideal model can only be resolved by modifying the assumptions made when developing the first conceptual model. This second conceptual model is further developed into ideal and simulation models.

This development process of simulation model can be interpreted as an analogy with software engineering. It includes the analogy with life cycle. The real world task is like software specification, the conceptual world task is like software analysis, the physical modelling task is like software design, the simulation model implementation is like software coding, debugging and testing, the simulation experiment is like software usage. The analogy can be followed in techniques used for programming, testing, debugging. Some analogies are missing in specification, analysis, design. The interpretation of simulation model development as a software development is very useful and enables to exclude many problems.

For the support of development process of simulation models specific knowledge tools have been developed within the project Clockwork (Zdrahal et al., 2007; Steinbauer et al., 2003). They support the capture of knowledge accumulated during the simulation model development. It can be used for the support of geographically distributed design teams, for the reuse of knowledge from previous cases or for the cautious simulation model development where companies must cooperate on the simulation model development but do not wish to share their whole know-how (Zdrahal et al., 2007).

5 Model based control of mechatronical systems

Multibody systems are the most often used models of mechatronical systems. The control tasks for multibody systems (MBS) can be summarized as (Valasek, 2003)

- The first fundamental problem is to stabilize the motion of multibody system around its equilibrium. It is the basic control problem. An important subproblem is just to check the stability of some proposed control.
- The next problem is to move the multibody system from one position into another. It is one of the most frequent control problems of MBS. This problem is transformed into other problems. The problem can be stated as the global stabilization of the system from the initial position around the final position as the stabilized equilibrium. By this way the problem is transformed into previous problem. The other problem statement is to stabilize the system around the given trajectory connecting both positions. It is well known trajectory tracking. For nonlinear multibody systems it can be a serious problem just to construct admissible trajectory connecting initial and final positions.
- The previous problem can be formulated as the determination of optimal trajectory with respect to some performance index.
- The problem could be just to find any admissible trajectory of the system connecting two positions of multibody system. It is a preparatory problem for control by trajectory tracking.

The stabilizing problems are addressed by several groups of methods:

- There were developed methods that exactly linearized the nonlinear system and thus enable after the linearizing transformation to apply standard control methods for linear systems.
- Another approach is to propose a control law based on some insight into the system and then to carry out the synthesis of control parameters for the nonlinear plant as control gain optimization by the technique of multiobjective parameter optimization (MOPO).
- Very promising method is to rewrite the nonlinear system into the form of linear system and then to apply the standard control methods for linear systems however for position (state) dependent system. It is another way for transforming the nonlinear system into linear-like one.

The problem of constructing suitable trajectory is formulated

- either directly as the optimal control problem that is generally not quite easy to be solved

- or in specifically difficult cases of under-actuated systems as time planning problem that requires first to construct the missing control actions in order to modify the system into equal-actuated one.

The role of models in MBS control is being increased. In terms of control requirements the model enables to predict the system behaviour. The system model enables to solve the inverse problem, i.e. what actions should be taken to achieve desired specifications from the current states of the system. If an accurate model exists and if its inverse exists, then system dynamics can be cancelled by the inverse model, i.e. the system output is always equal to the desired output. In other words, model based control design has the potential to provide perfect control. However, all models contain some degree of error and not every model is invertible, thus perfect control is very difficult to be realised. Nevertheless the model is an essential tool for control of mechatronical systems and the model based control, i.e. the control that systematically uses the model of the plant, achieves better results than other approaches.

For control purposes many different models are used. An overview of such models is given in Figure 20. In the following three models will be investigated in more detail. It is black box time series model CTU-LOLIMOT (Stefan et al., 2006) nonlinear predictive model, lumped parameter nonlinear model that is the multibody model, see, e.g. (Stejskal and Valasek, 1996), and distributed parameter nonlinear model that is flexible multibody model as the combination between MBS and FEM, see, e.g. (Stejskal and Valasek, 1996).

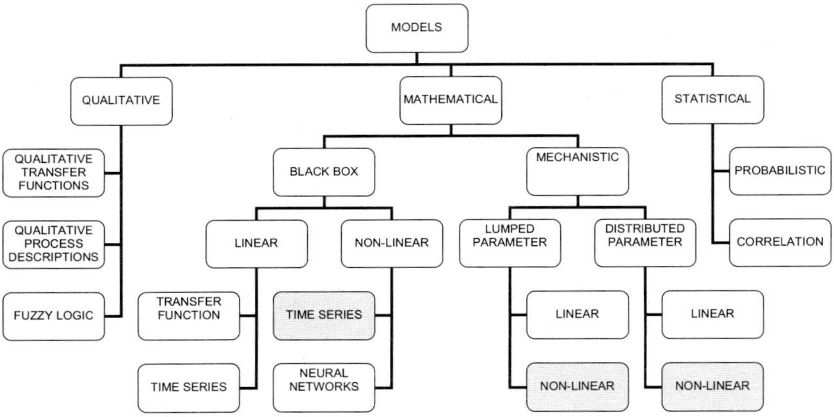

Figure 20. Overview of models used for control.

5.1 Main problems of control of multibody systems

Thus multibody systems are the most frequent models of mechanical systems being controlled. They concentrate the mechanical functionality of machines to be enough precise on one hand and simultaneously to be enough concise for control on the other hand. The main difficulties of control of mechanical (multibody) systems are the following (Valasek, 2003, 2004):

- Mechanical (multibody) systems are usually nonlinear systems. The well established control methods are developed for linear systems, only recently there have been developed methods for control of nonlinear systems, however still not being so general and powerful as for linear systems.
- Nonlinearity is sometimes essential for the control of multibody systems. There exist multibody systems (e.g. nonholonomic systems) that being linearized about operation point become uncontrollable.
- Multibody system undergo large displacements where linearization about operation point is impossible. The control problem is quite often not just stabilization around working (equilibrium) point, but the motion from one state into another one.
- Mechanical (multibody) systems in many cases have different number of inputs and outputs, or different number of inputs and degrees of freedom (DOFs). The inputs are operated by actuators (drives). Therefore it can be spoken about under-actuated, equal-actuated and over-actuated systems. This is causing especially problems in combination with nonlinearity of mechanical systems.
- Multibody systems are quite often suitably described by redundant coordinates. Then the models are described by differential-algebraic equations (DAE). DAE cannot be generally and easily transformed into ordinary differential equations (ODE). Only ODE can be described by state space models. The absolute majority of current control methods suppose that the dynamic system being controlled is described by ODE and thus the traditional control methods are difficult to be applied.

The above stated problems and further described solution methods can be demonstrated on the following simple problem. Despite its simplicity this example includes all serious problems of control of multibody systems.

The exemplary multibody system (Figure 21) consists of two bodies connected by revolute joints in plane. It has two DOFs. The first joint is always actuated by controlled action M_1. The second joint can be or cannot be actuated by the torque M_2. If it is actuated then the system is equal-actuated and methods for exact linearization can be applied. It is an example of a rigid planar robot manipulator. If the second joint is not actuated then the system is under-actuated and there are problems of its control. If this not-actuated joint is equipped with spring and damper then it is a model of flexible robotic arm. If there is no spring in this not-actuated joint then it is a model of mechatronic toy with different control challenges.

The multibody system can be described by the independent coordinates (e.g. φ_1 and φ_2) or by the dependent coordinates (e.g. the cartesian coordinates of centre of mass and angles of links with respect to the frame or natural Cartesian coordinates of ends of links). According to that the control is formulated as traditional control of ODE or control of DAE.

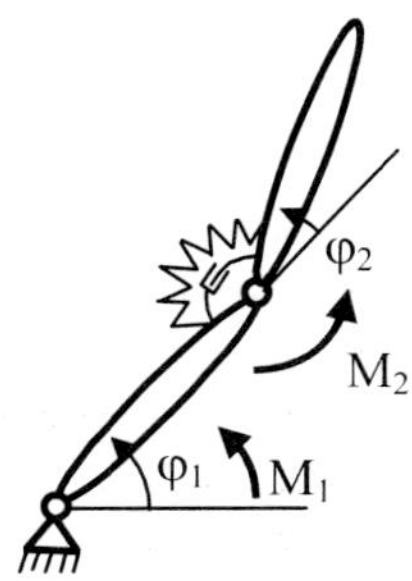

Figure 21. Example of simple multibody system with different control problems.

5.2 Description of Nonlinear Multibody Systems

The (nonlinear) multibody systems including the control action are generally described by Lagrange equations of mixed type, see, e.g. (Stejskal and Valasek, 1996)

$$\mathbf{M}(\mathrm{s})\frac{d^2\mathbf{s}}{dt^2} - \mathbf{\Phi}^T\boldsymbol{\lambda} = \mathbf{Q}(\mathrm{s}, \frac{d\mathbf{s}}{dt}) + \mathbf{T}(\mathrm{s}, \frac{d\mathbf{s}}{dt})\mathbf{u} \tag{12}$$

$$\mathbf{F}(\mathrm{s}, \frac{d\mathbf{s}}{dt}) = \mathbf{0} \tag{13}$$

where $\mathbf{s}$ are the coordinates describing the multibody system, $\mathbf{u}$ is the vector of input control variables, $\mathbf{M}$ is the mass matrix, $\mathbf{Q}$ are the generalized forces, $\mathbf{F}$ are the kinematic (holonomic or nonholonomic) constraints, $\mathbf{\Phi}$ is the Jacobian corresponding to the kinematic constraints, $\boldsymbol{\lambda}$ are the Lagrange multipliers, the control inputs (forces) $\mathbf{u}$ are acting through transmissions $\mathbf{T}$. This fact that the general description of multibody systems uses differential-algebraic equations (DAE) is one difficult problem of control synthesis for multibody systems.

These equations of motion for the purposes of control must be often transformed into the state space description. There are two possible ways how to do it. The first approach preserves the same variables $\mathbf{s}$ from original description and only eliminates the Lagrange multipliers $\boldsymbol{\lambda}$. If the kinematic constraints (13) do not depend on the velocities $\frac{d\mathbf{s}}{dt}$, i.e. the kinematic constraints (13) are holonomic ones, then these kinematic constraints $\mathbf{F}$ in (13) are twice differentiated with respect to time

$$\boldsymbol{\Phi}\frac{d^2\mathbf{s}}{dt^2} = -(\boldsymbol{\Phi}\frac{d\mathbf{s}}{dt})_s\frac{d\mathbf{s}}{dt} \tag{14}$$

where subscript s means partial differentiation with respect to the coordinates $\mathbf{s}$. In case of nonholonomic constraints, i.e. the kinematic constraints (13) depend on the velocities $\frac{d\mathbf{s}}{dt}$, then the kinematic constraints $\mathbf{F}$ in (13) are differentiated with respect to time just once and resulting into the same form (14). The accelerations are expressed from the equations of motion (12) by multiplying it by the inverse mass matrix $\mathbf{M}$ and they are substituted into (14). The equations for the Lagrange multipliers are thus derived. Solving for them and substituting back it is obtained the formulation of equations of motion of multibody system (12) in seemingly ordinary differential equations (ODE) only.

$$\begin{aligned} \frac{d^2\mathbf{s}}{dt^2} &= \mathbf{M}^{-1}(\mathbf{I} - \boldsymbol{\Phi}^T(\boldsymbol{\Phi}\mathbf{M}^{-1}\boldsymbol{\Phi}^T)^{-1}\boldsymbol{\Phi}\mathbf{M}^{-1})(\mathbf{Q} + \mathbf{T}\mathbf{u}) - \\ &\quad -\mathbf{M}^{-1}\boldsymbol{\Phi}^T(\boldsymbol{\Phi}\mathbf{M}^{-1}\boldsymbol{\Phi}^T)^{-1}(\boldsymbol{\Phi}\frac{d\mathbf{s}}{dt})_s\frac{d\mathbf{s}}{dt} \end{aligned} \tag{15}$$

Hence the state space description with state vector [$\mathbf{s}$, d/dt $\mathbf{s}$] in dependent coordinates $\mathbf{s}$ can be obtained. However, in such case the resulting equations are after linearization not controllable. It follows from the fact that in such case the dependent coordinates on the level of acceleration are related by the quadratic terms in velocities as follows from the second derivatives of kinematic constraints $\mathbf{F}$ in (14). Such equations are after linearization not controllable in usual positions with zero velocities. Nevertheless this simple formulation is sometimes very useful, e.g. for optimal control (Roubicek and Valasek, 2002).

The other approach is based on the description of multibody systems in independent coordinates $\mathbf{q}$. The coordinates must be independent, e.g. the independent coordinates $\mathbf{q}$ are selected (constructed) from $\mathbf{s}$ generally

$$\begin{aligned} \mathbf{s} &= \mathbf{r}(\mathbf{q}) \\ \frac{d}{dt}\mathbf{s} &= \mathbf{R}\frac{d}{dt}\mathbf{q} \\ \frac{d^2}{dt^2}\mathbf{s} &= \mathbf{R}\frac{d^2}{dt^2}\mathbf{q} + \frac{d}{dt}\mathbf{R}\frac{d}{dt}\mathbf{q} \end{aligned} \tag{16}$$

Substituting from (16) into the first time derivative of the kinematic constraint it is found out that this matrix $\mathbf{R}$ spans the null space of the Jacobian $\mathbf{\Phi}$

$$\mathbf{\Phi R} = \mathbf{0} \tag{17}$$

Thus multiplying the equations of motion (12) by $\mathbf{R}^T$ eliminates again the Lagrange multipliers. Then using the state vector $\mathbf{x}$=[$\mathbf{q}$, d$\mathbf{q}$/dt] the equations of motion can be transformed into the state space model

$$\begin{aligned} \frac{d}{dt}\mathbf{q} &= \dot{\mathbf{q}} \\ \frac{d}{dt}\dot{\mathbf{q}} &= (\mathbf{R}^T\mathbf{MR})^{-1}(\mathbf{R}^T\mathbf{g} - \mathbf{R}^T\mathbf{M}\dot{\mathbf{R}}\dot{\mathbf{q}}) + (\mathbf{R}^T\mathbf{MR})^{-1}\mathbf{R}^T\mathbf{Tu} \end{aligned} \tag{18}$$

The theory of multibody systems and the solution of corresponding DAE describes several methods for deriving and obtaining the null space matrix $\mathbf{R}$, see, e.g. (Stejskal and Valasek, 1996).

Now the dynamics of multibody systems can be described in the form of general nonlinear system in state space description

$$\frac{\mathrm{d}\mathbf{x}}{\mathrm{dt}} = \mathbf{f}(\mathbf{x}) + \mathbf{g}(\mathbf{x})\mathbf{u} \tag{19}$$

where $\mathbf{x}(n\times 1)$ is the state vector, $\mathbf{u}(m\times 1)$ is the control vector and generally $\mathbf{f(0)}=\mathbf{0}$. For the purpose of control it must be stated which variables $\mathbf{y}$ can be measured and used as the input to the control system. It is supposed their general dependence

$$\mathbf{y} = \mathbf{h}(\mathbf{x}) \tag{20}$$

Thus there is the output variable vector $\mathbf{y}(p \times 1)$.

5.3 Description of nonlinear predictive models

The nonlinear black box time series models are very suitable for model based predictive control. The considered concept CTU-LOLIMOT (LOcal LInear MOdel Tree implemented by CTU in Prague) consists of multiple linear dynamic models valid just in local sub-regions and smoothly summed into one resulting nonlinear model (Stefan et al., 2006). The identification algorithm is called LOLIMOT and was developed in (Nelles, 1999). The new implementation was done at CTU in Prague (Stefan et al., 2006).

The local models are described as linear difference models with inputs $\mathbf{u} = [\mathbf{u_1}, \mathbf{u_2}, \ldots, \mathbf{u_p}]$, $\mathbf{u}_1 = [u_1(k), u_1(k-1), \ldots, u_1(k-N)]$, $\mathbf{u}_2 = [u_2(k), u_2(k-1), \ldots, u_2(k-N)]$, ..., $\mathbf{u}_p = [u_p(k), u_p(k-1), \ldots, u_p(k-N)]$ of depth N, outputs y_i of depth n_y and coefficients $w_{i,j}$. These local models are summed using the Gaussian smoothening functions $\Phi_i(\mathbf{u})$ with centers c_{ij} and variances σ_{ij}

$$\Phi_i(\mathbf{u}) = \frac{\mu_i(\mathbf{u})}{\sum\limits_{j=1}^{M} \mu_j(\mathbf{u})},$$

$$\mu_i(\mathbf{u}) = \exp(-\frac{1}{2(N+1)}(\frac{\|\mathbf{u}_1 - c_{i1}\mathbf{j}\|^2}{\sigma_{i1}^2} + \frac{\|\mathbf{u}_2 - c_{i2}\mathbf{j}\|^2}{\sigma_{i2}^2} + \ldots + \frac{\|\mathbf{u}_p - c_{ip}\mathbf{j}\|^2}{\sigma_{ip}^2})) \tag{21}$$

where $\mathbf{j}(1 \times (N+1)) = [1, 1, \ldots, 1]$. The resulting CTU-LOLIMOT model description with M submodels is

$$\begin{aligned}
\tilde{y} = \sum_{i=1}^{M} \tilde{y}_i \Phi_i(\mathbf{u}) = \sum_{i=1}^{M} (w_{i,0} + w_{i,1} u_1(k) + w_{i,2} u_1(k-1) + \ldots \\
+ w_{i,N+1} u_1(k-N) + \ldots \\
+ w_{i,N+2} u_2(k) + w_{i,N+3} u_2(k-1) + \ldots + w_{i,2N+2} u_2(k-N) + \ldots \\
\ldots \\
+ w_{i,(p-1)N+p} u_p(k) + w_{i,(p-1)N+p+1} u_p(k-1) + \ldots \\
+ \ldots + w_{i,pN+p} u_p(k-N) + \ldots \\
+ w_{i,pN+p+1,y} y(k-1) + w_{i,pN+p+2,y} y(k-2) + \ldots \\
+ w_{i,pN+p+n_y,y} y(k-n_y)) \, \Phi_i(\mathbf{u})
\end{aligned} \tag{22}$$

The advantage of this description is that the model is locally linear, but globally nonlinear. The CTU-LOLIMOT model can be identified based on simulation model as well as based on experimental data from real system.

5.4 Model based control systems

The model based control of mechatronical systems can be divided into 4 groups: control with implicit or explicit usage of plant model for linear or nonlinear plant (Figure 22). The control approaches are combined with different kinds of plant models. Two model types are distinguished: analytical models based on physical principles and identified (black-box) models. Based on that particular control approaches are following:

- **MOPO (Multiple Objective Parameter Optimization):** it is a complete development methodology by design by simulation and optimization. It can be applied to any (linear/nonlinear) plant just having its detailed simulation model. Multiple objectives and requirements can be raised.
- **NQR-SDRE (Nonlinear Quadratic Regulator – State Dependent Riccatti Equation):** the majority of nonlinear systems can be stabilized and controlled by this approach. The nonlinear model is decomposed into linear one with nonlinear coefficients and state dependent Riccatti design is used. The decomposition can be computed for any system just having its simulation model.
- **LQ adaptive:** optimal LQR control is applied to linear plant with on-line adapted parameters.
- **MPC (Model Predictive Control):** it is a standard model predictive control.
- **NQR – MPC:** nonlinear plant model is decomposed into local linear one with nonlinear coefficients and MPC is applied.
- **CTU-LOLIMOT – MPC:** nonlinear plant model is decomposed into local linear ones and MPC is applied for local linear model. However, in the next horizon the used local linear model is different from the previous one. The controlled models are locally linear, but globally nonlinear.
- **Feedback exact linearization:** for the systems usually equal-actuated the input is modified by nonlinear transformation derived from the system model that the resulting model becomes a linear system.

	Linear plant	Nonlinear plant
Implicit plant model	MOPO	MOPO, NQR-SDRE
Explicit plant model	LQ adaptive, MPC	Feedback exact linearization, CTU-LOLIMOT – MPC, NQR – MPC

Figure 22. Model based control approaches.

These approaches are described further in details.

5.5 Feedback exact linearization of nonlinear systems

One of the fundamental approaches towards the control of nonlinear systems is the exact feedback linearization (Isidori, 1989; Slotine and Li,

1991; Fliess, 1992; Khalil, 1996). Its idea is simple. Considering the system (19) it is investigated the existence and the construction of a suitable (static) state feedback control

$$\mathbf{u} = \alpha(\mathbf{x}) + \beta(\mathbf{x})\mathbf{w} \tag{23}$$

and a transformation of variables

$$\mathbf{z} = \mathbf{T}(\mathbf{x}) \tag{24}$$

that together transform the original system (19) into a linear one

$$\frac{\mathrm{d}\mathbf{z}}{\mathrm{dt}} = \mathbf{Az} + \mathbf{Bw} \tag{25}$$

The same problem can be investigated for the existence of a suitable dynamic state feedback control

$$\begin{aligned} \mathbf{u} &= \alpha(\mathbf{x}, \mathbf{p}) + \beta(\mathbf{x}, \mathbf{p})\mathbf{w} \\ \frac{\mathrm{d}\mathbf{p}}{\mathrm{dt}} &= \phi(\mathbf{x}) + \eta(\mathbf{x})\mathbf{u} \end{aligned} \tag{26}$$

with a transformation of variables (24) that again transform the original system into a linear one (25). These problems are called input-state (exact feedback) linearization. And finally both these problems can be investigated for output feedback where instead of state variables $\mathbf{x}$ in feedbacks (23) or (26) only the output variables $\mathbf{y}$ from (20) are used. Then these problems are called input-output (exact feedback) linearization.

Then after transforming the original system into the equivalent linear one (25) there are applied the control techniques for linear systems like pole placement, LQR etc.

Several variants of feedback exact linearization can be distinguished. It is input-output exact feedback linearization that begins from the output dependence (20), then the input-state exact feedback linearization that investigates the construction of such suitable output dependence (20) that it is exact feedback linearizable

The solution of the problem of input-output exact feedback linearization is simpler. The construction of possible linearizing feedback for the system (19-20) is the recursive application of the time differentiation rule

$$\begin{aligned} \mathbf{y}^{(1)} &= \frac{\partial \mathbf{h}}{\partial \mathbf{x}}(\mathbf{f}(\mathbf{x}) + \mathbf{g}(\mathbf{x})\mathbf{u}) = \mathbf{f}_1(\mathbf{x}) + \mathbf{g}_1(\mathbf{x})\mathbf{u} = \\ &= L_{\mathbf{f}}\mathbf{h} + L_{\mathbf{g}}\mathbf{h}\mathbf{u} =_{\mathrm{def}} \dot{\mathbf{z}}_1 =_{\mathrm{def}} \mathbf{z}_2 \\ \mathbf{y}^{(k+1)} &= \frac{\partial \mathbf{z}_k}{\partial \mathbf{x}}(\mathbf{f}(\mathbf{x}) + \mathbf{g}(\mathbf{x})\mathbf{u}) = \mathbf{f}_{k+1}(\mathbf{x}) + \mathbf{g}_{k+1}(\mathbf{x})\mathbf{u} = \\ &=_{\mathrm{def}} \dot{\mathbf{z}}_{k+1} =_{\mathrm{def}} \mathbf{z}_{k+2} \end{aligned} \tag{27}$$

There is supposed besides necessary smoothness that $\mathbf{g}_i$=0 for $i = 1, \ldots, r-1$ and $\mathbf{g}_r \neq 0$ enabling to express

$$\begin{aligned} \mathbf{f}_r(\mathbf{x}) + \mathbf{g}_r(\mathbf{x})\mathbf{u} &= \mathbf{w} \\ \mathbf{u} &= \frac{\mathbf{w} - \mathbf{f}_r(\mathbf{x})}{\mathbf{g}_r(\mathbf{x})} \end{aligned} \tag{28}$$

and this leads to

$$\mathbf{y}^{(r)} = \mathbf{w} \tag{29}$$

This means that the original system is first transformed into the Brunovsky canonical form (27) and then into the Frobenius canonical form (29). These canonical forms prove the controllability of the system and enable to stabilize the system easily by pole placement.

The transformation (24) is

$$\mathbf{z} = \mathbf{T}(\mathbf{x}) = [(\mathbf{h}(\mathbf{x}))^T, (L_{\mathbf{f}}\mathbf{h}(\mathbf{x}))^T, ..., (L_{\mathbf{f}}^{r-1}\mathbf{h}(\mathbf{x}))^T]^T \tag{30}$$

Here and in previous expressions there is used the formalism of Lie algebra

$$\begin{aligned} &L_{\mathbf{f}}\mathbf{h}(\mathbf{x}) = \frac{\partial \mathbf{h}}{\partial \mathbf{x}}\mathbf{f}(\mathbf{x}), L_{\mathbf{f}}^{i+1}\mathbf{h}(\mathbf{x}) = L_{\mathbf{f}}L_{\mathbf{f}}^{i}\mathbf{h}(\mathbf{x}) \\ &[\mathbf{f}, \mathbf{g}](\mathbf{x}) = \frac{\partial \mathbf{g}}{\partial \mathbf{x}}\mathbf{f}(\mathbf{x}) - \frac{\partial \mathbf{f}}{\partial \mathbf{x}}\mathbf{g}(\mathbf{x}), ad_{\mathbf{f}}^0\mathbf{g}(\mathbf{x}) = \mathbf{g}(\mathbf{x}), \\ &ad_{\mathbf{f}}^1\mathbf{g}(\mathbf{x}) = [\mathbf{f}, \mathbf{g}](\mathbf{x}), ad_{\mathbf{f}}^{i+1}\mathbf{g}(\mathbf{x}) = [\mathbf{f}, ad_{\mathbf{f}}^{i}\mathbf{g}](\mathbf{x}) \end{aligned} \tag{31}$$

It can happen and it happens that $r < n$. In that case there is another dynamics in the system (19) besides (29). It is so called zero dynamics that is expressed from the original system stating the condition $\mathbf{y}=\mathbf{y}^{(1)}=\ldots=\mathbf{y}^{(r)}=\mathbf{0}$. The zero dynamics is not influenced by the transformation (30) and the control (28). The stability of this zero dynamics is decisive for the stabilizability of the whole system just by stabilizing the input-output dynamics (29) by the feedback (28) and pole placement.

The input-state exact feedback linearization is the problem whether there exists such function $\mathbf{y}=\mathbf{h}(\mathbf{x})$ for which the input-output exact feedback linearization gives $r = n$. This means that the system is completely linearized and there is no zero dynamics with the danger of destabilizing the system. The necessary and sufficient condition for its existence have been proved (Isidori, 1989). This is the full rank (equal n) of the vector field [$\mathbf{g}$, ad$_f\mathbf{g}$,...,ad$_f^{n-1}\mathbf{g}$] and the set [$\mathbf{g}$, ad$_f\mathbf{g}$,...,ad$_f^{n-2}\mathbf{g}$] is involutive (this means that if two vectors $\mathbf{k}_1$ and $\mathbf{k}_2$ are from this set then [$\mathbf{k}_1$,$\mathbf{k}_2$] is also from this set). However, the construction of such $\mathbf{h}(\mathbf{x})$ is not so easy as the computation of the input-output exact feedback linearization (27).

In all these derivations it is supposed that the dimension of $\mathbf{y}$ is equal to the dimension of $\mathbf{u}$ and $\mathbf{w}$. This is certainly often violated that prevents to apply this method for the control of under-actuated systems.

5.6 MOPO (Multiobjective Parameter Optimization)

This method of control synthesis is the least restrictive one regarding the requirement on the description of nonlinear (multibody) system and the form of control law. However, the strength of claims about the properties of the synthetized control is relatively low. The synthesis procedure is following:

The approach of Multi-Objective Parameter Optimization (MOPO) (Joos, 1992; Kortüm et al., 2002) is based on the design-by-simulaton. There is available the nonlinear simulation model of the controlled plant including the control law, the performance index usually in the form of integral performance index or maximum value and selected set of considered excitations. The control law is described in parametric form and its parameters are determined by the numerical optimization of the performance index evaluated by the simulation response of the plant to the considered excitations. Thus, by means of the MOPO approach the nonlinear models and models, which cannot be analytically treated, can be used. This approach enables not only to find parameters of nonlinear control of nonlinear plant, but it allows one also to find a satisfactory compromise among the performance criteria despite the fact that they conflict with each other. The MOPO approach is based on a search in the parameter space (Pareto optimality) by model simulation.

"Free system parameters", the tuning parameters, e.g. control coefficients, mass properties or installation positions, are varied within their given limits until an "optimal compromise" is found. In doing so the performance criteria (also called objective functions) $\mathbf{c}$ are weighted by user-defined weighting factors, the design parameters $\mathbf{d} \geqslant \mathbf{c}_{\text{start}}$. The optimization strategy is to minimize all weighted criteria $c_i(\mathbf{p})/d_i(\mathbf{p})$ in such a way that the currently *worst* criterion with the maximal value will be reduced: optimization tries to decrease the weighted criterion $c_i(\mathbf{p})/d_i(\mathbf{p}) \to 0$. The tuning parameters $\mathbf{p}$ are determined by solving a minmax-optimization problem with constraints and parameter restrictions

$$\alpha^* = \min_{\mathbf{p}}(\max_i c_i/d_i) \tag{32}$$

$$c_i(\mathbf{p}) \leqslant d_i \ for \ i \in \Psi \tag{33}$$

$$p_k^- \leqslant p_k \leqslant p_k^+ \tag{34}$$

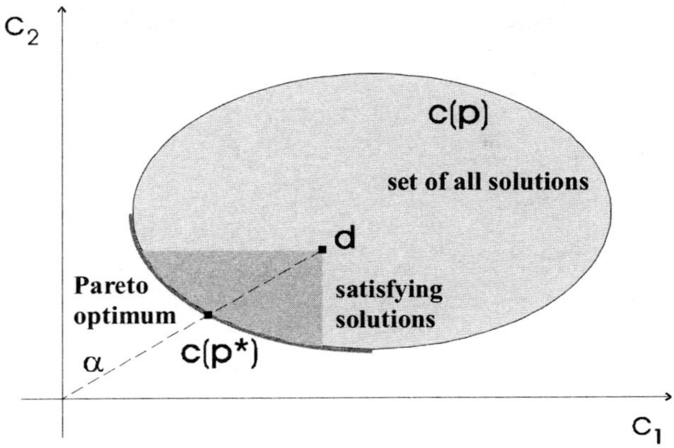

Figure 23. Pareto-optimum for two objective functions.

The performance criteria (objective functions) may be free to be optimized or may be partially limited by a set Ψ of performance constraints, e.g. mechanical, hydraulic or electronic restrictions like power, pressure, current and voltage. The weighting factors or design parameters enable the user to adapt the criteria to adequate sizes and to determine the direction of optimization process by weighting some criteria more important $(c_l(\mathbf{p})/d_l(\mathbf{p}) = 1)$, others less important $(c_j(\mathbf{p})/d_j(\mathbf{p}) < c_l(\mathbf{p})/d_l(\mathbf{p}))$. The parameter optimization is finished, when the maximum of all weighted criteria cannot be further decreased. The result is a point on the Pareto-optimal boundary (Figure 23).

The state-of-the-art of traditional design methodology of controlled vehicle suspensions has been based on restricted design models, linear control laws based on linear design models and limited experimental verification replaced usually by only experimental parameter tuning. On the contrary, the advanced design methodology is based on the full usage of realistic design and reference simulation models (taking into account the essential nonlinearities and important degrees of freedom) and on the control design taking into account the significant nonlinearities of the controlled plant.

The other problem arising from the nonlinear nature of models and/or control laws is the design procedure of control law. The current direct synthesis methods of control of nonlinear systems are limited by many necessary additional conditions of the controlled plant (Valasek, 2003, 2004). There is only one new approach for control synthesis with realistic requirements on the nonlinear plant (Valasek and Steinbauer, 1999). Nevertheless the control design with minimal requirements on the nonlinear plant that is combined with the new advanced control design methodology is the Multi-Objective Parameter Optimization (MOPO) approach.

The main steps of the MOPO design methodology are the following:

Models: The reference simulation model of the plant to be controlled (vehicle suspension system) is developed. It is as detailed as possible in order to cover the dynamical phenomena to be controlled. Besides this reference model the design model is also developed. It is simpler than the reference model but includes all essential features (DOFs, nonlinearities etc.).

Control law selection: Based on the design model the physical insight into the plant to be controlled is developed. Using this physical insight a suitable control law is proposed. It usually includes parameters to be designed for the appropriate function of the controlled plant.

Control law design: The control law is parametrized. The parameters are determined by the MOPO approach. A performance index usually in the integral form is selected. Important and representative excitation inputs are taken. The choice of inputs, the choice of performance indeces and the weights of their individual components are the most problematic parts of this approach. The design simulation model is integrated with these inputs and the performance index is evaluated. Then the performance index is just a function of the unknown control law parameters. They are determined by its optimization. It is design-by-simulation and design-by-optimization.

Verification: Firstly, the reference model is verified by comparison with experiments on real plant. Secondly, the design model is verified against the reference model. Then the designed control law is verified by the simulation of the reference model also using a larger set of input excitations. Finally the resulting control law is verified by experiments on real plant.

Example of application of MOPO synthesis for vehicle suspensions is in (Valasek et al., 1998).

5.7 NQR (Nonlinear Quadratic Regulator) and nonlinear linearization of nonlinear systems

This is another way of transforming a nonlinear system into a linear form where methods of linear control can be applied (Banks and Mhana, 1992; Cloutier, 1997; Langson and Alleyne, 1997; Valasek and Steinbauer, 1999; Cloutier et al., 1999; Erdem and Alleyne, 1999; Valasek and Kejval, 2000). The dynamics of the nonlinear system is described by the equations (19). The multibody system must be described by independent coordinates, otherwise further described decomposed system is uncontrollable. The quadratic performance index of the control which is to be synthesized is the infinite horizon control

$$J = \int_0^\infty (\mathbf{x}^T\mathbf{Q}\mathbf{x} + \mathbf{u}^T\mathbf{R}\mathbf{u})\,\mathrm{dt} \tag{35}$$

If there exists the decomposition of the system dynamics

$$\mathbf{f}(\mathbf{x}) = \mathbf{A}(\mathbf{x})\mathbf{x} \tag{36}$$

and if the following assumptions (Banks and Mhana, 1992) are valid

1. The matrices $\mathbf{Q}$ and $\mathbf{R}$ are positive definite.
2. The matrices $\mathbf{A}$ and $\mathbf{g}$ are analytic valued functions.
3. The control function $\mathbf{u}(\mathrm{t})$ exists and corresponding state trajectory $\mathbf{x}(t)$ for $t \in (0, \infty)$ which satisfy the system dynamics (19) and the performance index (35) is finite.
4. The pair of matrices $(\mathbf{A}(\mathbf{x}),\mathbf{g}(\mathbf{x}))$ is controllable and stabilizable for each $\mathbf{x}$ in the linear system sense, i.e.

$$\mathrm{rank}[\mathbf{g}(\mathbf{x}), \mathbf{A}(\mathbf{x})\mathbf{g}(\mathbf{x}), ..., \mathbf{A}^{n-1}(\mathbf{x})\mathbf{g}(\mathbf{x})] = n \tag{37}$$

5. The state vector $\mathbf{x}$ is fully measured.

then there exists the control

$$\mathbf{u} = -\mathbf{K}(\mathbf{x})\mathbf{x} \tag{38}$$

which minimizes the performance index (35). However, it has been shown that this control is only suboptimal.

The nonlinear gain matrix $\mathbf{K}(\mathbf{x})$ is determined as

$$\mathbf{K}(\mathbf{x}) = \mathbf{R}^{-1}\mathbf{g}^T(\mathbf{x})\mathbf{P}(\mathbf{x}) \tag{39}$$

where $\mathbf{P}(\mathbf{x})$ is the solution of the Riccati equations

$$\mathbf{A}^T(\mathbf{x})\mathbf{P}(\mathbf{x}) + \mathbf{P}(\mathbf{x})\mathbf{A}(\mathbf{x}) + \mathbf{Q} - \mathbf{P}(\mathbf{x})\mathbf{g}(\mathbf{x})\mathbf{R}^{-1}\mathbf{g}^T(\mathbf{x})\mathbf{P}(\mathbf{x}) = \mathbf{0} \tag{40}$$

This control approach is called Nonlinear Quadratic Regulator (NQR) (Valasek and Steinbauer, 1999) or State Dependent Riccati Equation (SDRE) control (Cloutier, 1997).

The key problem is the computation of the decomposition $\mathbf{A}(\mathbf{x})$. There have been derived some formulas for the computation of this decomposition, like (Langson and Alleyne, 1997)

$$\mathbf{A}(\mathbf{x}) = \int_0^1 \left.\frac{\partial \mathbf{f}}{\partial \mathbf{x}}\right|_{\mathbf{x}=\lambda\mathbf{x}} d\lambda \tag{41}$$

But the computation of this formula for the multibody systems is very difficult because it requires the symbolic manipulation with the equations of motion which might be very large and complex and might include rational and trigonometric functions. Therefore it is not realistic to suppose to compute (41) for multibody systems.

Therefore there have been developed efficient algorithms (Valasek and Steinbauer, 1999) for computation of this decomposition that moreover enables to use the non-uniqueness of the decomposition (36) for control synthesis. The simplest procedure among them is following. Its idea is based on the following decomposition described for scalar function f of three variables x_1, x_2, x_3

$$\begin{aligned} f(x_1,x_2,x_3) &= \frac{f(x_1,x_2,x_3)-f(0,x_2,x_3)}{x_1}x_1+ \\ &+\frac{f(0,x_2,x_3)-f(0,0,x_3)}{x_2}x_2+ \\ &+\frac{f(0,0,x_3)-f(0,0,0)}{x_3}x_3 \end{aligned} \tag{42}$$

This decomposition must certainly solve the division by zero and it is dependent on the order of variables. According to l'Hospital rule, the division by zero is in the limit equal to the corresponding derivative which for smooth functions exists and can be numerically computed by division for magnitudes of numerical accuracy of the computer.

The computation of this decomposition is not unique. It depends on the order of variables and besides that there are some free parameters for decomposition parametrization (Valasek and Steinbauer, 1999). The influence of this ambiguity can be described and investigated by the recursive decomposition (42) applied to each addend of the previous decomposition step. There are 2^n such addends at the end for n variables which is very natural for polynomial functions f. However, the problem is that any addend f_i in any level of decomposition can be split into convex sum $f_i\lambda + f_i(1-\lambda)$ where λ can be any function. This makes the decomposition fully ambiguitious. The level of recursive decomposition and-or the free functions λ can be selected in order to improve controllability of decomposed system (19) with (36).

The main problem with NQR is that first it is not the optimum control of general nonlinear system as it has been shown in (Banks and Mhana, 1992) and second there is still missing a proof of global stability of the control law (38) despite partial results (Erdem and Alleyne, 1999). There has not been found any case of unstable NQR control but general proof is missing. The practical experience is excellent and the stability can be checked for individual cases using the computed solution of the Riccatti equation (40). The Lyapunov function V is proposed as

$$V = \mathbf{x}^T \mathbf{P}(\mathbf{x})\mathbf{x} \tag{43}$$

It is positive definite according to the properties of the solution of Riccatti equation and the negativeness of its time derivative can be checked similarly in the same grid as the computed values of the gain $\mathbf{K}(\mathbf{x})$ deriving into

$$\frac{dV}{dt} = -\mathbf{x}^T(\mathbf{Q}+\mathbf{P}(\mathbf{x})\mathbf{g}(\mathbf{x})\mathbf{R}^{-1}\mathbf{g}^T(\mathbf{x})\mathbf{P}(\mathbf{x}) - \frac{d\mathbf{P}(\mathbf{x})}{d\mathbf{x}}(\mathbf{A}(\mathbf{x})-\mathbf{g}(\mathbf{x})\mathbf{K}(\mathbf{x}))\mathbf{x})\mathbf{x} \leqslant 0 \tag{44}$$

The first two terms are proving stability and the third term is usually small.

An important practical issue of NQR is the implementation of the synthetized control law. One approach can be the real-time computation of Riccatti equation and the whole control. It is certainly very demanding of computational power but for some small systems reasonable.

The other approach is based generally on the look-up tables. It means that the solutions of NQR control are computed for some values of states off-line and for the on-line application the pre-computed gains are either retrieved or interpolated. Such approaches were already used for gain scheduling approaches. The problem is certainly the efficient choice of suitable states for off-line computation and the on-line interpolation between them. This approach is applicable even for large multibody systems.

Another important advantage of NQR approach is that the control can be extended into incorporating different limitations of admissible control sets (e.g. asymmetric interval of control variable for semi-active systems (Valasek and Kejval, 2000)).

This approach of NQR is applicable to all under-actuated (certainly equal-actuated) systems that satisfy the condition that the linearized system about the equilibrium $\mathbf{x}=\mathbf{0}$ is controllable. Even this condition is violated for important classes of under-actuated systems (e.g. some nonholonomic multibody systems). For this class of multibody systems the control is very difficult. The stabilizing control was developed for such systems having $\mathbf{f}(\mathbf{x})=\mathbf{0}$, but for general systems with $\mathbf{f}(\mathbf{x})\neq\mathbf{0}$ it is still an open problem.

5.8 MPC (Model Predictive Control)

Model Predictive Control (MPC) has become very successful and widely accepted as advanced control methodology. It solves many traditional control problems by the direct application of fundamental principles. Instead of difficult and tedious finding the closed loop optimal feedback control it is based on the repeated solution of open loop optimal control using updated state and knowledge of the plant predictive model, see, e.g. (Rossiter, 2003; Camacho and Bordons, 2004).

The general formulation of MPC is following. The nonlinear system

$$\dot{\mathbf{x}} = \mathbf{f}(\mathbf{x}, \mathbf{u}) \tag{45}$$

is subjected to the optimal control problem with the cost function

$$\min_{\mathbf{u}(.)} J(\mathbf{u}(.),\mathbf{x}(t)) = \min_{\mathbf{u}(.)} \int_t^{T_p} \mathbf{F}(\mathbf{x}(\tau), \mathbf{u}(\tau))d\tau \tag{46}$$

and possible nonlinear constraints

$$\mathbf{h}\ (\mathbf{x}, \mathbf{u}) \leqslant \mathbf{0} \tag{47}$$

This problem is solved on-line for the finite prediction horizon T_p (Figure 24) but the optimal control input $\mathbf{u}$ is used just for the interval to the next sampling period $\tau \in [t, t+\delta]$ where δ is sampling time and the optimal control is solved for the moved prediction horizon $[t+\delta,\ T_p+\delta]$ (Figure 25).

The MPC is applied for linear systems as well as nonlinear ones. Further investigated variant is based on the CTU-LOLIMOT predictive models (Valasek et al., 2005). They enable to consider the plant model locally linear and globally nonlinear. If the predictive control can be based on linearized model even only locally it simplifies the problem significantly. Otherwise the optimisation choice of control cannot be solved easily and becomes computationally very difficult task.

The formulation of LOLIMOT predictive model (22) enables to compute the linear system matrices $\mathbf{A}$, $\mathbf{B}$, $\mathbf{C}$ by direct way. The response is computed as weighted sum of responses of all subsystems. Instead of computing the response, the parameters of linear subsystems are added by weighted sum. The following formula is used for computation of discrete transfer function instead of the predicted output

$$\frac{y}{u} = \frac{(\sum\limits_i w_{i,0}\Phi_i(\mathbf{u}), \sum\limits_i w_{i,1}\Phi_i(\mathbf{u}), \ldots, \sum\limits_i w_{i,m}\Phi_i(\mathbf{u}))}{(1, -\sum\limits_i w_{i,1,y}\Phi_i(\mathbf{u}), \ldots, -\sum\limits_i w_{i,n,y}\Phi_i(\mathbf{u}))} \tag{48}$$

Then the transfer function is transformed into the classical discrete state description with locally constant state matrices $\mathbf{A}$, $\mathbf{B}$ and $\mathbf{C}$

$$\begin{aligned} \Delta\mathbf{x}_{k+1} &= \mathbf{A}\Delta\mathbf{x}_k + \mathbf{B}\mathbf{u}_k \\ \mathbf{y}_k &= \mathbf{C}\Delta\mathbf{x}_k \end{aligned} \tag{49}$$

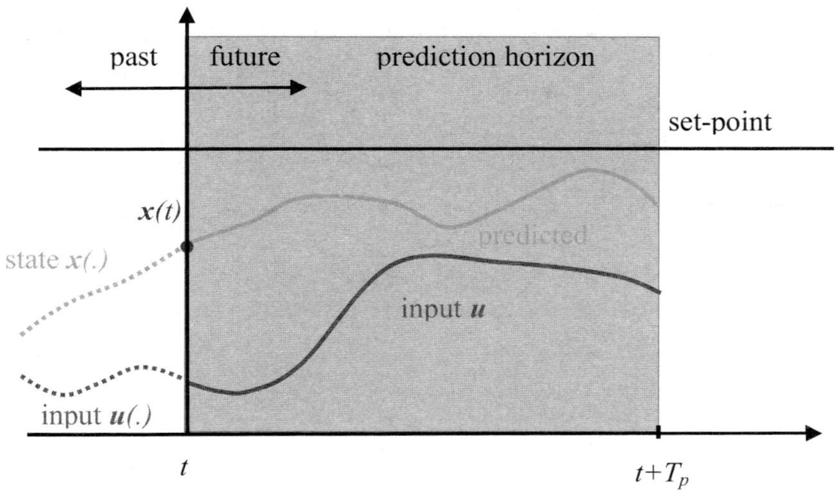

Figure 24. Model predictive control for the prediction horizon.

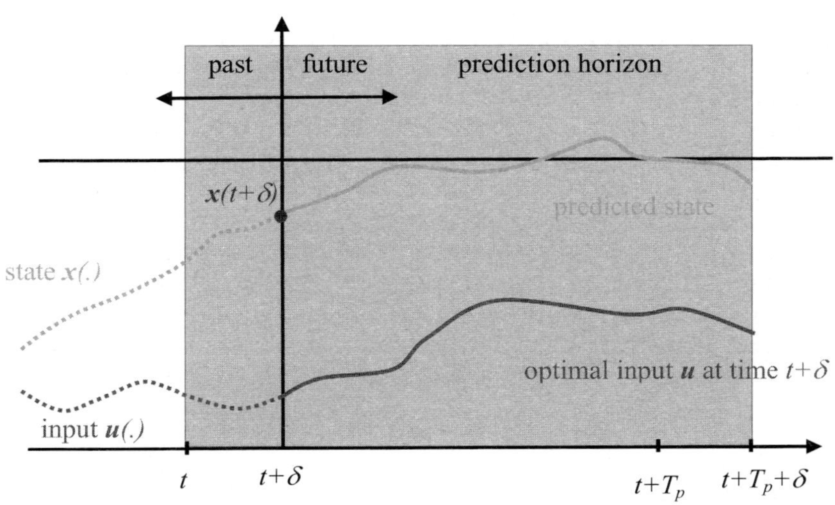

Figure 25. Model predictive control for the moved prediction horizon.

where $\Delta \mathbf{x}$ are system states and $\mathbf{y}$ are investigated outputs. The predictive control can be applied to the model created using the above described procedure. Writing the relation (49) for N subsequent steps as follows

$$\begin{aligned}
\hat{\mathbf{y}}_k &= \mathbf{C}\Delta\mathbf{x}_k \\
\Delta\hat{\mathbf{x}}_{k+1} &= \mathbf{A}\Delta\mathbf{x}_k + \mathbf{B}\mathbf{u}_k \\
\hat{\mathbf{y}}_{k+1} &= \mathbf{C}\mathbf{A}\Delta\mathbf{x}_k + \mathbf{C}\mathbf{B}\mathbf{u}_k \\
&\vdots \\
\Delta\hat{\mathbf{x}}_{k+N} &= \mathbf{A}^N\Delta\mathbf{x}_k + \mathbf{A}^{N-1}\mathbf{B}\mathbf{u}_k + \ldots + \mathbf{B}\mathbf{u}_{k+N-1} \\
\hat{\mathbf{y}}_{k+N} &= \mathbf{C}\mathbf{A}^N\Delta\mathbf{x}_k + \mathbf{C}\mathbf{A}^{N-1}\mathbf{B}\mathbf{u}_k + \ldots + \mathbf{C}\mathbf{B}\mathbf{u}_{k+N-1}
\end{aligned} \tag{50}$$

and rewriting these equations into the matrix form, the prediction of the outputs is obtained

$$\hat{\mathbf{y}} = \mathbf{f} + \mathbf{G}\mathbf{u} \tag{51}$$

where

$$\mathbf{f} = \begin{bmatrix} \mathbf{CA} \\ \vdots \\ \mathbf{CA}^N \end{bmatrix} \Delta\mathbf{x}_k,\ \mathbf{G} = \begin{bmatrix} \mathbf{CB} & \ldots & \mathbf{0} \\ \vdots & \ddots & \vdots \\ \mathbf{CA}^{N-1}\mathbf{B} & \ldots & \mathbf{CB} \end{bmatrix} \tag{52}$$

The control is derived from the optimization of a quadratic performance index. The performance index J_k is optimized in the step k using the prediction $\hat{\mathbf{y}} = \begin{bmatrix} \hat{\mathbf{y}}_{k+1} & \ldots & \hat{\mathbf{y}}_{k+N} \end{bmatrix}^T$ for $\mathbf{y}$.

$$\begin{aligned}
J_k &= \sum \left\{ (\widehat{\mathbf{y}} - \mathbf{w})^T \mathbf{Q} (\widehat{\mathbf{y}} - \mathbf{w}) + \mathbf{u}^T \lambda \mathbf{u} \right\} = \\
&= \sum \left\{ (\mathbf{G}\mathbf{u} + \mathbf{f} - \mathbf{w})^T \mathbf{Q} (\mathbf{G}\mathbf{u} + \mathbf{f} - \mathbf{w}) + \mathbf{u}^T \lambda \mathbf{u} \right\}
\end{aligned} \tag{53}$$

where the summation is over the prediction horizon, $\mathbf{y}$ is the output vector, $\mathbf{w}$ is the desired output vector, $\mathbf{Q}$ is a penalization matrix of the outputs, $\lambda\mathbf{I}$ with identity matrix $\mathbf{I}$ is a penalization of the inputs, and $\mathbf{u} = \begin{bmatrix} \mathbf{u}_k, & \ldots & \mathbf{u}_{k+N-1} \end{bmatrix}^T$ is the input vector. From the requirement of the minimization of the performance index

$$J_k = \min_{\mathbf{u}} J_k(\mathbf{u}) \tag{54}$$

the control law is derived

$$\mathbf{u} = (\mathbf{G}^T\mathbf{Q}\mathbf{G} + \lambda\mathbf{I})^{-1}\mathbf{G}^T\mathbf{Q}(\mathbf{w} - \mathbf{f}) \tag{55}$$

However, only the first element of the vector $\mathbf{u}$ is used for the nearest control action.

The control problem very often includes the constraints of the values of different variables. In order to incorporate these constraints, the direct computation procedure (53) must be generalized. Quadratic programming appears to be a good tool for this purpose (Valasek et al., 2005). The unconstrained optimization problem (54) is reformulated as a quadratic optimization problem

$$\min_{\mathbf{u}} \left(\frac{1}{2} \mathbf{u}^T \mathbf{H}_{\mathrm{QP}} \mathbf{u} + \mathbf{f}_{\mathrm{QP}}^{\mathrm{T}} \mathbf{u} \right) \tag{56}$$

with a constraint condition

$$\mathbf{A}_{\mathrm{QP}} \mathbf{u} \leqslant \mathbf{b}_{\mathrm{QP}} \tag{57}$$

The particular matrices used for quadratic programming follow from equations (51)-(55)

$$\begin{aligned} \mathbf{H}_{\mathrm{QP}} &= \mathbf{G}^T \mathbf{Q} \mathbf{G} + \lambda \mathbf{I} \\ \mathbf{f}_{\mathrm{QP}} &= \mathbf{G}^T \mathbf{Q} (\mathbf{w} - \mathbf{f}) \end{aligned} \tag{58}$$

This optimization problem is solved in every sampling instant. Again, only the first element of vector $\mathbf{u}$ is used for the nearest control action.

6 Applications

6.1 Mechatronic and X-Friendly Suspension

The mechatronic suspension can be called a vehicle suspension in which the properties can be controlled. The important feature of mechatronic suspension is the synergy between the used technologies. This means the synergy between the mechanical, hydraulic, electrical, electronic, control and computer components, where the drawbacks are compensated and advantages are strengthened, as well as the synergy between the suspension and the function of other subsystems like ABS. For classification (Valasek and Kortüm, 2001) it is important to consider the time scale of changes of suspension properties. The adaptive suspensions with the change of the suspension setting in time longer than 1 sec and rapidly controlled suspensions with the change of the suspension setting below 1 sec can be distinguished. This classification is based on the control strategy, not on the actuator's dynamic capabilities. There can be several variants of actuators. Either an active servodrive (e.g. hydraulic cylinder) acts between sprung and unsprung masses or semi-active actuator is considered. Such semi-active actuator either changes the damping constant of the damper or the stiffness of the spring. The difference between the active and semi-active actuators is that the active one can add energy into the system but semi-active one can only dissipate the energy from the system.

The control of vehicle suspension is implemented in order to minimize (reduce) certain performance index. It is spoken about certain suspension friendliness like road-friendliness for reduction of road-tyre forces and thus for reduction of road damage being caused by dynamic peaks of road-tyre forces. This terminology can be generalized into the concept of X-friendly vehicle suspension for the range of passenger-friendly, driver-friendly, load-friendly, road-friendly, bridge-friendly, brake-friendly, soil-friendly. This list can be further extended. The suspension controller should be a complex X-friendly controller containing all the mentioned functionalities. The appropriate control strategy will be switched according to the current conditions. The priority must to be the vehicle safety e.g. during emergency stopping. The second level of importance is then the road/bridge/soil-friendliness, which is particularly important for heavier vehicles. The lower level will usually be the passenger/driver/load comfort.

The further described different controlled vehicle suspensions were designed by the MOPO design methodology. The vehicle system to be controlled is usually significantly nonlinear. Thus the traditional control design approaches based on linear models provide only limited results. Therefore an advanced methodology for control design in the case of semi-active vehicle suspensions has been developed (Kortüm et al., 2002; Valasek et al., 1998; Valasek and Kortüm, 2001), which can also be considered as a methodology for the design of mechatronic systems for vehicles in general.

The different concepts of X-friendly vehicle suspension are further described. They were designed by the design methodology described above for currently realistic actuators. The influence of actuators from point of view of dynamics and energy consumption is described in the next section.

6.2 Comfort oriented Driver/Passenger/Load Friendly Suspension

The concept of controlled suspension for comfort is obvious. The comfort or the interaction with the driver or load has been investigated since the first studies. A lot of references exist (Karnopp et al., 1974; Sharp and Crolla, 1987; Venhovens, 1993; Elbeheiry et al., 1995). The performance index is equivalent to minimization of filtered sprung mass acceleration with respect to the inertial space according to ISO 2631. Hence Karnopp's idea of Sky-Hook comes (Karnopp et al., 1974). Sky-Hook is a fictitious damper between the sprung mass and the inertial frame (fixed in the sky) (Figure 26).

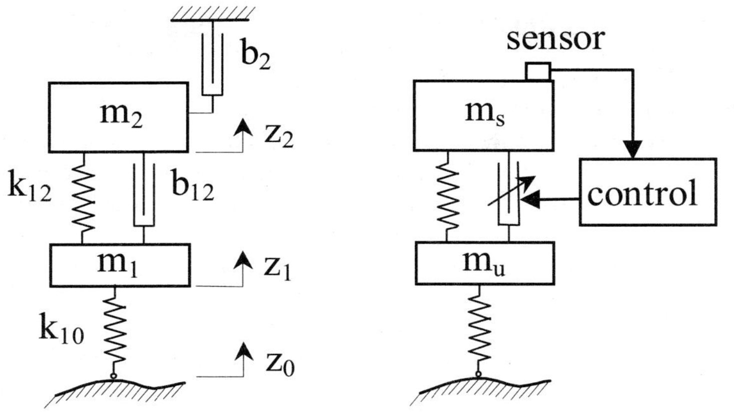

Figure 26. Sky-Hook – ideal concept (a) and realization (b).

The further design considerations are based on the simple quarter-car model (Figure 26). Despite its simplicity, it covers the fundamental suspension dynamic properties of the real vehicle. For the description of the control concept a linear quarter-car model is used; however, the nonlinearities of the force elements are taken into account in further descriptions. The damping force of this fictitious damper reduces the sprung mass vibration. Its equations of motion are

$$\begin{aligned} &m_1\,\ddot{z}_1 + k_{10}\,(z_1 - z_0) + b_{10}\,(\dot{z}_1 - \dot{z}_0) - k_{12}\,(z_2 - z_1) + F_d = 0 \\ &m_2\,\ddot{z}_2 + k_{12}\,(z_2 - z_1) - F_d = 0 \end{aligned} \tag{59}$$

where m_1 is the unsprung mass, m_2 is the sprung mass, k_{12} is the stiffness of the main spring, k_{10} is the stiffness of the tyre, b_{10} is the tyre damping constant and F_d is the force of the passive or semi-active damper or active element. The meaning of the kinematical quantities z_2, z_1 and z_0 is also clear from this figure. Figure 26a represents the ideal fictitious case. The parallel fictitious damper b_2 tries to add the higher damping to the sprung mass. Figure 26b represents the realization of this concept. The fictitious force computed from added Sky-Hook damper is applied by the actuator F_d. The actuator can be a fully active element (active force generator) or a semi-active element (variable shock absorber). The ideal active force F_d of this element, according to the control law of the Sky-Hook, is

$$F_d = -b_2 \dot{z}_2 \tag{60}$$

If this force is directly applied by active force generator then it is the active Sky-Hook suspension. For semi-active suspension this force must be limited to the range of forces applicable by the semi-active devices. For the ideal semi-active damper this transformation from the required (active) force F_{act} to the setting of the damping rate $\text{b}_{semi-active}$ such that the damping force is nearest to the desired one. For an ideal linear variable shock absorber the damping rate $b_{\text{semi-active}}$ is set for the interval $(b_{\text{min}}, b_{\text{max}})$ as

$$b_{\text{semi-active}} = \begin{array}{ll} b_{\text{max}} & \text{if} \quad b_{\text{max}} < b_{\text{act}} \\ b_{\text{act}} & \text{if} \quad b_{\text{min}} < b_{act} < b_{\text{max}} \\ b_{\text{min}} & \text{if} \quad b_{\text{act}} < b_{\text{min}} \end{array} \tag{61}$$

$$b_{\text{act}} = \frac{F_{\text{act}}}{(\dot{z}_2 - \dot{z}_1)} \tag{62}$$

The semi-active Sky-Hook suspension is then realized by taking $F_{act} = F_d$ from (60) into (61)-(62) and (59). The real variable shock absorber is, however, nonlinear and time-dependent with internal dynamics and the transformation (61)-(62) must take it into account.

Such control concept of semi-active suspension for comfort can achieve based on the simulation studies that the riding comfort is improved by about 10% in almost whole frequency range. Further improvement can be achieved by force feedback control loop of the semi-active damper (Valasek et al., 2002a) as well as by speeding up the damper response, i.e. reducing of the time constant of the damper as indicated in Figure 27.

The simulation study on Figure 27 has been done for the truck on Figure 29 passing a good stochastic road at the speed 72 km/h with the damper dynamics characterized by the triple of parameters (time delay of damper response [ms], time constant for decreasing force [ms], time constant for increasing force [ms]). A good stochastic road means the stochastic simulation of a new highway road (Kortüm and Valasek, 1998).

6.3 Road-Friendly Suspension

One of the objectives of the SADTS project was to investigate the potential of improvement of road-friendliness by the semi-active suspension (Yi and Hedrick, 1989; Kortüm and Valasek, 1998; Valasek et al., 1998). There have been investigated many different control concepts. The newly developed one is so called Ground-Hook suspension control (Novak and Valasek, 1996; Valasek et al., 1997a,b).

The further design considerations are based on the simple quarter-car model. Despite its simplicity, it covers the fundamental suspension dynamic properties of the real vehicle. For the description of the control concept a

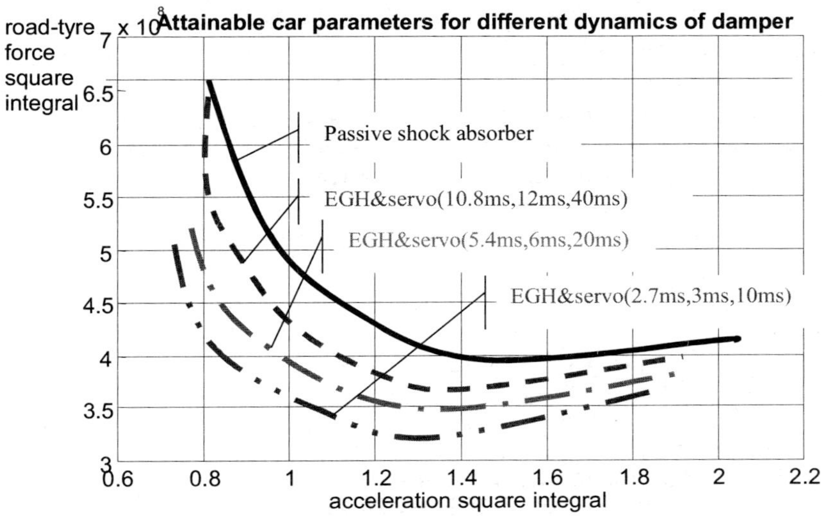

Figure 27. Pareto set of attainable vehicle properties (comfort – handling).

linear quarter-car model is used; however, the nonlinearities of the force elements are taken into account in further descriptions.

The basic idea of the Extended Ground-Hook (EGH) is depicted in Figure 28.

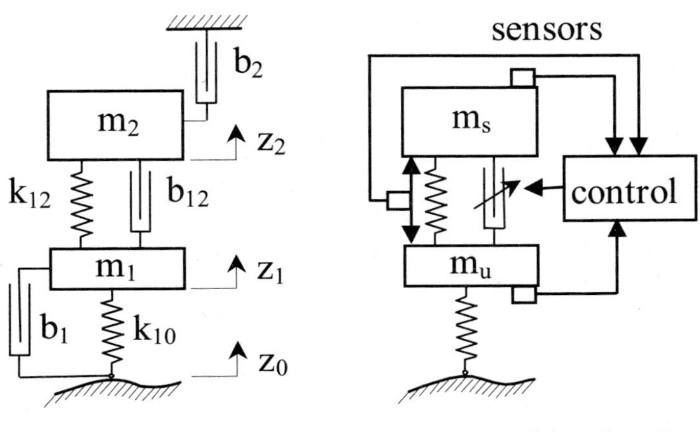

Figure 28. Extended Ground-Hook - ideal concept (a) and realization (b).

Figure 28a represents the ideal fictitious case. The real damping of the tyre b_{10} has a very low value, and the parallel fictitious damper b_1 tries to add the higher damping value b_1 to it. The preservation of low accelerations of the sprung mass has been reached by the combination with a Sky-Hook. Figure 28b represents the realization of this concept. The fictitious force computed from added Ground-Hook and Sky-Hook dampers is applied by the actuator F_d. The actuator can be a fully active element (active force generator) or a semi-active element (controllable shock absorber). The ideal force F_d of this element, according to the control law of the Extended Ground-Hook, is

$$F_d = b_1 (\dot{z}_1 - \dot{z}_0) - b_2 \dot{z}_2 - b_{12} (\dot{z}_2 - \dot{z}_1) + \Delta k_{10} (z_1 - z_0) - \Delta k_{12} (z_2 - z_1) \quad (63)$$

where Δk_{10} and Δk_{12} are further additional terms for the fictitious stiffness cancellation, which are mainly important in the case of a fully active actuator (Valasek et al., 1997b). This final variant contains all the terms of the full state feedback, with the exception of one absolute position term, such as z_0, z_1 or z_2. The advantage of the last equation is that all the terms are either directly measurable or reconstructable from other measurements.

By changing the parameters b_1, b_2, b_{12}, Δk_{10} and Δk_{12} a variety of modified control laws for the system can be obtained. For the systematic determination of these parameters, the MOPO approach was applied. The multi-objective parameter optimization (MOPO) (Kortüm and Valasek, 1998; Valasek et al., 1997a,b; Valasek and Kortüm, 1998) allows one to find a satisfactory compromise among the performance criteria despite the fact that they conflict with each other. The MOPO approach is based on a search in the parameter space (Pareto optimality) by model simulation. MOPO requires one to define the criteria, the parameter set and the representative reference inputs. Thus, by means of the MOPO approach the nonlinear models and models, which cannot be analytically treated, can be used. The single performance index was the time integral of the dynamic tyre force

$$I = \int_0^{T_{\text{large}}} F_{10}^2 dt \quad (64)$$

The parameters were determined by means of different optimization procedures both deterministic and genetic. The parameters of the Extended Ground-Hook were originally considered constant for the whole shock absorber velocity interval (Kortüm and Valasek, 1998; Valasek and Kortüm, 1998). Because the characteristics of the shock absorber are nonlinear, the nonlinear Extended Ground-Hook control with state-dependent gains (gain scheduling) is used. The strong nonlinearity of the controlled shock absorber, especially its asymmetry, can be taken into account for the determination of control-law parameters. Therefore the nonlinear EGH version, which enables the state-dependent coefficients (gains) of the EGH control law (63), was developed with excellent results (Valasek and Kortüm, 1998). Their dependence on the relative velocity is determined by the MOPO optimization approach.

Figure 29. Experimental truck: tractor M.A.N. with semi-trailer.

This concept was implemented for two truck prototypes SKODA-LIAZ and M.A.N. (Figure 29). The used controllable dampers are in Figure 30. The final results from real experiments on the prototypes are promising. Road friendliness was improved in all cases. For very good stochastic road the measurements have shown (Figure 31) that standard deviation of road-tyre dynamic force (DLC) was reduced by 20%. The same effect in time dependence can be seen on Figure 32 for cosine obstacle. This means that the measures for the road damage (DLSF95%) was reduced by 5%. The detailed definition of DLSF95% is in (Yi and Hedrick, 1989; DIVINE, 1997; Kortüm and Valasek, 1998; Valasek et al., 1998). For deterministic obstacle of a ramp standard deviation of road-tyre dynamic force (DLC) was reduced by 26%. This means that the measure for the road damage (DLSF95%) was reduced by 39% or the aggregated forces by 8-70%. The other interpretation

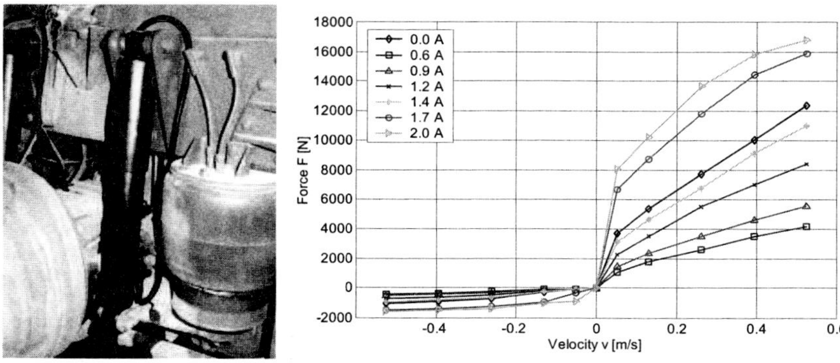

Figure 30. Controllable damper and its controllable characteristics.

of these results maybe that the allowed useful payload can be increased by 1-1.6 ton on 10 tons.

6.4 Bridge-Friendly Suspension

Studies (Valasek and Kortüm, 1999; Valasek et al., 2002b) deal with the investigation whether the road friendly suspension is also the bridge friendly one. The developed semi-active suspension was investigated within the bridge-truck interaction. The comparison between the response of bridge - truck interaction for passive and semi-active suspension is on the Figure 33. The maximum interaction force under the drive axle is lower for semi-active suspension compared with passive one. The conclusion based on the simulation study is that if the truck is on the entrance to the bridge excited which usually occurs then the semi-active suspension can reduce the dynamic road-bridge forces and bridge deflection by about 5%.

6.5 Soil-Friendly Suspension

Similarly as the interaction between truck and road can be influenced by the controlled suspension also the interaction between the truck (agriculture tractor) and the soil (off road) can be influenced. The control concept is based on the Extended Ground-Hook. The initial investigation has shown the potential of about 5% soil deformation reduction. The Pareto set between comfort and soil damage for both passive optimized and semi-active suspensions is depicted on Figure 34a. The Figure 34b presents the detail of the resulting soil deformation after multi-pass loading based on the

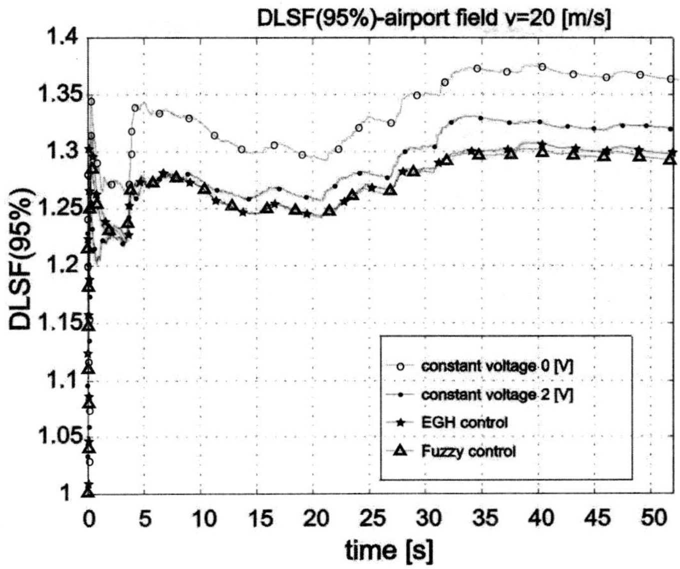

Figure 31. Comparison of the measurement of improvement of road-friendliness.

simulation study. The figure describes the development of the obstacle in the form of the initial bump on the soil road (upland sandy loam) after 7 passes with passenger off-road with passive, semi-active or limited-active suspension, (Valasek et al., 2006b). The usage of the passive suspension leads to the development of typical wavy road, the usage of the controlled suspension (semi-active or active) can prevent such road deterioration.

6.6 Brake-friendly Suspension

It was realized that the function of ABS is strongly influenced by the road unevenness and resulting fluctuation of road-tyre forces. The consequence of this consideration is the concept of brake-friendly suspension (Valasek et al., 2004). It is based on the control of suspension in order to keep the road-tyre forces maximal for optimal brake function.

The influence of fluctuation of road-tyre forces is twice. First, the longitudinal force coefficient between the tyre and the certain given road depends on the vertical load and the longitudinal slip in a nonlinear way. As a result of this nonlinearity, the oscillations of the vertical force result in an extension of the stopping distance. The unloading of a wheel in some time instant and

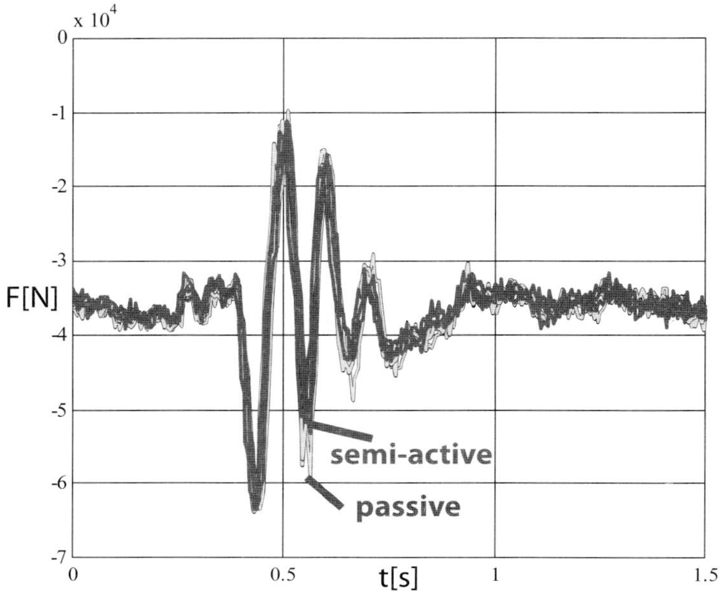

Figure 32. Time behaviour of road-tyre forces on cosinus bump. Comparison between conventional and controlled damping from real measurements.

the additional load on the wheel in the next time instant leads in the sum to less effort of longitudinal braking force than in the case of mean value of wheel vertical load. Second, the controlled suspension is in any case capable of decreasing the situation of losing contact between tyre and road (short wheel hop) when the braking is completely interrupted. These undesirable situations are even worse in the case of ABS. The ABS algorithm is avoiding wheel blocking which happens exactly after losing tyre-road contact while the wheel blocking in other situations during tyre-road contact is prevented by the ABS. The consequence of the ABS algorithm and especially its sampling time is that after losing and gaining the contact, the wheel is for some time interval not being braked. This is the time interval between the time

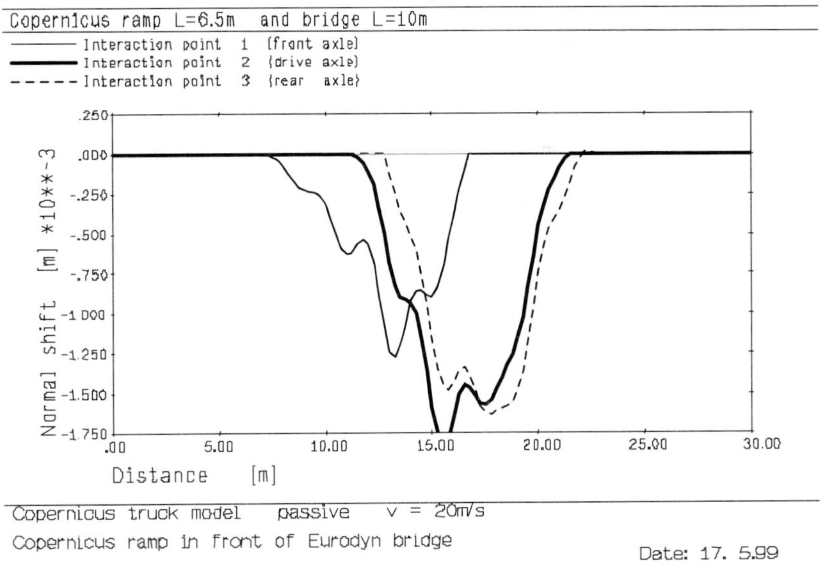

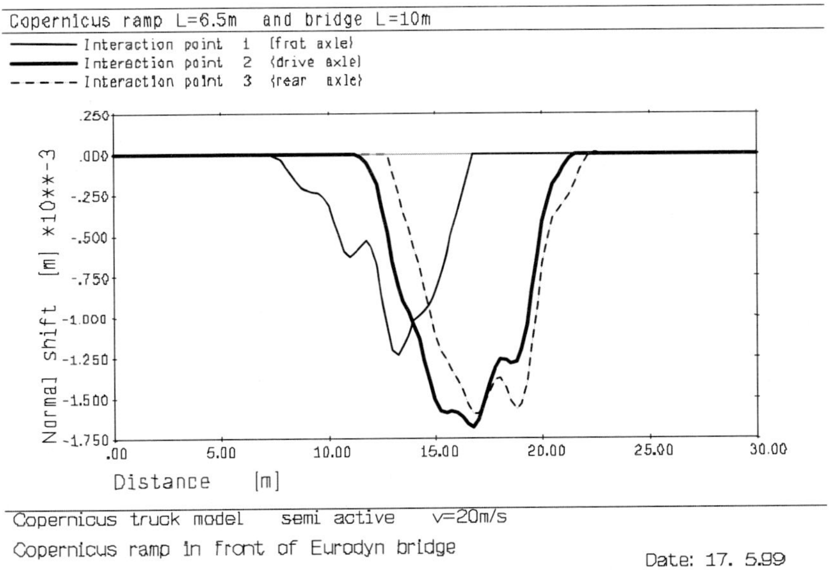

Figure 33. Comparison between passive and semi-active suspension on a bridge (bridge deflection).

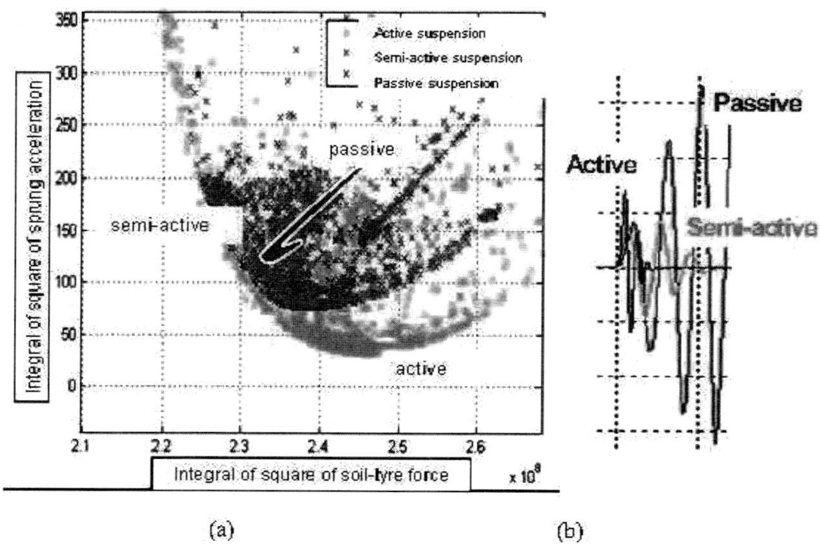

Figure 34. Pareto set of soil-friendly suspension (a) and detail of resulting deformation after repetitive loading by passive, semi-active or limited active suspensions (b).

instant of the losing of the wheel contact and the first sampling time instant after the gaining of the wheel contact. It can be tens of milliseconds. The problem is the sum of such drop-outs of braking.

A dynamically controlled vehicle suspension, where the controller is optimized to the braking manoeuvre, can suppress the undesirable vibrations and shorten the braking distance. This is called brake-friendly suspension. The main goal during the stopping manoeuvre is to reduce the fluctuation of the vertical forces between road and tyre. This is the control objective to be provided. Initially it is the same objective as the road-friendliness but whether some specific brake-friendliness significantly different from road-friendliness exists is an open question.

A wide range of simulation experiments has been conducted. There were different passive settings of suspensions, semi-active suspensions with different configurations (with non-symmetric and symmetric characteristics and different time constants) and limited-active suspensions. The results are summarized in Figure 35.

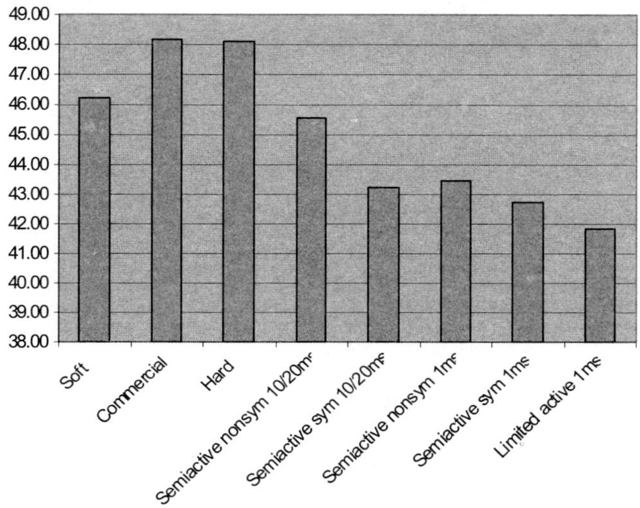

Figure 35. The stopping distance for different suspension concepts on a good stochastic road.

The vehicle has an initial velocity of 100 km/h. The emergency braking manoeuvre is simulated on a road excitation representing a good stochastic road. The simulation results presented in Figure 35 indicate a shortening of the vehicle stopping distance by 2.5 to 6.3 meters. Besides the influence of the symmetric characteristics of the semi-active damper, the contribution of the reduced time constant is observable. The limited active concept brings further reduction of the stopping distance by 0.8 m compared to the best semi-active concept. All these results are based on the simulation study (Valasek et al., 2004).

In order to validate the simulation results a Hardware-in-the-Loop (HiL) test rig has been designed and manufactured, see Figure 36a, (Steinbauer et al., 2005). The quarter car is equipped with a magnetorheological damper. The test rig represents the vertical vehicle dynamics of the vehicle during the stopping manoeuvre. The longitudinal dynamics is modeled in a real-time computer environment together with the damper control algorithm. The results (Figure 36b) achieved on the test rig indicate that for the given road conditions the stopping distance has been decreased by about 1.3 meters for the brake-friendly suspension.

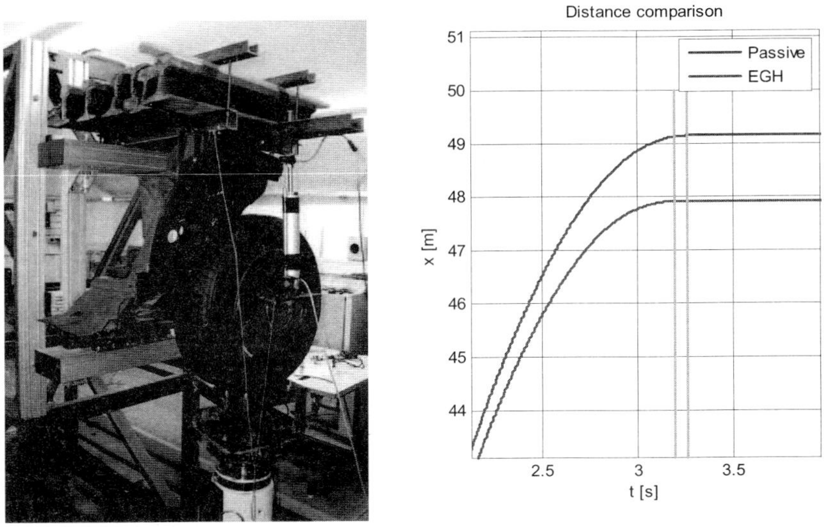

(a) (b)

Figure 36. Quarter car test rig with a magnetorheological damper for the HiL simulation (a) and the comparison of the stopping distance for passive and brake-friendly (EGH) vehicle.

6.7 Model based predictive control of combustion engine

Current combustion engines in vehicles are equipped with sophisticated electronic control. The global engine control (motor management) is usually based on partial characteristics measured on experimental engine stands and stored as look-up tables. The advantage of such approach is that it can accumulate a lot of know-how experience. The drawback is that the control is based on quasistatic characteristics, it is not capable to fully utilize the engine dynamic capabilities and to be adjusted to modified engine properties. Moreover the overall control of powertrain is being developed and there is necessary to coordinate the engine dynamic capabilities and vehicle dynamic requirements. Therefore the recent attention has been paid to the engine and complete powertrain control using the knowledge of their models (Pischinger et al., 2004; Winsel et al., 2004).

This section describes the development of an alternative approach towards combustion engine control. It is based on the model based predictive control. The basis is the development of predictive model that is capable to predict the engine states for several future injection time instances. Using this model the future engine behaviour in dependence on possible control inputs (fuel injections, supercharger blade position etc.) are parametrized and using optimization of performance index the control input is chosen.

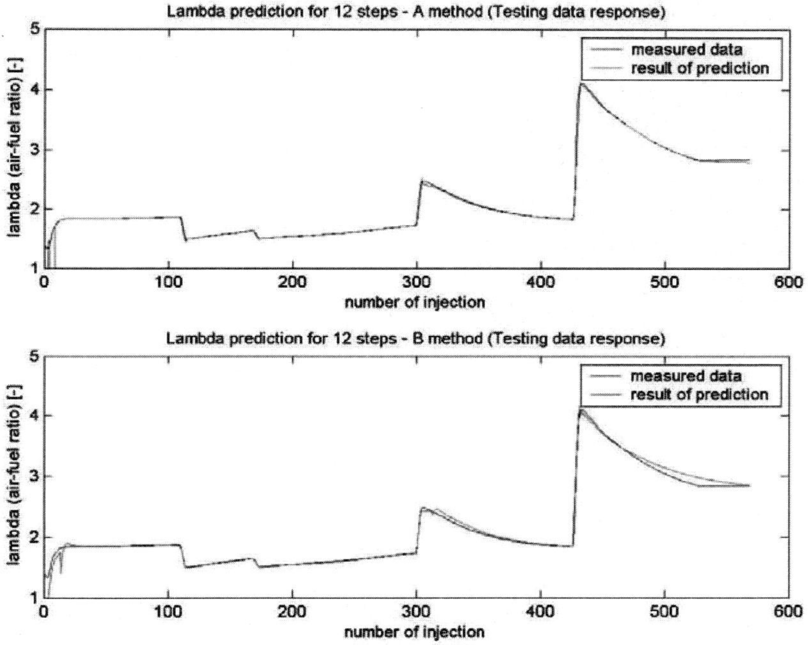

Figure 37. Testing of prediction by identified model.

The whole approach described in this section has been developed and verified in three phases. Firstly, the simulation model with reasonable ratio between accuracy and computational complexity is developed. Secondly, the engine simulation model is identified by LOLIMOT approach and corresponding engine predictive model is developed (see the previous section dealing with model based control). The problem of predictive control is the solution of computational complexity of control input selection during optimization. It is especially complicated by nonlinearity of the predictive model. This problem can be solved by using locally linearized dynamic models enabling efficient choice of optimized control input and preserving the global nonlinear behaviour of the engine (CTU-LOLIMOT – MPC approach). Thirdly, the engine predictive control concept based on LOLIMOT identification is implemented to a combustion engine model and overall investigated. There was a serious problem how to describe the complicated dynamic model of combustion engine with many time delays and possible control inputs only in small time interval of fuel injection. This has been solved by using the discrete instances of fuel injection as time discretization of the dynamic model. The resulting dynamic model is a discrete dynamic model describing the engine behaviour from one fuel injection until the next one. The simulation results indicate a good predictive capability of the applied approach (see Figure 37) and control capability of predictive control method.

Simulation model of combustion engine The engine simulation model has been developed in order to create a realistic, but computationally efficient simulation model (Macek et al., 2006). The physical simulation model uses all special features of combustion engine simulation. The model is based on acceptably simplified simulation model of diesel combustion engine (Figure 38).

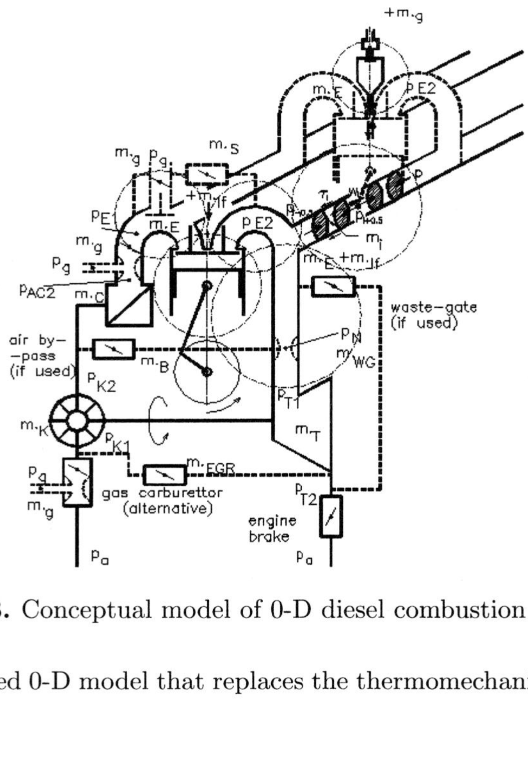

Figure 38. Conceptual model of 0-D diesel combustion engine.

It is a so-called 0-D model that replaces the thermomechanical processes with pure time delays of appropriate variables. The simulation model has 7 state variables

α −crankshaft angle [$^\circ$]
n_M −crankshaft revolutions per minute [rpmin]
n_{TD} – revolutions of turbocharger [rpmin]
m_{SP} – air mass in inlet tube [kg]
m_{VP} – air mass in exhaust pipe [kg]
T_{SP} – air temperature in inlet tube [K]
T_{VP} – air temperature in exhaust pipe [K]
and 2 control variables
m_p −mass of injected fuel into 1 cylinder [kg]
A -reference area of turbocharger [m^2].

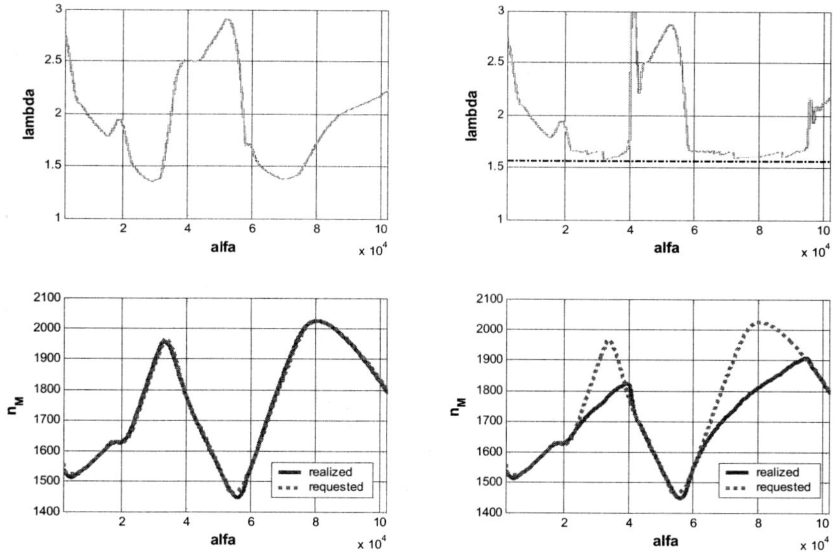

a) without constraint b) constrained lambda

Figure 39. Model predictive control of engine rpm – simulation results (lower constraint of lambda).

Model based engine predictive control The model based predictive control of combustion engine is based on the CTU-LOLIMOT predictive model (22) that enables to compute the system matrices **A**, **B**, **C** for MPC by direct way. Then the transfer function is transformed into the state space description. Based on that the predictive control can be applied. The solution procedure is generalized using quadratic programming in order to easily incorporate various constraint conditions. The algorithm of model predictive control based on identified CTU-LOLIMOT models and quadratic programming can be briefly summarized as follows (Valasek et al., 2005) :

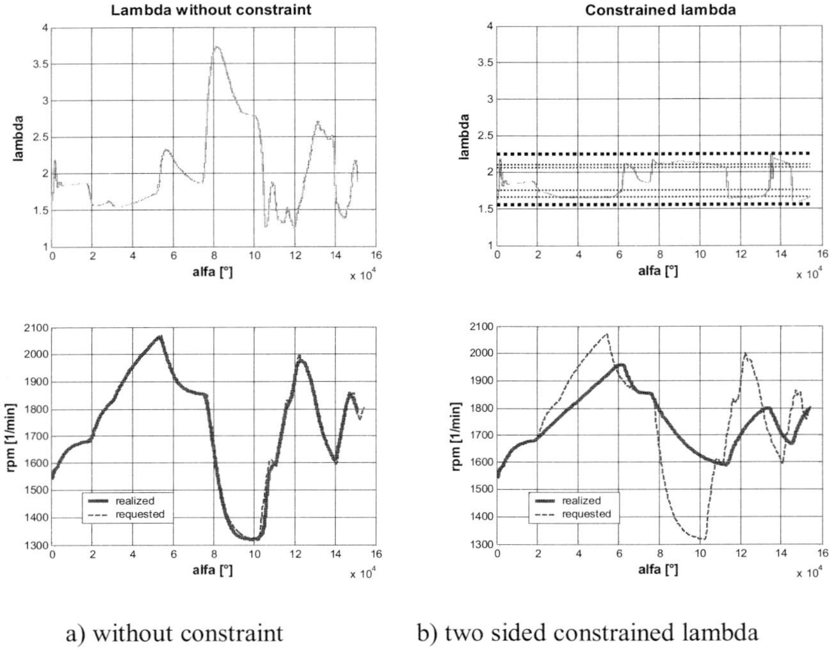

Figure 40. Model predictive control of engine rpm – simulation results (two sided constraint of lambda).

- The nonlinear predictive CTU-LOLIMOT models of motor (and/or complex simulation model of motor) are identified, assembled and tested by CTU-LOLIMOT program. The measurable quantities of system are chosen as outputs of individual CTU-LOLIMOT models. The additional CTU-LOLIMOT models are assembled and tested for controlled quantities (for example lambda subjected to constraints during control).
- The nonlinear predictive CTU-LOLIMOT models are incorporated into controller (simulation model of controller). The outputs of main

CTU-LOLIMOT models (measurable quantities) are considered as states of substitutional linear model used for predictive control. That means, the CTU-LOLIMOT models are locally (for measured actual values of states) linearized and inserted into matrix form (58). The locally linearized additional CTU-LOLIMOT models are assembled to equality and non-equality constraints. Actually the upper and lower limits of lambda have been considered.

- The control input is computed by solution of quadratic programming problem (56). Only the first element of vector $\mathbf{u}_{\mathrm{con}}$ is used for the nearest control action.

The quality of CTU-LOLIMOT predictive models is demonstrated in Figure 37. The simulation results of control are on Figure 39. The numerical experiments show, that the used prediction horizon should be truncated in the vicinity of lambda limits (see the line on Figure 39b). The application of limits affects attainable values of primarily controlled quantity (rpm). Similar results are for two sided constraint of lambda in Figure 40. The real experiment with implemented MPC with CTU-LOLIMOT predictive models is in Figure 41. It is the MPC control of supercharged gas engine held on constant speed where the engine torque is controlled by means of gas flap operation.

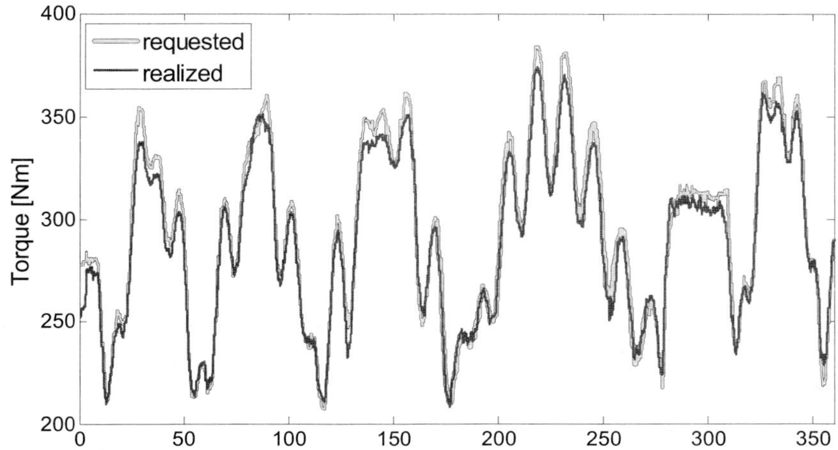

Figure 41. Implementation of MPC control on supercharged gas engine.

This section describes the method for creating efficient predictive model and the predictive control as control approach that uses the knowledge of engine models. The simulation indicates high quality of prediction and control processes. The control input is computed by solving the quadratic programming problem. The simulation experiments were carried out on diesel combustion engine and they have proven the capability of this approach to control the combustion engine even in dynamic states. The long-term goal is to develop model based predictive control of combustion engines based on on-line identified nonlinear predictive models.

7 Conclusion

In this contribution the ways of efficient modeling, simulation and model based control design have been described. They are essential for the design and development of mechatronical systems. The described methods were applied for the design of integrated chassis and combustion engine control of vehicles. The applications have demonstrated that the proposed methods are fully applicable in practice and very efficient.

Bibliography

20-SIM. http://www.20sim.com/.

ACSL. http://www.aegistg.com/.

M. Arnold. Numerical methods in the simulation of vehicle-guideway interaction. In A. Buikis, R. Ciegis, and A.D. Fitt, editors, *Progress in Industrial Mathematics at ECMI 2002.* pages 115–120. Springer–Verlag, Berlin Heidelberg, 2004.

M. Arnold, A. Carrarini, A. Heckmann, and G. Hippmann. Simulation techniques for multidisciplinary problems in vehicle system dynamics. In M. Valasek (ed.). *Computational Mechanics in Vehicle System Dynamics, Supplement to Vehicle System Dynamics.* 40:17-36, 2004.

S.P. Banks and K.J. Mhana. Optimal control and stabilization for nonlinear systems. *SIAM Journal of Mathematical Control and Information.* 9:179-196, 1992.

E.F. Camacho and C. Bordons. *Model Predictive Control.* Springer Verlag, London 2004.

F.E. Cellier. *Continuous System Modeling.* Springer Verlag, 1991.

J.R. Cloutier. State-Dependent Riccati Equation Techniques. *Proc. of 1997 American Control Conference.* Albuquerque, pages 932-936, 1997.

J. R. Cloutier, D. T. Stansbery, and M. Sznaier. On the recoverability of nonlinear state feedback laws by extended linearization control techniques. *Proc. of the 1999 American Control Conference.* San Diego, pages 1515-1519, 1999.

DIVINE (1997). *Proceedings of DIVINE Concluding Conference.* Rotterdam 1997.

DYMOLA. http://www.dynasim.se/.

E.M. Elbeheiry, D.C. Karnopp, M.E. Elaraby and A.M. Abdelraaouf. Advanced Ground Vehicle Suspension Systems - a Classified Bibliograhy. *Vehicle System Dynamics.* 24:231-258, 1995.

E.B. Erdem and A. Alleyne. Globally stabilizing second order nonlinear systems by SDRE control. *Proc. of the 1999 American Control Conference.* San Diego, pages 2501-2505, 1999.

M. Fliess (ed.). *Nonlinear Control System Design*, Pergamon Press 1992.

G. Franklin, J.D. Paell and Abbas Emami-Naeini. *Feedback Control of Dynamic Systems.* Edison-Wesley 1993.

A. Guran and H.S. Tzou (eds.). *Structronic Systems. Smart Structures, Devices and Systems, Part I, II.* World Scientific, Singapure 1998.

J. Hewit. *Lecture Notes for Mechatronics Advanced School*, CISM, Udine 1992.

A. Isidori. *Nonlinear Control Systems*, Springer-Verlag, Berlin 1989.

H. Janocha. *Adaptronics and Smart Structures. Basics, Materials, Design, and Applications.* Springer, Berlin 1999.

H.D. Joos. *Informationstechnische Behandlung des mehrzieligen optimierungsgestutzten regelungstechnischen Entwurfs.* PhD Thesis, Univ. of Stuttgart 1992.

D.C. Karnopp, M.J. Crosby and R.A. Harwood. Vibration Control Using Semi-Active Force Generators. *Journal of Engineering for Industry.* Transactions of ASME, 96:619-626, 1974.

D.C. Karnopp, D.L. Margolis and R.C. Rosenberg. *System Dynamics. A Unified Approach.* John Wiley & Sons Inc., N.Y., 1990.

A. Kelly and C. Zweben. Smart Composite Materials Systems. In *Comprehensive Composite Materials.* Elsevier Science Ltd., 2000.

H.K. Khalil. *Nonlinear Systems*, Prentice Hall, New Jersey 1996.

W. Kortüm, R.M. Goodall and J.K. Hedrick. Mechatronics in Ground Transportation - Current Trends and Future Possibilites. *Annual Reviews in Control*, Pergamon Press, pages 133-144, 1998.

W. Kortüm and M. Valasek (Eds.). *SADTS Final Technical Report.* DLR, Wessling 1998.

W. Kortüm, M. Valasek, Z. Sika, W. Schwartz, P. Steinbauer and O. Vaculin. Semi-Active Damping in Automotive Systems. Design by Simulation. *International Journal of Vehicle Design.* 28:103-120, 14, 2002.

R. Kübler and W. Schiehlen. Two Methods of Simulator Coupling. *Mathematical and Computer Modelling of Dynamical Systems.* 6:93-113, 2, 2000.

W. Langson and A. Alleyne. Inifinte horizon optimal control of a class of nonlinear systems. *Proc. of the IEEE American Control Conference 1997.* Albuquerque, pages 3017-3022, 1997.

J. Macek, M. Polasek, Z. Sika, M. Valasek, M. Florian and O. Vitek. Transient Engine Model as a Tool for Predictive Control, *SAE Paper.* 2006-01-0659, 2006.

MATLAB. http://www.mathworks.com/.

MODELICA. http://www.modelica.org/.

O. Nelles. Nonlinear system identification with local linear fuzzy-neuro models. *Automatisierungstechnik.* Shaker Verlag, Aachen 1999.

R. Neugebauer, B. Denkena and K. Wegener. Mechatronic systems for machine tools, *CIRP Annals.* 56:657-686, 2, 2007.

M. Novak and M. Valasek (1996). A New Concept of Semi-Active Control of Truck's Suspension, In *Proc. of AVEC 96.* Aachen, pp. 141-151, 1996.

S. Pischinger, C. Schemus, G. Luekemeyer, H. J. Theuerkauf and T. Winsel. Investigation of Predictive Models for Application in Engine Cold-Start Behaviour. *SAE Paper.* 2004-01-0994, 2004.

J.A. Rossiter. *Model-Based Predictive Control. A Practical Approach.* CRC Press, Boca Raton 2003.

T. Roubicek and M. Valasek. Optimal Control of Causal Differential-Algebraic Systems. *Journal of Mathematical Analysis and Its Applications.* 26:616-641, 269, 2002.

R.S. Sharp and D.A. Crolla. Road Vehicle Suspension System Design - a Review. *Vehicle System Dynamics.* 16:167-192, 1987.

J.J. Slotine and W. Li. *Applied Nonlinear Control*, Prentice Hall, New Jersey 1991.

M. Stefan, Z. Sika, M. Valasek and V. Bauma. Neuro-Fuzzy Identification of Nonlinear Dynamic MIMO Systems, *Engineering Mechanics.* 13:223-238, 3, 2006.

P. Steinbauer, Z. Sika, M. Valasek, T. Skopalik and P. Mikolas. HIL Experiments with Quarter Car. *Engineering Mechanics.* 12:99-106, A1, 2005.

P. Steinbauer, M. Valasek, Z. Zdrahal, P. Mulholland and Z. Sika. Knowledge Support of Virtual Modelling and Simulation, In *Virtual Nonlinear Multibody Systems.* Dordrecht. Kluwer Academic Publishers, , vol. 1, pages 233-242, 2003.

V. Stejskal and M. Valasek. Kinematics and Dynamics of Machinery, Marcel Dekker, New York 1996.

M. Tomizuka. Mechatronics. From the 20th to 21st Century, In R. Isermann (ed.). *Proc. of 1st IFAC Conference on Mechatronic Systems.* Darmstadt, pages 1-10, 2000.

M. Valasek. Simulation of mechatronic systems as systems with different physical nature, In M. Acar, J. Makra and E. Penney (eds.). *Mechatronics.* Computational Mechanics Publ., Southampton, pp. 749-754, 1994.

M. Valasek. Mechatronic System Design Methodology - Initial Principles Based on Case Studies, In. J. Adolfsson and J. Karlsen (eds.). *Mechatronics 98, Proc. of Mechatronics 98 Conf.* Pergamon Press, Amsterdam, pages 501-506, 1998.

M. Valasek. Design of Nonlinear Control of Nonlinear Multibody Systems, In *Virtual Nonlinear Multibody Systems.* Dordrecht. Kluwer Academic Publishers, vol. 1, pages 263-278, 2003.

M. Valasek. Design and Control of Under-Actuated and Over-Actuated Mechatronical Systems - Challenges of Mechanics and Mechatronics. *Supplement of Vehicle System Dynamics.* 37:37-50, 40, 2004.

M. Valasek. Simulation Model Development in Analogy with Software Engineering, In *Proceedings 5th MATHMOD Vienna Symposium on Mathematical Modelling.* (CD-ROM). Vienna. ARGESIM, pages 140, 2006.

M. Valasek, M. Babic, Z. Sika and L. Magdolen. Development of Semi-Active Truck Suspension, In *Proc. of IFAC Symposium on Transportation Systems.* Chania, pages 470-475, 1997a.

M. Valasek, P. Breedveld, Z. Sika and T. Vampola. Software Tools for Mechatronic Vehicles. Design Through Modelling and Simulation. *Supplement Vehicle System Dynamics.* 33:214-230, 1999.

M. Valasek and J. Kejval. New Direct Synthesis of Nonlinear Optimal Control of Semi-Active Suspensions, In *Proc. of Advanced Vehicle Control AVEC 2000.* Ann Arbor, pages 691-697, 2000.

M. Valasek, J. Kejval and J. Maca. Control of truck-suspension as bridge-friendly, In H. Grundmann and G. I. Schueller (eds.). *Structural Dynamics Eurodyn 2002.* Munich, Balkema, pages 1015-1020, 2002b.

M. Valasek and W. Kortüm. Nonlinear Control of Semi-active Road-friendly Truck Suspension, In *Proc. of AVEC 98.* Nagoya, pages 275-280, 1998.

M. Valasek and W. Kortüm. Road-friendly Trucks, In L. Fryba and J. Naprstek (Eds.). *Structural Dynamics-Proc. of Eurodyn '99.* Balkema, Rotterdam, pages 855-860, 1999.

M. Valasek and W. Kortüm. Semi-active Suspension Systems II, In *Mechanical Design Handbook.* CRC Press, 2001.

M. Valasek, W. Kortüm, Z. Sika, L. Magdolen and O. Vaculin. Development of Semi-Active Road-Friendly Truck Suspensions. *Control Engineering Practice.* 6:735-744, 1998.

M. Valasek, J. Maca, P. Steinbauer, J. Spurny, V. Smilauer and O. Vaculin. Coupling of Simulation Tools for Simulation of Multiphysical Systems. *Engineering Mechanics.* 13:213-222, 3, 2006a.

M. Valasek, M. Novak, Z. Sika and O. Vaculin. Extended Ground-Hook - New Concept of Semi-Active Control of Truck's Suspension. *Vehicle System Dynamics.* 27:289-303, 5-6, 1997b.

M. Valasek, Z. Sika, J. Kejval, J. Honcu and P. Steinbauer. Force Control of Variable Shock Absorber. In *Proceedings of AVEC'02.* Tokyo. Society of Automotive Engineers of Japan, vol. 1, pages 141-146, 2002a.

M. Valasek, Z. Sika, M. Florian, J. Macek and M. Polasek. Model Based Predictive Control of Combustion Engine with Constraints. *Review of Automotive Engineering JSAE.* 26:349-356, 3, 2005.

M. Valasek, J. Sveda and Z. Sika. Soil-friendly Off-road Suspension. *Supplement Vehicle System Dynamics.* 44:479-488, 2006b.

M. Valasek and P. Steinbauer. Nonlinear Control of Multibody Systems, In J. Ambrosio, W. Schiehlen (eds.). *Proc. of Euromech Colloquium 404, Advances in Computational Multibody Dynamics.* Lisbon, pages 437-444, 1999.

M. Valasek, O. Vaculin and J. Kejval. Global Chassis Control. Integration Synergy of Brake and Suspension Control for Active Safety, In *Proc. of AVEC 2004.* Arnhem, pages 74-78, 2004.

VDI 2206 2003. *Design methodology for mechatronic systems.* VDI, Beuth-Verlag, Berlin 2003.

A. Veitl, T. Gordon, A. van de Sand, M. Howell, M. Valasek, O. Vaculin and P. Steinbauer. Methodologies for Coupling Simulation Models and Codes in Mechatronic System Analysis and Design. *Supplement Vehicle System Dynamics.* 33:231-243, 1999.

P.J. Venhovens. *Optimal Control of Vehicle Suspensions.* PhD Thesis, Delft University of Technology, 1993.

VHDL-AMS. http://www.eda.org/vhdl-ams/wwwpages/language_ proposal/intro.html.

G.G. Wallace. *Conductive Electroactive Polymers. Intelligent Materials Systems.* CRC Press, Boca Raton 2003.

T. Winsel, H. J. Theuerkauf, S. Pischinger, C. Schernus, G. Lütkemeyer and M. Ayeb. HiL-Calibration of SI Engine Cold Start and Warm-Up Using Neural Real-Time Model. *SAE Paper* 2004-01-1362, 2004.

K. Yi and J.K. Hedrick. Active and Semi-Active Heavy Truck Suspensions to Reduce Pavement Damage. *SAE Paper* 892486, 1989.

Z. Zdrahal, P. Mulholland, M. Valasek and A. Bernardi. Worlds and Transformations. Supporting the Sharing and Reuse of Engineering Design Knowledge. *International Journal of Human-Computer Studies.* 65:959-982, 12, 2007.

Port-based modelling of multidomain physical systems in terms of bond graphs

Peter Breedveld *

* University of Twente Faculty of Electrical Engineering, Mathematics and Computer Science, P.O. Box 217, 7500 AE Enschede, Netherlands
p.c.breedveld@utwente.nl

1 Introduction

1.1 Port-based modeling versus input-output modeling

The key idea in any modeling approach that starts from a priori knowledge about the physical behavior of the system to be modeled is that a conceptual separation is made between various *fundamental behaviors* that are (considered to be) relevant in a given problem context. Although this step is always present, it is often preceded by a step in which the system to be modeled is subdivided into subsystems on the basis of aspects of *function* and/or *configuration*. Note that within the context of dynamic modeling, and thus in the context of this text, the adjective 'physical' (physical system, physical model, physical modeling, etc.) is not used in the sense of 'concrete', but in the sense of 'obeying the laws of physics'. Consequently, a physical model in this sense can be rather abstract. A pump drive system, for example, can be seen as an interconnection of an electric power source (an amplifier), an electric motor, a transmission (a gear box) and a mechanism for displacement of a fluid (a load). This is the first conceptual level of observable, *functional components*. The total behavior consists not only of the observable dynamic interaction between these subsystems, but also of the conceptual interaction between the fundamental behaviors that constitute the behavior of these subsystems themselves. Already when modeling at this level, it is useful to consider both forms of interaction (observable and conceptual) from the point of view of *bilateral* relations (Figure 1), instead of the unidirectional input-output relations that are often used, thus implicitly assuming no 'back-effect'. This bilateral relation and its relation to the concept of energy is the key idea behind a port-based modeling approach, as will be one of the main messages of this treatise.

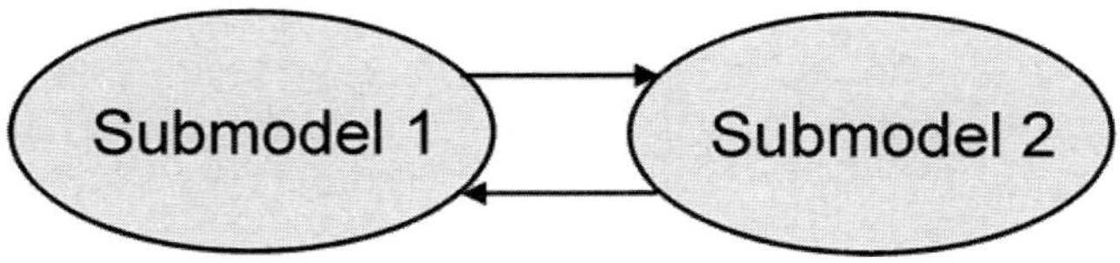

Figure 1. Bilateral relation

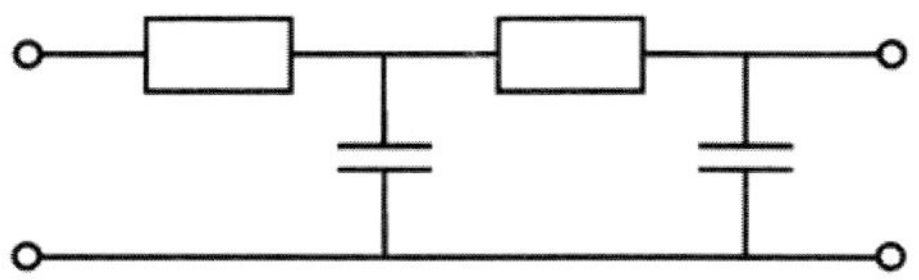

Figure 2. Double RC network

To give a rather simple example: when modeling the behavior of a double RC network (Figure 2) assuming connection to a fit source, e.g. a sine generator, the first step is to identify the components (two resistors and two capacitors). Unlike many other domains of physics, the fabrication of components applied in electrical circuits is given so much attention that, in a context of normal operation (i.e. neglecting radiation and high frequencies), the components map during modeling in a one-to-one way to the conceptual elements with the same name (capacitor C and resistor R) that represent two basic physical concepts, viz. electrical storage (of charge & electrostatic energy) and Ohm's law for dissipation respectively. It is discussed later that the use of this conceptual resistor implies an implicit assumption that the variations of the temperature are such that they may be neglected in the given context, such that 'dissipatable' free energy can be used instead of globally conserved 'true' energy. If one would identify each RC network as a subsystem and consider the transfer function between the input and the output voltages, the total transfer function is not equal to the product of these transfer functions, unless one includes a so-called separation amplifier that suppresses the back-effect by means of modulating an external power supply. This is due to the bilateral relation between all these elements and these two subsystems in particular. In most classical modeling approaches, most of the relations are thought to be unilateral (input-output), in other words: the back-effect is implicitly considered negligible. In those cases where the back-effect cannot be neglected, it is thought of as a separate, unilateral relation, often addressed as the 'feedback'.

The next step in the modeling process commonly is that all conceptual behaviors are linked at the conceptual level into a set of differential and algebraic equations using common variables and balance equations. In principle, the algebraic equations can be used to eliminate a number of variables, such that a set of (partial) differential equations remains. In some simple cases (e.g. linear systems), such a set may have an analytical solution. However, in most cases such an analytical solution does not exist and numerical integration (simulation) is used to obtain an approximate solution. However, if the model is written in terms of equations an important change in the nature of the model representation is implicitly made: from a *simultaneous*, by definition graphical, representation of a system model in terms of interconnected subsystems and interconnected conceptual elements, the representation becomes *sequential*, in the sense that the equations can only be read one after the other. As the parts of the actual system that is being described are present simultaneously and take part in the behavior simultaneously, simultaneous representations generally provide more insight, in particular during the conceptual phase of the modeling process, as well as during model debugging and interpretation of simulation results. Furthermore, when algebraic equations are eliminated, not only some physically relevant variables are lost, but also the (conceptual) structure of the relations, such that local changes in the model structure are not possible anymore, even though such elimination may be very useful for analysis or computational purposes. Several examples of domain-dependent, simultaneous model representations exist and are frequently used, in particular in engineering, although at the same time they represent the wish to take the structure apart for a local analysis per submodel or element. For example, a free-body diagram is used to compute the net contribution to the rate of change of momentum ($\mathrm{d}p/\mathrm{d}t$) on the basis of the forces acting on a body. For the analysis of electrical circuits similar techniques exist, like (modified) nodal analysis, mesh analysis, incidence matrices, see any standard textbook on circuit analysis, e.g. (Hayt, Kemmerly, and Durbin, 2002).

The bilateral interaction between conceptual elements requires the concept of a port, introduced in the context of electrical networks by Harold Wheeler (Wheeler and Dettinger, 1949) and extended by Henry Paynter to all relevant physical domains (Paynter, 1961). Some authors have taken this concept even beyond physical systems as any relation can be considered bilateral. As a bilateral interaction can be seen as a combination of two opposite unilateral interactions or signals, *two* relevant quantities are required to characterize such a port. This aspect is discussed later in more detail.

Important at this point is that port-based modeling can be distinguished from input-output modeling techniques not only by its use of bilateral relations a priori in a simultaneous representation (i.e. graphical), but in particular by the fact that no decisions are made about the input-output nature of the constituting signals of this relation. The relations between the ports are said to be acausal, a terminology to be explained later and somewhat different from the commonly used meaning of causality and acausality related to the 'arrow of time', even though it is related in case of so-called storage ports (cf. the concept of preferred causality, section 3.3). In case this difference in meaning can cause confusion, the concept of causality addressed herein is addressed as 'computational causality', as the computations in digital simulation commonly require an input-output structure of the relations.

This property of acausality of a port not only enables easy reuse of model parts within other contexts, but also gives immediate feedback to the modeler about his modeling decisions as the so-called causal properties of ports to be connected not only determine the final computational structure, i.e. computational causality, but may also indicate that certain physical phenomena are not taken into account. This feedback stimulates the modeler in taking a deliberate decision about what he should include in his model and what not.

1.2 Dynamic models of engineering systems

In order to be precise, the aspects of reality to be modeled are 'limited' herein to behavior in time, i.e. dynamics, of systems that obey the basic principles of classical physics, in the sense that quantum effects play no role. This means that the dynamic behavior is modeled of macroscopic phenomena that are all supposed to obey the basic principles of thermodynamics, viz. (global) conservation of energy and positive entropy production, independent of some specific domain. The adjective 'macroscopic' refers to the *nature* of the scope, viz. a thermodynamic approach, rather than the *scale* of the phenomena: phenomena at the nano scale may for instance still be modeled using a macroscopic viewpoint, which combines the kinetic energy of the random motions into thermal energy using entropy not only as an energy state, but also as a property to express the increase of random motion in terms of the positive entropy production principle. By contrast, the microscopic view does not collect any motion into thermal energy and this kind of view may even be applied to objects at a relatively large scale, like queuing behavior.

The particular problem context in which a model is made determines the criteria for the selection of basic phenomena that may determine the relevant behavior of each of the functional or tangible components. As discussed earlier, the structured set of physical *components* that constitutes a model, has to be further abstracted into a conceptually interconnected set of basic behaviors, like storage, transformation, etc. that will be discussed in more detail later in the form of ideal elements (section 3.2). It will turn out that this process may be supported by the use of multiport generalizations of the most basic conceptual elements, in particular if the relevant phenomena take place in various physical domains. Such systems are called 'multidomain', 'multidisciplinary', 'multi-physics', 'mechatronic', etc.

A clear distinction is required between intended behavior and realized behavior, where intended behavior is analyzed by functional modeling during design, while during design evaluation as well as during troubleshooting the behavior that will be or is actually realized should be analyzed by physical modeling of all realized aspects, in other words: including the undesired behaviors from a functional point of view. It can be often observed during troubleshooting that the functional models that served their purpose well in the design phase, fail to provide the required insight for troubleshooting. This means that when a designer is involved in either concept validation or troubleshooting, he has to make the difficult step of reducing the priority of what he envisioned on a functional basis, while trying to capture all physical behaviors (to be) actually realized, even when there is no direct link with the intended function. Making a clear separation between functional modeling and physical modeling supports this step considerably. For instance, in many cases mechanisms are designed on the basis of their configuration properties in order to generate a specific spatial path. However, in analyzing its behavior in time, the mass and flexibility of its constituting parts can often not be neglected.

2 Structured systems: physical components & interaction

In this section a special focus of the port-based approach to modeling is put on structured systems, i.e. systems that can be seen as a set of physical and/or functional components in a structure. Key element of such a structure is that it represents the relations between the constituting parts or subsystems. As these relations are bilateral, the (virtual) 'points' where these relations can be 'attached' to a subsystem have a bilateral nature too: the *ports*. In other words: the relations interconnect subsystems into a system via the ports of the subsystems. The properties of ports related to behavior in terms of classical physics are discussed, as well as the basic concepts that are required to describe the subsystem behavior: *state* and *change* (of state). In contrast with domain-specific simultaneous representations like free-body diagrams, the analysis of the model in the port-based approach is not based on tearing the structure apart and performing local analyses: the structural elements are considered subsystems (nodes of a graph) as well and called *junctions* (cf. section 3.2).

Within the context of dynamic systems, where the core of the dynamics itself lies in the process of 'storage', the two relevant quantities or variables in a bilateral relation discussed later in this section are strongly related to the concept of equilibrium, the balance between two parts in which a (locally) conserved quantity or so-called extensive state can be stored, like (angular) momentum, charge, magnetic flux (linkage), matter, entropy, etc.: they can either *establish* the equilibrium by changing the stored quantity ('equilibrium establishing variable' or 'rate of change of extensive state'), like electric current, flow of matter, etc., or they *determine* the equilibrium by forming a balance independent of the extent of the communicating parts ('equilibrium determining variable' or 'intensive state'), like voltage, temperature, etc. For example, in case of diffusion processes, the flows of matter of the participating species are the equilibrium-establishing variables (called generalized fluxes in non-equilibrium thermodynamics) that become zero when the equilibrium is reached. The concentrations of the participating species are the equilibrium-determining variables (called generalized forces in non-equilibrium thermodynamics) that become equal (zero difference or gradient) when equilibrium is reached.

The conjugation between these two types of variables that characterize a port is called *dynamic conjugation.* Diffusion phenomena are commonly studied at constant pressure and temperature, thus allowing energy exchange with the environment that is not considered relevant for the dynamics of the diffusion process. However, as soon as such assumptions cannot be made and more than one domain (i.e. type of conserved quantity) needs to be considered, the energy bookkeeping becomes an important means for consistent modeling and the dynamic conjugation has to be further constrained to *power conjugation* in the sense that the relation between two ports that dynamically interact, describes the power, in other words: the energy exchange linked to the relation between two connected *power ports* (in the sequel 'port' will refer to a 'power port' unless otherwise indicated). In other words: the power of a port has to be a function of the two power conjugated variables. The shape of this function is not dictated by nature, but a common choice is that the *product* of the two power-conjugate variables forms the power of the interconnection between two power ports, called *bond.* This is a natural consequence of the mathematical property

Figure 3. Word bond graph of a pump driven by a DC motor

that a change of energy E, being a function of one or more stored quantities q_i:

$$E\left(q_1, ..., q_i, ..., q_n\right) \tag{1}$$

can be written as the inner product of the partial derivatives with respect to the stored quantities and the changes of those stored quantities:

$$\mathrm{d}E\left(q_1, ..., q_i, ..., q_n\right) = \sum_{i=1}^{n} \frac{\partial E}{\partial q_i} \mathrm{d}q_i \tag{2}$$

Domain-dependent examples of power conjugate variables are: voltage and current, force and velocity, pressure and volume flow, temperature and entropy flow, etc. The domain-independent terminology for the equilibrium-determining variable is 'effort' and 'flow' is used for the equilibrium-establishing variable. In case of common (power) bonds the product of effort and flow is assumed to represent the power, in case of pseudo bonds effort and flow may be related to the power in another manner. For instance, in case of the common variables used to model thermal behavior, viz. temperature and heat flow, the heat flow itself is the (thermal) power already. However, it only represents a rate of change of the stored energy as long as the dynamics of other domains play no role; otherwise the amount of heat cannot play the role of a proper state variable, as discussed at length in many introductions to thermodynamics, e.g. (Callen, 1960). Figure 3 demonstrates how specific functional or material parts of an actual system can be represented in the form of submodels by a short text enclosed by some closed line and identifying them in an insightful way (often an ellipse is used to distinguish them from the common 'blocks' of a block diagram of which the interface consists of inputs and outputs and where the blocks themselves represent mathematical operations). These ellipses form the nodes of a graphical representation, a so-called *word bond graph*, of which the edges represent the so-called *bonds* between the ports of these subsystems (Paynter, 1961). The word bond graph can be seen as the highest level port-based description of a dynamic model, close to the functional and structural level. The direction of the edges in the form of the *half-arrow* that is typical for a bond represents the positive orientation of both flow and power related to that bond.

Figure 4. Causal stroke

Another aspect that can be added to a bond is its so-called causal stroke, a little stroke attached to one end of the bond and perpendicular to it. It represents the computational direction or *computational causality* (when chosen) of the effort signal and thus also the computational direction of the flow signal as the signals of the two conjugate variables are opposite by definition (Figure 4).

3 Bond graphs

Readers who have been exposed previously to analog models and the bond graph notation in particular are warned that many introductions start from the misconception that analogies are merely based on similarity of the underlying differential equations. In this treatise this is considered an exchange of cause and effect. The actual source of analogies that lead to the domain independence of both differential equations and bond graph notation is the mere fact that in human reasoning about dynamic behavior only a rather limited set of elementary concepts is exploited. This set of ideal concepts is chosen as the backbone of section 3.2 and corresponds directly to the bond graph symbolism. In section 3.3 it is shown how the concept of computational causality that augments the representation of the physical structure by a bond graph with a representation of the computational structure, leads to the formulation of a computable mixed set of differential and algebraic equations.

In models that are solved analytically, it is commonly preferable to eliminate the algebraic equations as much as possible, as only the structure of the differential equations determines the nature of the described behavior. In mathematics this is often taken one step further by scaling or normalizing the variables (sometimes called 'lifting'), which takes the model a step further away from its physical interpretation. In the days of analog computing scaling and reduction to a minimal form was a necessity and in the early days of digital computing 'superfluous' computations were to be avoided, due to the hardware limitations. However, the algebraic part describes the (conceptual) interconnection structure and maintaining this model information, particularly in a graphical representation, allows rapid model modification, quick interpretation of simulation results and thus enhances insight, in particular if the variables continue to represent common physical concepts. Due to the increase of computational power and robustness of modern computing maintaining this information that is not essential for solving the dynamics is less costly, in particular due to automatic symbolic manipulation.

Starting from the word bond graph, this section explains the basic structure of a bond graph and the most important semantics and grammar of this graphical language.

3.1 Appearance

The mathematical structure of a word bond graph is that of a labeled, directed graph or di-graph of which the directed edges are the bonds with their orientation (half-arrows) and of which the labeled nodes (ellipses with text) represent the subsystems characterized by some (functional) description (Figure 3). The transition to a regular bond graph is relatively simple: basic behaviors with respect to energy are represented by a mnemonic code (acronym) consisting of up to a few capital letters. In this chapter these basic behaviors will be discussed at different levels, which lead to different categorizations. However, the key elementary behaviors of macroscopic physics, viz. *conservation* (in fact expressing the need for any useful model to satisfy *time translation symmetry*) and *positive entropy production* (expressing the need for most macroscopic models to satisfy *time reflection asymmetry*, in other words: the 'arrow of time') will play a key role at all times.

3.2 Elementary behaviors

Storage We start with a bottom-up discussion that assigns all reversible storage to those nodes of the graph that are called storage elements. Note that the adjective 'reversible' is in fact a tautology when identifying this ideal process and will be omitted in the sequel. In principle, these storage nodes may have an arbitrary number of ports and are called multiport storage elements in that generic case. For reasons of simplicity, the one-port version will be described first. The storage element or capacitive element stores a specific, (locally) conserved quantity that, by definition, is extensive in nature, i.e. it is proportional to the spatial and/or material extent of the object to be modeled. Examples are electric charge, amount of moles, entropy, momentum, magnetic flux (linkage), etc. Some of these quantities, like charge and momentum, are also globally conserved, but others, like entropy, elastic displacement and flux linkage, are not. Such a locally conserved quantity q is by definition a state variable, in the sense that its cyclic integral is zero:

$$\oint \mathrm{d}q = 0 \tag{3}$$

which means that it describes the state of the system in a unique way, independent of the history of the system. One could also say that the history of the system in uniquely reflected in its current state, in other words: the states can be considered the 'memory' of the system.

The locally conserved quantity also determines the nature of the so-called 'physical domain' and will require some reference value. The process of storage of such a locally conserved quantity can be represented by the time integral of its rate of change, which is the already identified equilibrium establishing variable or flow variable f:

$$\int_{q(t_0)}^{q(t)} \mathrm{d}q\,(t) = \int_{t_0}^{t} \frac{\mathrm{d}q\,(\tau)}{\mathrm{d}\tau}\mathrm{d}\tau = \int_{t_0}^{t} f\,(\tau)\mathrm{d}\tau = q\,(t) - q\,(t_0) \tag{4}$$

and where $q(t) - q(0)$ is the increase of the stored quantity in a time interval $t - t_0$. Later it is shown that the powerful –in particular in the conceptual modeling phase– feature of the relation of the effort and flow variables with the concept of equilibrium is commonly lost due to a symmetrization of the role of the power conjugate variables. However, the concept of the symplectic gyrator to be discussed later, allows an alternative approach that maintains this asymmetry and can be converted into the traditional form.

Independent of the domain, the storage element is represented by the label **C** that stands for '*capacitor*'. The conjugate effort or equilibrium determining variable has to be an intensive state, i.e. a state that is independent of the extent of the described system, as equilibrium between two systems does not depend on their individual extents. Obviously, intensive states are related to the extensive states. In case of physical properties with a spatial orientation, like force, electrical current, etc., it is clear that the concept of extent has to be related to the nature of the spatial dependence of the physical quantity, so the role of the intensity is played by the linear vector and the role of the extensity by the axial vector. Without going in more detail, it is noted that this creates an additional constraint on how these vectorial extensities can be combined in case of equilibrium.

The relation between the intensive state and extensive state is described by the *constitutive relation* of the **C**-type element, which depends both on constituting matter and geometry. In the common, linear, 1-port case the constitutive relation reduces to a proportionality constant, e.g. the spring constant in Hooke's law, its inverse is called compliance.

In the generic form of the linear constitutive relation of a 1-port **C**

$$e(t) = \frac{q(t)}{C} = \frac{1}{C}\left(\int_{t_0}^{t} f(\tau)\,\mathrm{d}\tau + q(t_0)\right) \tag{5}$$

the inverse of this proportionality constant is called the *capacitance* C, which, as a natural consequence of the above discussion, should be extensive in nature. Examples are the *c*apacitance of an electric capacitor, the *c*ompliance of a spring, etc. Such constitutive parameters (also in the nonlinear case) are always dependent on material parameters that are not related to the extent, like mass density, and geometric parameters like length, width and height that are by definition related to the spatial extent. Sometimes they also contain global natural constants. Sometimes the spatial extent is combined with the mass density into a material extent expressed in amount of moles or in kilogram.

For example, an electric flat-plate capacitor of which fringing and losses can be neglected can be described by a linear constitutive relation:

$$u(t) = \frac{q(t)\,d}{A\varepsilon_0\varepsilon_r} = \frac{q(t)}{C} = \frac{1}{C}\left(\int_{t_0}^{t} i(\tau)\,\mathrm{d}\tau + q(t_0)\right) = \frac{1}{C}\int_{t_0}^{t} i(\tau)\,\mathrm{d}\tau + u(t_0) \tag{6}$$

where the capacitance is $C = \frac{A\varepsilon_0\varepsilon_r}{d}$, A (plate area) and d (distance between plates) are geometrical parameters, ε_r (relative permittivity or dielectric constant) is the material parameter and the global natural constant is ε_0 (permittivity).

Another example of a linear storage element is an ideal, non-relativistic mass m (product of mass density ϱ and volume V), storing momentum p:

$$v(t) = \frac{p(t)}{m} = \frac{p(t)}{\rho V} = \frac{1}{m}\int_{t_0}^{t} F(\tau)\,\mathrm{d}\tau + v(t_0) \tag{7}$$

Although the association with a capacitance may seem somewhat awkward at first sight, it is emphasized that so far only one type of storage has been identified on the basis of the asymmetric role of effort and flow. The kinetic equilibrium determining variable (kinetic effort) is the velocity v while the equilibrium establishing variable is the rate of change of momentum (kinetic flow) or net force F. The resulting paradox with common modeling approaches that treat a force as an effort and a velocity as a flow, will be resolved later.

Environment, sources, boundary conditions, constraints If the system to be modeled cannot be isolated in an energetic sense from the rest of the world, its so-called environment, this environment should 'store' too as this allows exchange of the stored conserved quantity with this environment. However, the criterion for selecting the system boundary that separates the system from its environment is that the environment does not influence the dynamic characteristic of the system. In other words: the intensive state of the environment does not depend on its extensive state, such that the extent of the environment becomes irrelevant, which coincides with the common understanding of the concept of an environment. This means that the intensive state or effort is also independent of the rate of change of its extensive state, i.e. its conjugate flow. This kind of imposed intensity at the boundary is often called a *source*, a *constraint* or a *boundary value*. In bond graph terminology such an environmental influence becomes a *s*ource of *e*ffort with acronym **Se**. Not only from a mathematical point of view it can be considered a **C**-type element of which the capacitance has become infinite, such that it becomes independent of a change in extent and just depends on its initial value (or 'initial condition'), but also from a modeling point of view, since storage in the environment can only be considered as such when the capacitances of all capacitors in the system itself are relatively small compared to the capacitance of the environment.

The energy of a system is the sum of all partial energies stored in the capacitors. In other words: the energy is a state that is a function of all stored quantities (independent extensive states) and as such an extensive state too. Since the environment is not related to the extent, the reference of each conserved state is also made at the intensive level and can be either absolute or relative in nature: for instance, temperature and pressure have absolute zero points, but voltage and velocity (the effort of the kinetic domain) have no absolute zero point.

In many cases these types of boundary conditions, viz. sources of effort, are combined with certain types of transducers that are considered ideal in such a way that the constant effort imposed to the transducer (ideal pump or motor) is translated into an imposed flow variable in another domain. Such combinations of effort sources with interdomain transducers are called flow sources with acronym **Sf**. Both types of sources can also be deliberately approximated by means of external energy supply and feedback control. In those cases the conceptual connection to some 'infinite' storage becomes much less elucidating. One might also say that the use of feedback control and external power supply lifts the requirement that the environment should be (infinitely) large with respect to the system to be modeled.

In the degenerate case of zero-valued sources the power is zero too (also called Dirichlet and von Neumann type of boundary conditions), which means that the modeled system is energetically isolated and its energy becomes an invariant. Many analyses of systems described by partial differential equations (such as finite-element analyses) are in principle based on these kinds of boundary conditions. If such models are considered submodels to be combined with other submodels the description of this interaction is not straightforward (Ligterink, Breedveld, and van der Schaft, 2006).

Powercontinuous structure Now that all storage in the system (**C**-type ports) and in the environment (**Se**- and **Sf**-type ports) has been described, the principle of energy conservation means that all other conceptual elements in the system need to be *power continuous* accordingly. In other words: they require an instantaneous power balance between all ports:

$$\sum_i \varepsilon_i P_i = \sum_i \varepsilon_i e_i f_i = 0 \tag{8}$$

where $\varepsilon_i = \pm 1$, depending on the orientation of the half arrow of the connecting bond.

This condition means that all other elements need to have at least two ports. This seems a quite unnatural conclusion, because most readers will immediate think of the resistor as a basic one-port element. In order to resolve this paradox, the ideal resistor is treated first.

Resistor Indeed, this basic resistive element in its general form is a two-port, an irreversible transducer, where one of the ports is by definition a thermal port with the flow of irreversibly produced entropy as conjugate flow variable of the thermal effort, the temperature. However, in most problem contexts it is *implicitly* assumed that the temperature variations can be neglected at the time scale of interest. This means that this thermal port can be considered to be connected to an effort source and that its conjugate flow is irrelevant for the dynamics of the system. In fact, this is also the case for the storage elements and, as will be explained later in more detail, this means that the stored energy is replaced by one of its Legendre transforms (cf. section 4.1), the free energy, as a generating function of the constitutive relations. The free energy does not satisfy a conservation principle and can thus be *dissipated* by a one port **R**-element.

Many constitutive relations of these resistive elements, like Ohm's law in the electric case, are considered linear algebraic relations between the two conjugate variables of the remaining port:

$$u - Ri = 0 \tag{9}$$

The power into the electric port is the product of effort and flow, voltage u and current i, and is considered dissipated when the (absolute) temperature T is considered constant:

$$P_{\text{diss}} = ui = Ri^2 = \frac{u^2}{R} = Tf_{S_{\text{irr}}} = P_{\text{thermal}} \tag{10}$$

By contrast, if the temperature cannot be considered constant, the thermal port needs to be explicitly modeled. Due to the power continuity constraint the flow of irreversibly produced entropy can always be computed, independent of the form of the constitutive relation of the resistive port, by dividing the input power which equals the thermal output power, i.e. the heat flow, by the absolute temperature:

$$f_{S_{\text{irr}}} = \frac{P_{\text{thermal}_{\text{out}}}}{T} = \frac{P_{\text{in}}}{T} \tag{11}$$

This relation cannot be inverted as the flow of produced entropy is zero in equilibrium.

Furthermore, it shows that the irreversible transducer that represents the positive entropy production in the system is by definition nonlinear, even if the constitutive relation of the non-thermal port is linear, like in the case of an ohmic resistor. From another perspective to be discussed later, it will become clear that all linear power continuous two-ports are *reversible* transducers, which confirms the conclusion that an *irreversible* transducer needs to have at least one *nonlinear* constitutive relation. In order to distinguish the two-port irreversible transducer from the one-port dissipator of free energy **R**, the symbol **RS** is used, emphasizing the explicit representation of the production (*s*ource) of entropy.

Any nonlinear constitutive relation of the non-entropy producing port of the **R(S)** should be a relation that lies in the first and third quadrant due to the second law of thermodynamics. This means that all resistive constitutive relations cross the origin and that all linear resistors are positive. Only differential resistances in an operating point that is kept out of equilibrium by external energy input can be considered 'negative'. Living systems are in such a state with local entropy reduction (information generation) at the cost of global entropy production by the metabolic system.

Generalized junction structure: transduction and interconnection
Now that all energy storage and irreversible production of entropy (dissipation of free energy) has been given a conceptual location in the **C** & **Se** (**Sf**) and **R(S)** respectively, the rest of the model can be seen as a power continuous multiport that is called *generalized junction structure* (**GJS**). However,

this **GJS** can be usefully substructured into elements with specific properties that can be globally split into *transducers* and *interconnections*.

The common choice for the relation between the two power conjugate variables and the power is a product operation, which can be observed from the common choices for effort and flow. However, as a simple linear transformation transforms effort- and flow-type variables into so-called *(wave) scattering variables* of which the relation with the power is a sum operation, this illustrates that this is merely the modeler's choice (Paynter and Busch-Vishniac, 1988). The transformation to scattering variables provides quite some insight if the properties of power continuous n-ports ($n \geq 2$) are studied, thus providing a means to create a substructure in the generalized junction structure.

When the *interconnection* aspect of power continuous n-ports is studied, the key property is that one should be able to make *arbitrary connections*, in other words: all ports should behave in the same way and cannot be mutually distinguished. This means that an interconnection n-port is not only power continuous, but also port-symmetric, which can be easily translated into the form of the scattering matrix that represents the constitutive relations in terms of scattering variables. By means of this scattering approach it is relatively straightforward to prove that there are only two possible solutions when these two constraints are applied (Hogan and Fasse, 1988). These two solutions are described by linear constitutive relations without any characteristic parameter and can be seen as a combination of a generalized form of a Kirchhoff's law for one of the conjugate variables combined with the identity of the other conjugate variable:

$$\sum_{i=1}^{n} \varepsilon_i e_i = 0 \text{ with } \varepsilon_i = -1, +1 \text{ and } f_i = f_j \; \forall i, j : i \neq j \tag{12}$$

or

$$\sum_{i=1}^{n} \varepsilon_i f_i = 0 \text{ with } \varepsilon_i = -1, +1 \text{ and } e_i = e_j \; \forall i, j : i \neq j \tag{13}$$

In all cases these kinds of analyses start from the assumption that the inbound orientation is positive in principle, which means that $\varepsilon_i = +1$ for inbound port orientations and ε_i = -1 for outbound port orientations. These interconnection multiports are called *junctions*, where the **0**-junction describes the instantaneous balance of the flow variables combined with the identity of the effort variable and where the dual case, i.e. interchanged role of efforts and flows, is called a **1**-junction. No other assumption about the form of the constitutive relations was made than power continuity and port symmetry, such that this linear, parameter-free result holds for any domain in which power and ports are meaningful concepts.

It should be noted that the **0**-junction virtually represents its common effort, whereas the **1**-junction virtually represents its common flow. From a mnemonic point of view the labels **E** and **F** respectively seem more appropriate than the rather meaningless **0** and **1**, even though Paynter has claimed that the **0** resembles a node and the **1** resembles a line representing a flow. The reason that this is not the case is historic: upon first introduction of bond graphs by Paynter in the late fifties and early sixties (Paynter, 1961), he pursued a minimalist representation with only one character per ideal concept and chose **E** and **F** for the effort- and flow-source respectively. Soon after, his students Karnopp and Rosenberg (1968) formalized the notation further and adapted some of the mnemonics in order to achieve more expressive power.

Application of scattering variables to power continuous two-ports without the port symmetry constraint in order to get more insight in the *transformation* aspect of the **GJS**, demonstrates that these two-ports are characterized by *multiplicative* constitutive relations $e_1 = -\varepsilon n(.)\, e_2$ and $f_2 = \varepsilon n(.)\, f_1$ with $\varepsilon = -1, +1$ that in the linear case reduce to one parameter: $e_1 = -\varepsilon n e_2$, $f_2 = \varepsilon n f_1$ and $e_1 = -\varepsilon r f_2$, $e_2 = \varepsilon r f_1$, both with $\varepsilon = -1, +1$. If the two port orientations are chosen unequal (one inbound and one outbound) these relations reduce to the common relations for a transformer **TF** and a gyrator **GY** respectively: $e_1 = n e_2$, $f_2 = n f_1$ and $e_1 = r f_2$, $e_2 = r f_1$ (originally called **T** and **G** by Paynter). The nonlinear case (including the time-variant case) is constrained to *modulation* by the multiplier n or r, which is expressed by adding the letter **M** to the acronym: **MTF**: $e_1 = n(.)\, e_2$, $f_2 = n(.)\, f_1$ and **MGY**: $e_1 = r(.)\, f_2$, $e_2 = r(.)\, f_1$, where (.) stands for any modulating signal. If these signals are external, this does not require additional attention. By contrast, internal signals mean that system variables or functions of system variables may become part of the constitutive relation, which may result in constitutive relations of other types. An example is the gravitational force on a pendulum mass (Figure 5a). In principle, in the common context of distances from the earth surface that are small with respect to the earth radius, this force may be considered constant and independent of the velocity (**Se**-type source). However, the kinematic constraint has the nature of an **MTF** between the translational domain and the rotational domain with the angular position as a modulating signal (Figure 5b). Due to this state modulation the **Se**-port acts via the **MTF** as a **C**-type port on the rotational domain, for small deflections even as a linear **C** that results in an oscillator when combined with the kinetic storage in the mass (Figure 5c&d).

Both the **MTF** and **MGY** can be generalized to modulated multiport versions, where the generic multiport **MTF** (multiport modulated trans-

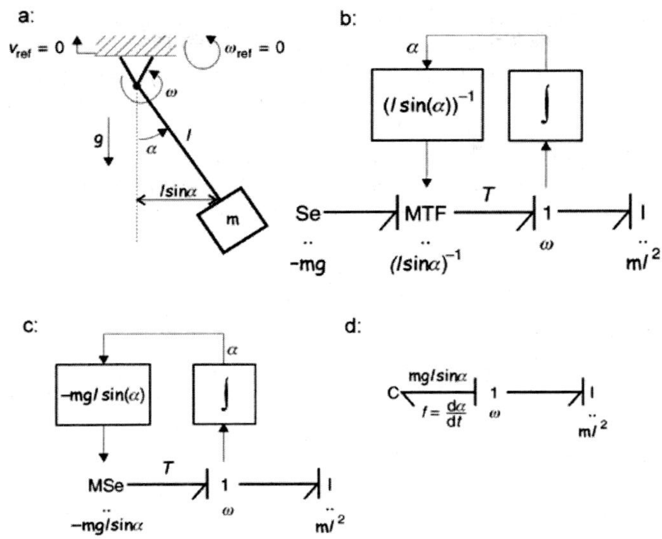

Figure 5. Pendulum: **Se** via position-modulated **MTF** results in **C**-type storage

former) is an $m+n$–port, characterized by a $m \times n$–matrix $\mathbf{T}(.)$:

$$\mathbf{e}_1 = \mathbf{T}^t\,(.)\,\mathbf{e}_2,\ \ \mathbf{f}_2 = \mathbf{T}\,(.)\,\mathbf{f}_1 \tag{14}$$

where $\mathbf{e}_1$ and $\mathbf{f}_1$ are the power conjugate variables of an m-dimensional inbound multiport and $\mathbf{e}_2$ and $\mathbf{f}_2$ are the power conjugate variables of an n-dimensional outbound multiport. The transposition of the matrix follows from the power continuity constraint:

$$\left.\begin{array}{c}\mathbf{f}_2 = \mathbf{T}\,(.)\,\mathbf{f}_1\\ P_1 = P_2 : \mathbf{e}_1^t\mathbf{f}_1 = \mathbf{e}_2^t\mathbf{f}_2\end{array}\right\}\mathbf{e}_1^t\mathbf{f}_1 = \mathbf{e}_2^t\mathbf{T}\,(.)\,\mathbf{f}_1 \tag{15}$$

$$\mathbf{e}_1^t = \mathbf{e}_2^t\mathbf{T}\,(.) \tag{16}$$

$$\mathbf{e}_1 = \mathbf{T}^t\,(.)\,\mathbf{e}_2 \tag{17}$$

The generic multiport **MGY** is an arbitrary n–port characterized by an $n \times n$–matrix $\mathbf{G}(.)$:

$$\mathbf{e} = \mathbf{G}\,(.)\,\mathbf{f} \tag{18}$$

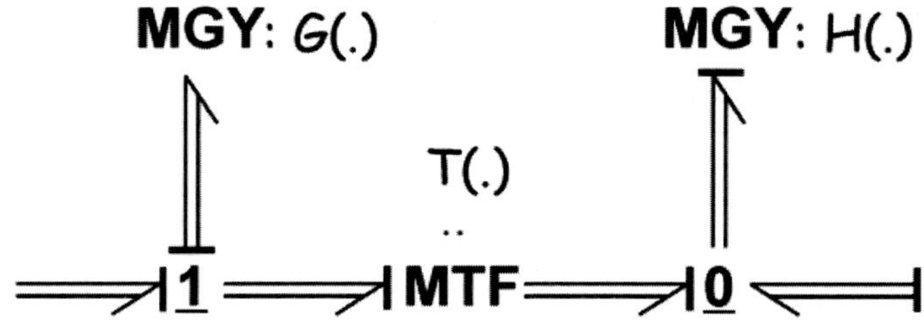

Figure 6. GJS decomposition

The power continuity constraint results in this case in the antisymmetry of the matrix **G:**

$$P = \mathbf{e}^t\mathbf{f} = \mathbf{f}^t\mathbf{G}^t\,(.)\,\mathbf{f} = 0,\ \mathbf{G}\,(.) = -\mathbf{G}^t\,(.) \tag{19}$$

By mixing efforts and flows in the input and output vectors and by taking all ports inbound again ($-\mathbf{f}_2$), the multiport **MTF** can be written in a similar manner:

$$P = \mathbf{e}^t\mathbf{f} = \begin{bmatrix}\mathbf{e}_1^t & -\mathbf{f}_2^t\end{bmatrix}\begin{bmatrix}\mathbf{f}_1\\ \mathbf{e}_2\end{bmatrix} = \begin{bmatrix}\mathbf{f}_1^t & \mathbf{e}_2^t\end{bmatrix}\begin{bmatrix}0 & -\mathbf{T}^t\,(.)\\ \mathbf{T}\,(.) & 0\end{bmatrix}\begin{bmatrix}\mathbf{f}_1\\ \mathbf{e}_2\end{bmatrix} = 0 \tag{20}$$

The generalized junction structure is a combination of all possible forms and thus leads to an antisymmetric matrix too. The quadratic form of an arbitrary matrix representing a multiplicative, algebraic relation between power conjugate efforts and flows of an arbitrary multiport represents the power generated or absorbed by such a multiport. The power continuity constraint thus requires this quadratic form to be zero. Indeed, this corresponds to the antisymmetry of the constitutive matrix:

$$P = \mathbf{e}^t\mathbf{f} = \begin{bmatrix}\mathbf{e}_1^t & -\mathbf{f}_2^t\end{bmatrix}\begin{bmatrix}\mathbf{f}_1\\ \mathbf{e}_2\end{bmatrix} = \begin{bmatrix}\mathbf{f}_1^t & \mathbf{e}_2^t\end{bmatrix}\begin{bmatrix}\mathbf{G}\,(.) & -\mathbf{T}^t\,(.)\\ \mathbf{T}\,(.) & \mathbf{H}\,(.)\end{bmatrix}\begin{bmatrix}\mathbf{f}_1\\ \mathbf{e}_2\end{bmatrix} = 0 \tag{21}$$

where $\mathbf{G} = -\mathbf{G}^t, \mathbf{H} = -\mathbf{H}^t$ and $\mathbf{H}$ is an 'inverse' gyration matrix ($\mathbf{f}_2{}^{\mathrm{GY}} = \mathbf{H}\,(.)\,\mathbf{e}_2$, where $-\mathbf{f}_2 = -\mathbf{f}_2{}^{\mathrm{TF}} + \mathbf{f}_2{}^{\mathrm{GY}}$), not necessarily generated though by taking some inverse, in other words: **H** does not need to have an inverse. This results in the **GJS** decomposition in Figure 6, where the double lined bonds represent multibonds, i.e. arrays of bonds, similar to the common notation for signal arrays in block diagrams.

The categorization of the junction structure elements used here initially is not based on the asymmetry between effort and flow pointed out earlier on a physical basis as this type of categorization corresponds to the conventional viewpoint provided by the existing mathematical analyses for engineering systems like electrical networks. Even though a voltage (potential difference) or a current (rate of change of charge / flow of moving electrons) are physically quite different, they are treated in a symmetric, dual way in Kirchhoff's equations.

In the sequel, we return to the physical differences, in which an effort can be considered a point-like variable (scalar potential, intensive state) when related to storage and a line vector (negative gradient) when related to a potential difference. By contrast, a flow can be seen as a rate of change of an extensive stored quantity (proportional to available volume or amount of matter) or one of the axial vectors that contribute to this rate of change in the balance equation of the stored quantity (divergence).

Given this distinction it is straightforward to conclude that the discrete versions of the (negative) gradient and divergence operations are represented by the summing port relations of the 1- and 0-junction respectively, keeping in mind that the junctions themselves represent the respective conjugate flow and effort, which are common at al ports.

When elaborating this distinction based on operators in a continuum approach, it should be noted here already that a unit gyrator, called *symplectic gyrator* (**SGY**) for reasons to be discussed later, represents a spatially discretized version of two rotation operators between the efforts at the one hand and the two non-conjugate flows at the other hand. This **SGY** is indeed similar to the junctions in the sense that it is linear without any constitutive parameter. Later, this **SGY** will turn out to be useful for explicitly representing two interdomain couplings that are not self-evident, despite the fact that they are often considered that way and structurally eliminated by partial dualization accordingly. Dualization means here the inversion of the roles of effort and flow in the constitutive relations and 'partial' means that not all ports are dualized. Allowing dualization immediately leads to a full symmetrization of the roles of effort and flow in the constitutive relations, thus not only going back to dual **0**- and **1**-junctions, but also giving every other type of port its dual, viz. **Se** and **Sf**, a storage element in which the effort is integrated into a conserved quantity (generalized momentum) called **I**-type port (generalized 'inertia' or 'inductance'). Inverting the roles of effort and flow in a constitutive relation of an **R**-type port (e.g. in Ohm's law) does not change its nature, although the relation is inverted and a resistance parameter becomes a conductance parameter. Finally, it should be noted that **TF** and **GY** are each others *partial* dual, i.e. they are obtained from each other by dualization of only one of the two ports. However, the **S**-type port of the **RS** cannot be dualized, as temperature and entropy production are by definition asymmetric. In the thermal domain dual storage is not a viable concept as it would violate the second law of thermodynamics (Breedveld, 1982).

The full symmetrization of effort and flow thus leads to the common *nine* basic port types:

C & **I,** **(M)Se** & **(M)Sf**, **(M)R(S)**, **(M)TF** & **(M)GY**, **(X)0** & **(X)1** where the letter **M** is added in case the constitutive relations can be modulated and the combination of the letter **X** with a junction means that the junction can be switched by a (Boolean) condition:

$$\begin{aligned}\mathbf{X1}: & e_j = \text{if condition then } -\varepsilon_j \sum_{i \neq j}^{n} \varepsilon_i e_i \text{ else 0 end;} \\ & \text{where } \varepsilon_i = -1, +1, \text{ and} \\ & f_i = \text{if condition then } f_j \ \forall i, j : i \neq j \text{ else 0 end;}\end{aligned} \tag{22}$$

or

$$\begin{aligned}\mathbf{X0}: & f_j = \text{if condition then } -\varepsilon_j \sum_{i \neq j}^{n} \varepsilon_i f_i \text{ else 0 end;} \\ & \text{where } \varepsilon_i = -1, +1, \text{ and} \\ & e_i = \text{if condition then } e_j \ \forall i, j : i \neq j \text{ else 0 end;}\end{aligned} \tag{23}$$

Modulation of storage ports is omitted in principle as it violates energy conservation and contradicts the concept of storage (change of content without a rate of change). At a level where it can be accepted that the energy exchange via one of the ports of a multiport **C** can be neglected with respect to the other ports, one might choose to represent this kind of so-called 'bond activation' by a modulated **C**. However, it should be clear that this condition exists and it should be checked that it is satisfied.

These are the nine basic node types in common bond graphs that are frequently used to describe electric circuits and simple mechanical systems. Trying to fit other domains in this approach often leads to complications although the hydraulic domain can be relatively easily incorporated as long as incompressibility is assumed, which allows description of a mass flow by a volumetric flow.

This is not a coincidence: in the descriptions of both electrical networks and mechanical systems some implicit assumptions are made by default that lead to a simplification of a coupling that is an interdomain coupling in principle. In case of electrical circuits containing capacitors and coils both the electrical domain (storage & conservation of charge, dielectric displacement) and the magnetic domain (storage & conservation of magnetic flux, more particular flux linkage of a coil) play a role. In its most general form the coupling between these domains is described by Maxwell's equations, in particular the third equation (Faraday's law of induction)

$$\operatorname{rot}\mathbf{E} = \nabla \times \mathbf{E} = -\frac{\partial \mathbf{B}}{\partial t} \tag{24}$$

and the fourth equation (Ampere's law)

$$\operatorname{rot}\mathbf{H} = \nabla \times \mathbf{H} = \frac{\partial \mathbf{D}}{\partial t} + \mathbf{J} \tag{25}$$

where $\mathbf{E}$ is the electric field strength, $\mathbf{D}$ the dielectric displacement, $\mathbf{B}$ the magnetic induction, $\mathbf{H}$ the magnetic field strength and $\mathbf{J}$ the electric current density.

However, if one assumes that electrical networks do not radiate, which is identical to assuming that changes in the EM fields take place quasi-stationary and if simple geometries are considered (e.g. an ideal coil with n windings), such that these relations can be written in terms of standard efforts and flows, this coupling (based on two rotation operators, cf. Equations (24) & (25)) reduces to two simple identities in terms of scalar quantities:

$$E.'l' = u = \frac{nd(B.'A')}{dt} = \frac{nd\Phi}{dt} = \frac{d\lambda}{dt}, \text{ i.e. } u = e_{\mathrm{elec}} = \frac{d\lambda}{dt} = f_{\mathrm{mag}} \tag{26}$$

the identity between the electric effort (voltage u) and the magnetic flow (rate of change of flux, for one current loop) or flux linkage λ (for a coil with n windings; change of sign due to choice of orientation) and

$$H.'l' = MMF = n(J.'A') = ni = ne_{\mathrm{mag}}, \text{ i.e. } e_{\mathrm{mag}} = i = \frac{dq}{dt} = f_{\mathrm{elec}} \tag{27}$$

the identity between the magnetic effort and the electrical flow (electric current). The quotes refer to the fact that the configuration is chosen such that the required spatial integrations are simplified. These two identities are exactly represented by a unit gyrator called **SGY** between the two domains, which can be eliminated next by dualization of the magnetic storage into the common **I**-type storage element. Magnetic dissipation is commonly not assumed relevant for electric circuits and neither are the magnetic sources. The only element that is commonly considered relevant, the magnetic storage in a coil or solenoid, is thus considered part of the electric domain as a *dual* type of storage. This explains the common modeling difficulties that occur if permanent magnets play a role.

Although this may be harder to appreciate, a similar situation exists between the potential (or elastic) domain and the kinetic domain that are commonly considered one domain, viz. the mechanical domain. In the mechanical domain, the implicit **SGY** coupling between the potential domain (storage of displacement) and the kinetic domain (storage of momentum) is commonly considered unconditionally present. However, the implicit assumptions that are made by default here are that motions are described with respect to an *inertial* reference frame and in *inertial* coordinates.

Only in that case Newton's second law states that the rate of change of momentum (kinetic flow) is an effort of the potential domain (force). Obviously, in such a situation the velocity (kinetic effort) serves as a rate of change of elastic or gravitational displacement (potential flow).

In recent work (Golo et al., 2000) the link between the spatially discrete, quasistationary network approach and the continuum approach, viz. **1**-junction related to gradient, **0**-junction related to divergence and **SGY** related to 2 rotations (Breedveld, 1984a), has been generalized into the so-called 'Dirac TransFormer' or **DTF**, which can be seen as a condensed notation for the fact that for numerical solution (simulation) of the underlying partial differential equations a non-trivial spatial discretisation has to be performed that commonly cannot be considered to be independent from the required time discretisation. The **DTF** represents the so-called Dirac-structure (Maschke, van der Schaft, and Breedveld, 1995). An extension that includes the rotation operators unconditionally is called the Stokes-Dirac structure, which is represented by the acronym **SDTF** or **SDS** (Maschke and van der Schaft, 2001).

Before making the relation between the physical structure of the bond graph and the computational structure that can be added via causality assignment, the example of the pump system in Figure 3 will be converted into an initial bond graph by translating the components into their most dominant elementary behavior. From this perspective the electrical power supply, e.g. the power grid voltage, can be considered a voltage source (**Se**). The dominant behavior of the electric motor, i.e. the Lorentz force that relates the torque with the same ratio to the motor current as the voltage (rate of change of flux linkage) to the angular velocity, can be considered a gyrator (**GY**). The transmission basically relates two angular velocities with the same ratio as the two torques (**TF**). The momentum balance in a centrifugal pump relates the pressure difference to the angular velocity and the volume flow to the torque. Although these relations will generally be nonlinear, this dominant behavior can be captured by a gyrator (**GY**).

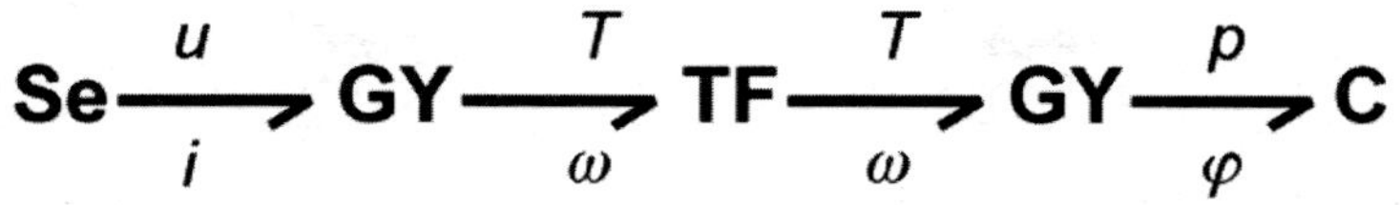

Figure 7. Initial bond graph of the pump system

Finally, the water tank primarily stores water, where the pressure at the bottom of the tank (assuming that the inlet is there too), depends on the stored volume, i.e. the integral of the net volume flow. This means that Figure 3 can be converted into the initial bond graph in Figure 7, assuming that the environmental pressure can be taken as the zero reference pressure.

3.3 Causality

Causal port properties Each of the nine basic elements (**C**, **I**, **R**(**S**), **TF**, **GY**, **Se**, **Sf**, **0**, **1**) introduced above has its own causal port properties, that can be categorized as follows: fixed causality of the first kind, fixed causality of the second kind, preferred causality, arbitrary causality and causal constraints between ports. The representation of (computational) causality by means of the causal stroke has been introduced already (cf. Figure 4).

Fixed causality of the first kind It needs no explanation that a source of effort always has an effort as output signal, in other words, the causal stroke is attached to the end of the bond that is connected to the rest of the system (Figure 8, Figure 9a). Mutatis mutandis the causal stroke of a flow source is connected at the end of the bond connected to the source (Figure 9b). These causalities are called 'fixed causalities' accordingly.

Fixed causality of the second kind Apart from the fundamentally fixed causalities of the first kind, all ports of elements that may become nonlinear and non-invertible, i.e. all but the regular (i.e. non-switched) junctions, may become computationally fixed due to the fact that the constitutive relation may only take one form that cannot be inverted.

Preferred causality A less strict causal port property is that one of the two possibilities is, for some reason, preferred over the other. Commonly this kind of property is assigned to storage ports, as the two forms of the constitutive relation of a storage port require either differentiation with

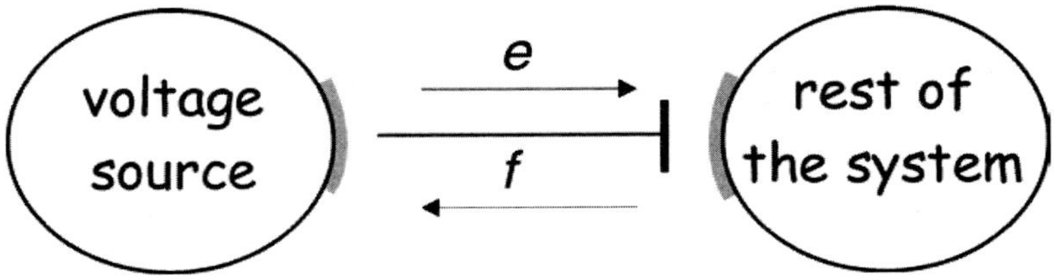

Figure 8. Fixed effort-out causality of an effort (voltage) source

Figure 9. Fixed causality of sources

respect to time or integration with respect to time (Figure 10). Due to the amplification of numerical noise by numerical differentiation the integral form is preferred from a computational perspective, but there are more fundamental arguments as well. A first indication is that the integral form requires an initial condition, while the differential form does not. Obviously, an initial state or content of some storage element is a physically relevant property that supports the statement that 'integration exists in nature, whereas differentiation does not'. Although one should be careful with the concept of 'existence' when discussing modeling, this statement seeks to emphasize that differentiation with respect to time requires information about future states in principle, in contrast with integration with respect to time. The discussion of causal analysis will make clear that violation of a *preferred integral causality* gives important feedback to the modeler about his modeling decisions. Some forms of analysis, e.g. finding the rank of the system matrix, may require that the differential causal form is preferred at some point in the analysis too, but this requirement is never used as a preparation of the equations for numerical simulation.

Arbitrary causality The expected next possibility in the sequence is that the causality of a port is neither fixed nor preferred, thus arbitrary. Examples of arbitrary port causality are linear, resistive ports that are consequently invertible and algebraic respectively. The acausal form of the constitutive relation of an ohmic resistor is $u - Ri = 0$, the effort-out causal form is $u = Ri$, while the flow-out causal form is $i = u/R$ (cf. Figure 11).

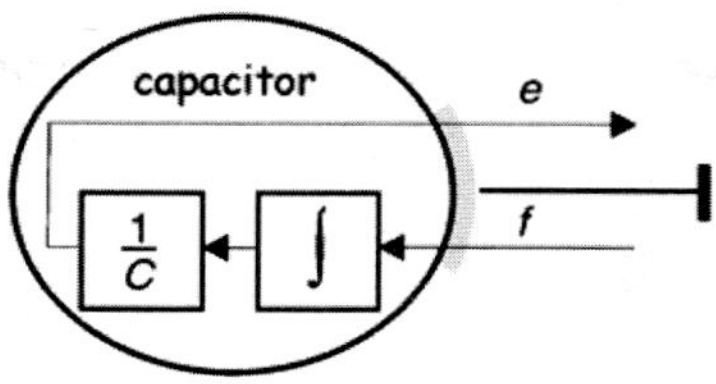

Figure 10. Preferred integral causality of a capacitor

Causal constraints Causal constraints only exist for basic multiports, i.e. elements with two or more ports like the transducers (**TF**, **GY**) and the junctions (**0**, **1**). For instance, if the constitutive relation of the two-port transducers is linear (the junctions are intrinsically linear), the first port to which causality is assigned is arbitrary, but the causality of the second port is immediately fixed. For instance, the two-port transformer always has one port with effort-out causality and one with flow-out causality. By contrast, the causalities of the ports of a two-port gyrator always have the same type of causality. In graphical terms: a **TF** has only one causal stroke directed to it, while a **GY** has either both causal strokes directed to it or none.

The fundamental feature of the junctions that either all efforts are common (**0**) or all flows are common (**1**) shows that only one port of a **0**-junction can have 'effort-in causality', i.e. flow-out causality, viz. the result of its flow-balance. By contrast, only one port of a **1**-junction can have 'flow-in causality', i.e. effort-out causality, viz. the result of its effort-balance. In graphical terms: only one causal stroke can be directed towards a **0**-junction, while only one open end can be directed towards a **1**-junction.

Multiport generalizations of the **TF** and the **GY** as well as junction structures consisting of junctions and bonds may be characterized by non-invertible matrices that add causal constraints.

Causal analysis: feedback on modeling decisions *Causal analysis*, also called *causality assignment* or *causal augmentation*, is the algorithmic process of putting the causal strokes at the bonds on the basis of the causal port properties induced by the nature of the constitutive relations. Not only the final result, but also the assignment process itself provides immediate feedback on modeling decisions.

Fixed causality of the first kind The first step is to assign fixed causalities of the first kind and immediately propagate them via the causal constraints, as the latter cannot be violated. For instance, if a flow source is

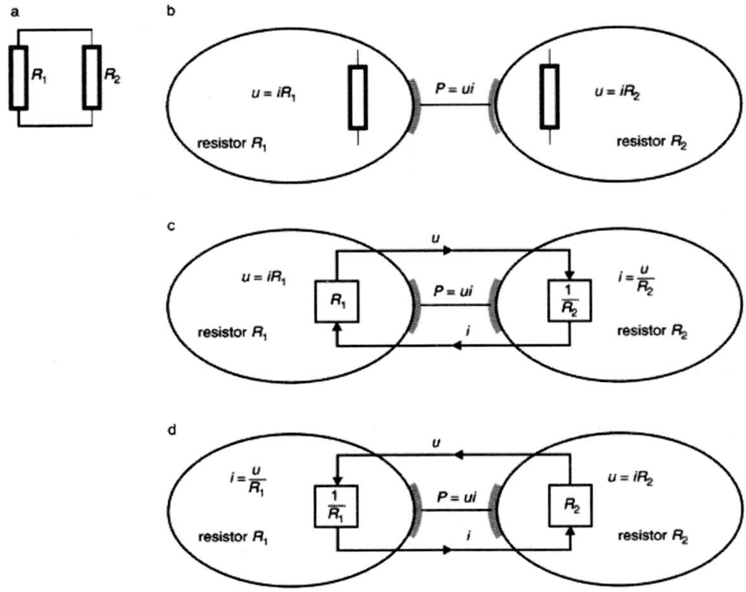

Figure 11. Arbitrary causality of two resistors causing an algebraic loop

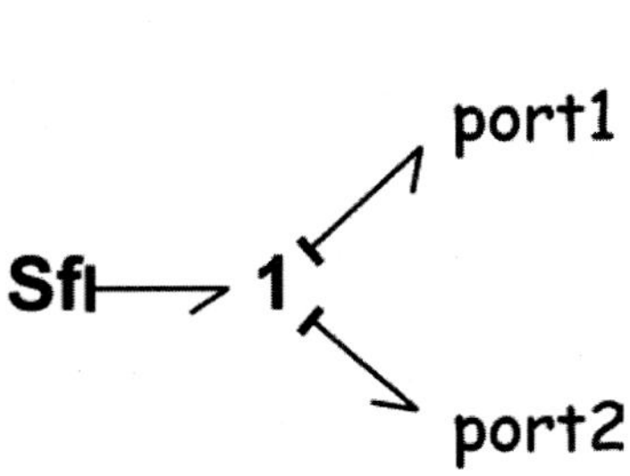

Figure 12. Propagation of a fixed causality via a 1-junction

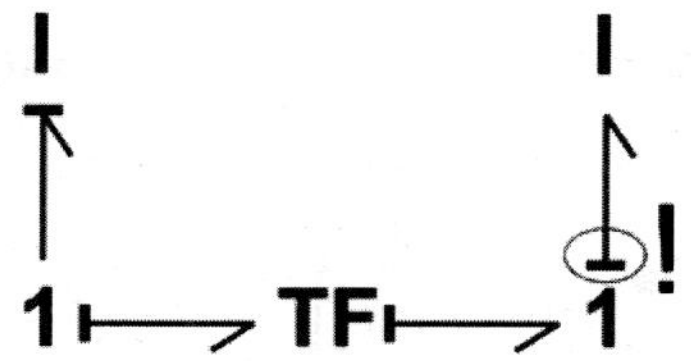

Figure 13. Dependent inertias via the causal constraints of **1**-junctions and transformer

connected to a **1**-junction, the source-port has flow-out causality by definition, which in turn means that the corresponding port at the **1**-junction gets flow-in causality, which means that all other ports of the **1**-junction get flow-out causality, etc. (Figure 12), unless one can apply numerical iteration to impose the same flow at the junction from more than one of its ports, but this is not anticipated in regular causal analysis that is focused on feedback nor is it possible if only sources are involved. Conflicts between fixed causalities of the first kind via the constraints indicate that the problem is ill-posed, e.g. two voltage sources in parallel or two force sources trying to impose a different value to the same force. The causality propagation may lead to violation of preferred causalities, e.g. a voltage source in parallel to a capacitor or a velocity source on a mass. This violation tells the modeler that no independent state is related to the storage element as its content is imposed by a source, which also means that it is dynamically inactive as its rate of change will only influence the source, which is insensitive to it by definition. Not only storage ports, but also resistive ports that get their causality imposed by a source are dynamically inactive, as they cannot form signal loops (so-called causal paths) via other ports. A causal path can be found by following either a causal stroke or an open end from one **R**-, **C**- or **I**-type port to another. At a **GY** the causality reverses, so if one follows the stroke one should continue with the open end, vice versa.

The adjective 'dynamically inactive' does not mean that the variables of the port cannot change, but that the port does not contribute to the dynamic characteristics (like time constants and eigen frequencies) of the system, which will become more clear after the discussion of preferred causality below. Conflicts with fixed causalities of the second kind inform the modeler that the model has to be changed in such a way that either the constitutive relation is changed such that it can be inverted or the structure is changed such that the conflict disappears. If this is impossible, numerical iteration has to be applied in order to solve the non-invertible relation.

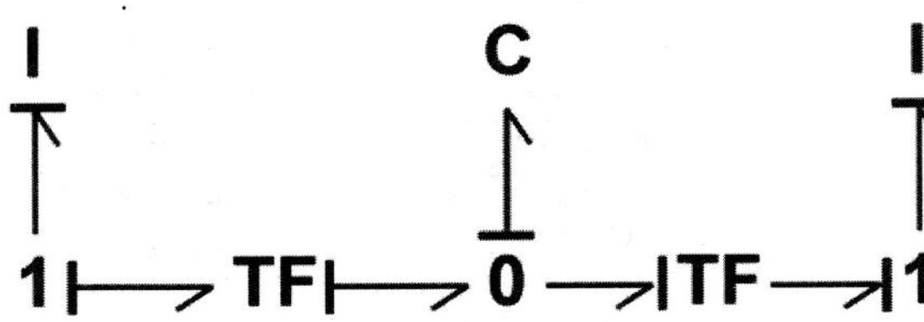

Figure 14. Independent inertias by adding the elasticity of the transmission (e.g. belt-drive)

Preferred causality After assignment of all fixed causalities of the first kind the preferred causalities are similarly assigned and immediately propagated via the causal constraints. Conflicts at this stage indicate that a port may get differential causality as a result of another port getting preferred integral causality. Figure 13 shows the bond graph of two rigidly linked inertias, e.g. the motor inertia and the load inertia in a servo system model, including a transmission (**TF**), but without any compliance. The causality of this bond graph shows the modeler that he has chosen a model in which two storage ports depend on each other and form a signal loop (causal path) with an integration that is compensated by a differentiation, i.e. a net algebraic loop. The computational problem may be solved

- by the application of implicit numerical integration,
- by changing the model (the sequence of putting the causal strokes hints the modeler where a model change should be made, e.g. adding the compliance of the transmission between the two rigid bodies; Figure 14), or
- by symbolic manipulation (either manually or automatically) of the model.

Preferred integral causalities that impose other preferred integral causalities give immediate feedback on the existence of *second-order* signal loops by identifying the *causal paths* in the bond graph, i.e. loops containing two integrations that lead to behavior as described by second-order differential equations, viz. potentially oscillatory behavior. A similar kind of feedback on the dynamic properties of the model is obtained by a port with arbitrary causality that is assigned its causality via propagation of an integral causality. The resulting *first-order causal path* informs the modeler about relaxation type of behavior. Finally, if a preferred causality creates a conflict with a fixed causality of the second kind, then the cause of this fixation has to be reconsidered and the problem solved by either changing the constitutive relation or the model structure or by implementing numerical iteration.

Fixed causality of the second kind As discussed above, fixed causalities of the first kind and preferred causalities are given a higher priority than fixed causalities of the second kind, unless a physical meaning can be assigned to the non-invertibility, in which case the model needs reconsideration in the sense that the model becomes ill-posed or that the number of free to choose initial conditions may be reduced. Accordingly, fixed causalities of the second kind can only propagate to ports with arbitrary causality. In that case mostly an algebraic loop will occur and, if not solved symbolically beforehand, this requires numerical iteration during simulation too. This shows that an explicit ODE model can only be obtained if fixed causalities of the second kind obtain their proper causalities via propagation of fixed causalities of the first kind and preferred causalities. The modeler should consider adapting the model if this is not the case.

Arbitrary causality Commonly all ports in a bond graph are causal after assigning and propagating fixed and preferred causalities, but if this is not the case, it means that at least two ports with arbitrary causality are present. If an arbitrary choice is made for one of these ports, this means that at least one other port will obtain its causality as a result of propagation via the causal constraints (cf. Figure 11). The dual choice would have the same effect. This shows the modeler that this situation always results in an algebraic loop (or its reverse form corresponding to the dual choice of causality) that requires numerical iteration during simulation. Similar to other causal conflicts, e.g. generated by differential causality, the assignment procedure itself hints the modeler how to change the model in order to prevent the algebraic loop. The causality assignment process is completely algorithmic and more advanced variations on this algorithm exist and are implemented that can handle all possible situations in an automated way (van Dijk and Breedveld, 1991; Golo, 2002). As a result, it can be used without using the notation itself, e.g. by replacing the bond graph with the more common iconic diagram representation or the linear graph notation. However, this largely reduces the amount of feedback that can be given to the modeler about his modeling decisions and the effect of model modifications becomes less obvious. Nevertheless, if one is merely interested in converting simple iconic diagrams into code ready for simulation, this is a powerful option.

The support offered by causal analysis in the common trade-off between conceptual and computational complexity of a model is illustrated by the simple example of a rigid constraint between two rigid bodies. Conceptual simplicity of this model leads to a causal problem (a so-called dependent inertia with differential causality) and consequently to numerical complexity when directly converted into computable equations. The example in Figure 13 already showed that a loop emerges containing an integration and a differentiation that cancel each other, i.e. a 'net' algebraic loop, similar to the situation where two resistors form such an algebraic loop (Figure 11). Direct equation generation does not generate a set of ordinary differential equations (ODE), but a mixed set of differential and algebraic equations (DAE). A DAE cannot be solved straightforwardly by means of explicit numerical integration (e.g. with the common Runge-Kutta 4th-order method). Nevertheless, an implicit DAE solver can be used to numerically solve the resulting DAE. However, the way in which the causal problem emerges in the model during causal analysis of the bond graph clearly suggests how the model can be modified in order to prevent the causal problem. In this example the rigid constraint can be replaced by an elastic element, i.e. a finite rigidity. Although this gives the model some more conceptual complexity, the numerical complexity is reduced, due to the fact that the resulting equations are a set of ordinary differential equations (ODE) that can be solved in a straightforward manner by explicit numerical integration schemes (Ascher and Petzold, 1998).

The resulting model needs a rather stiff constraint and thus introduces dynamics at a time scale that may or may not be of interest. This does not only mean that both options to formulate the model can be a solution depending on the problem context, the available tools, etc., but also that a third solution can be obtained, viz. a symbolic transformation of the model as to eliminate the dependent inertia. In other words: two rigidly connected rigid bodies may be considered as one rigid body. This possibility is directly induced by the causal analysis of the bond graph model too.

Example of causal analysis If causal analysis (also called 'causality assignment' or 'causal augmentation') is applied to the bond graph in Figure 7, the fixed causality of the **Se** propagates all the way through the graph via the causal constraints of the two-ports and imposes differential causality on the storage element (Figure 15).

It shows that the state of the system is imposed by the source, which means that the model cannot be used to capture any dynamic behavior. At best, it may serve to analyze a stationary situation, although it is not likely in a practical situation that none of the components has any losses. This immediate feedback on modeling decisions is obtained without writing any equation. If dynamic behavior is to be captured, one needs to modify the

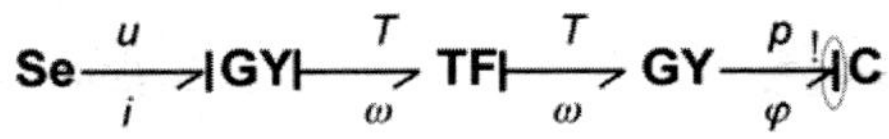

Figure 15. Causally augmented initial bond graph of the pump system: no dynamics

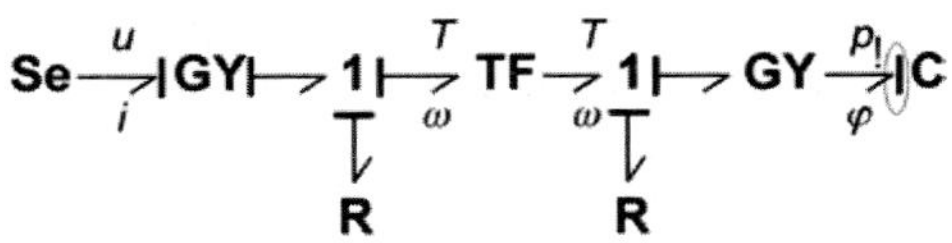

Figure 16. Addition of friction to the pump model: still no dynamics

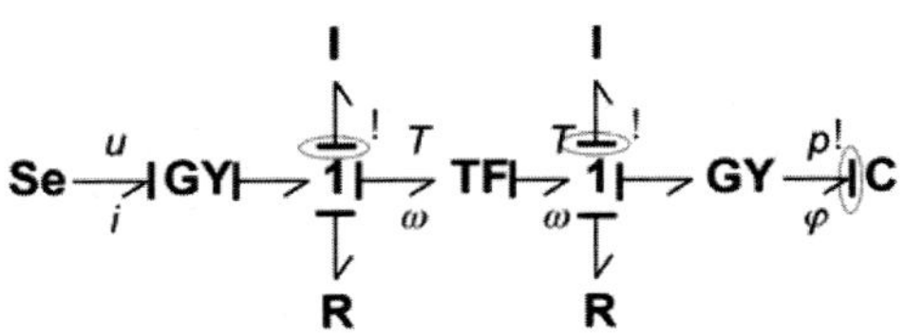

Figure 17. Addition of inertias to the pump model: still no dynamics

graph in such a way that it contains at least one independent state, e.g. the content of the water tank. To that end it will need to have integral causality. For instance, if one would decide that the friction in the bearings in the mechanical part should be modeled, the causal graph immediately shows that this does not change the causality of the storage element (Figure 16).

The same holds for adding the inertias of the rotating parts (Figure 17). The differential causality of the **I**-ports demonstrates that the imposed voltage not only imposes the hydraulic pressure, but also all velocities.

The bond graphs show that the only way to 'break' the causal propagation between voltage source, velocities and pressure is to add a junction that breaks the causal path. This can be either a **0**-junction in the mechanical connections, i.e. allowing some relative speed or a non-stiff connection, or a **1**-junction in the electrical or hydraulic domain, i.e. allowing a voltage difference between source and motor or a pressure difference between pump and tank. For instance, if the transmission is a belt drive, the elasticity of the belt can be modeled (Figure 18).

However, this only gives the tank and the belt independent states: the two **C**'s form a second-order loop via the **GY**-action of the pump. Furthermore, it forms a first order loop (causal path) via this **GY** with the

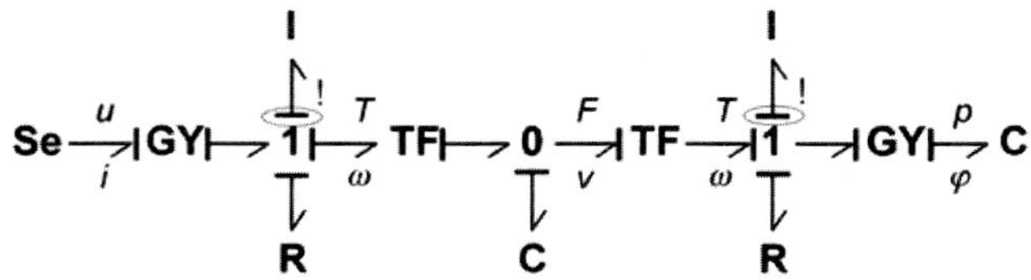

Figure 18. Addition of inertias to the pump model: some dynamics, but not the expected dominant behavior

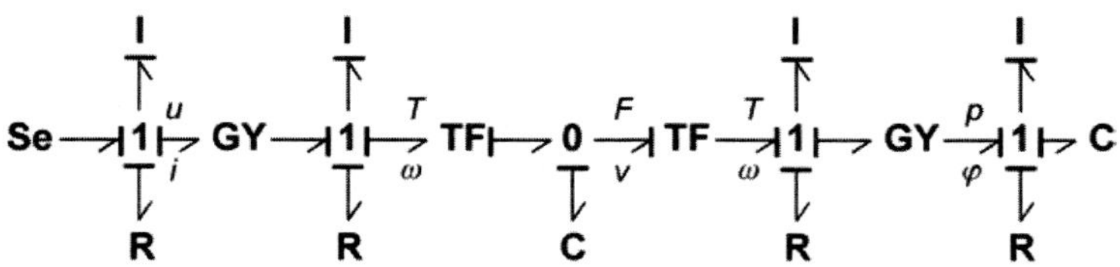

Figure 19. Addition of motor inductance and fluid inertia solves the causal conflicts: ample dynamics

mechanical **R** at the right-hand side. The first inertia still gets its velocity imposed by the source, while the state of the second depends on the state of the tank. The only way to resolve the first dependency is to add an electric **1**-junction connected to the resistance and/or the inductance of the current loop. Similarly, the second dependency can be resolved by adding a hydraulic **1**-junction connected to the resistance and/or the inertia of the hydraulic line (Figure 19).

The process of eliminating all dependencies resulted in a relatively high-order system, viz. a 6th order system (4x **I**, 2x **C**). If we assume for instance at this point that the elasticity of the belt can be neglected after all, the two mechanical **I**'s become dependent again, but these can be symbolically combined into one. If the hydraulic lines are relatively short, the hydraulic **I** can be omitted too (Figure 20), such that a third-order model (2x **I**, 1x **C**) remains to capture the observed dynamics. Further modeling iterations require a more detailed study of problem context and available measurement data. Note that this modeling approach also indicates in an early stage which measurements need to be performed in order to make additional modeling decisions.

The above example demonstrates that it is possible to make far-reaching modeling decisions quickly, because no detailed equations have to be written yet. Causality assignment thus proves to be not only a means to create a

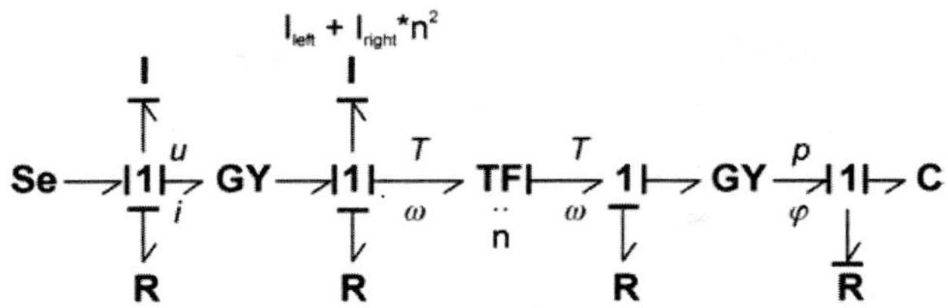

Figure 20. Simplification of the model by symbolic manipulation and maintaining key behavior

model in a form ready for numerical simulation, but also provides direct feedback on modeling decisions even before the exact form of the constitutive relations is known. The nine basic elements can be used to model many different dynamic problems. However, in the next section it will be shown how the generalization of the basic elements can lead to even more powerful modeling concepts.

4 Multiport generalizations

All basic elements can be generalized to a multiport form and put into a multibond notation of which the most important aspects are quite intuitive, cf. Figure 6 (Breedveld, 1985). Sources just form arrays without any coupling due to their nature, i.e. no relations between their conjugate variables. Junctions already are arbitrary multiports. The multiport forms of the (modulated) **TF** and **GY** where already introduced in section 3.2. Herein, it is merely mentioned that position modulated multiport transformers (**MTF**) are a powerful tool in modeling planar and spatial mechanisms (Tiernego and van Dixhoorn, 1979; Bos and Tiernego, 1985; Bos, 1986; Stramigioli, 1998).

The generalization of a resistor, or rather the resistive port of an irreversible transducer, is also straightforward as it concerns an algebraic relation between the port variables. Assuming that the generic algebraic form $\Phi(\mathbf{e}, \mathbf{f}) = 0$, where $\mathbf{e}$ and $\mathbf{f}$ are the conjugate effort and flow vectors of the resistive multiport, can be written $\mathbf{e} = \mathbf{e}(\mathbf{f})$ (or $\mathbf{f} = \mathbf{f}(\mathbf{e})$) with variations

$$de_i = \sum_j \frac{\partial e_i(\mathbf{f})}{\partial f_j} df_j \tag{28}$$

where $\frac{\partial e_i}{\partial f_i}$ is the 'self resistance' of the ith port (diagonal element of the Jacobian) and $\frac{\partial e_i}{\partial f_j} \forall j \neq i$ are the non-diagonal coupling terms. If the Jacobian is constant the relations are linear

$$e_i = \sum_j \frac{\partial e_i}{\partial f_j} f_j = \sum_j B_{ij} f_j \tag{29}$$

or $\mathbf{e} = \mathbf{Bf}$. In that case the irreversibly produced entropy flow $f_{S_{\rm irr}}$ conjugate to temperature T can be written

$$f_{S_{\rm irr}} = T^{-1}\mathbf{e}^t\mathbf{f} = T^{-1}\mathbf{f}^t\mathbf{B}^t\mathbf{f} \tag{30}$$

i.e. a quadratic form in the flows. As only the symmetric part $\mathbf{R} = \frac{\mathbf{B}+\mathbf{B}^t}{2}$ of the constitutive matrix contributes to the entropy production, only this part $\mathbf{R}$ is commonly considered to describe the resistive multiport. Its symmetry corresponds to what is called Onsager symmetry or reciprocity in irreversible thermodynamics (Callen, 1960), as the multiport $\mathbf{R}$ can be considered to represent a domain-independent form of description common in irreversible thermodynamics of the relations between generalized forces and fluxes. The antisymmetric part $\mathbf{G} = \frac{\mathbf{B}-\mathbf{B}^t}{2}$ is readily identified as a power continuous multiport gyrator as introduced already from another perspective in section 3.2. It corresponds to Casimir's extension of the Onsager relations (Breedveld, 1982). Although it is quite interesting to see that this extension only occurs between ports of domains that were identified earlier as having a symplectic interdomain coupling on different grounds, further discussion goes beyond the scope of this treatise. Finally it is noted that the multiport resistor, just like the elementary one-port can be modulated without changing its fundamental nature and that, in order to satisfy the positive entropy production principle at all times, the symmetric matrix $\mathbf{R}$ has to be positive-definite where not only all diagonal elements have to be positive, but also the determinants of all subblocks in this matrix. Finally, only the multiport generalization of a storage element adds a new behavior, viz. transduction. For this reason, multiport storage will be analyzed in more detail in the following subsection. For all the multiports mentioned above (canonical) decompositions (Breedveld, 1984b) into the nine basic elements can be found that not only enhance insight into the potential dynamic behavior of a model, but also allow recognition of substructures as specific multiports (composition), of which an alternative decomposition may improve numerical solution properties, as is the case for the so-called Euler Junction Structure (Karnopp and Rosenberg, 1968; Breedveld, 1999).

4.1 Multiport storage elements

The generalization of a 1-port storage element into an n-port storage element, where the ports may belong to different domains, requires an approach based on the commonly conserved property, the energy. It turns out that a new type of behavior will emerge from this extension, viz. transduction by a cycle process. In other words, power can be transformed from one port to another and thus from one domain into another by performing a cycle process with a multiport storage element with at least two ports. As much of the literature is rather inaccurate about the use of the concept of energy, the distinction between co-energies (often incorrectly considered energies when they have the same numerical value as the energy) and the globally conserved energy will be made clear first by a discussion of Legendre transformations of homogeneous functions.

Co-Energy & Legendre Transforms A concise overview of homogeneous functions and Legendre transforms is given and applied to energy functions, thus introducing the concept of co-energy. Various properties of energy and co-energy related to multiport storage are shortly discussed and related to a physical interpretation. Finally some domain specific forms of co-energy are discussed.

Mathematical background: Homogeneous Functions and Euler's Theorem A function $F(\mathbf{x})$ with $\mathbf{x} = (x_1, \ldots, x_k)^T$ is *homogeneous* of order n if

$$F(\alpha\mathbf{x}) = \alpha^n F(\mathbf{x}) \tag{31}$$

If $y_i(\mathbf{x}) = \frac{\partial F}{\partial x_i}$, then

$$y_i(\alpha\mathbf{x}) = \frac{\partial F(\alpha\mathbf{x})}{\partial \alpha x_i} = \frac{\alpha^n}{\alpha} \cdot \frac{\partial F(\mathbf{x})}{\partial x_i} = \alpha^{n-1} y_i(\mathbf{x}) \tag{32}$$

in other words: $y_i(\mathbf{x})$ is homogeneous of order $(n-1)$. For such homogeneous functions *Euler's theorem* holds:

$$\sum_{i=1}^{k} \frac{\partial F}{\partial x_i} \cdot x_i = n \cdot F(\mathbf{x}) \text{ or } F(\mathbf{x}) = \frac{1}{n} \sum_{i=1}^{k} y_i \cdot x_i = \frac{1}{n} \cdot \mathbf{y}^T \cdot \mathbf{x} \tag{33}$$

By definition the variation of F can be written:

$$\mathrm{d}F = \sum_{i=1}^{k} \frac{\partial F}{\partial x_i} \cdot \mathrm{d}x_i = \sum_{i=1}^{k} y_i \mathrm{d}x_i = \mathbf{y}^T \cdot \mathrm{d}\mathbf{x} \tag{34}$$

but also using Equation (33):

$$\mathrm{d}F = \mathrm{d}\left(\frac{1}{n}\mathbf{y}^T \cdot \mathbf{x}\right) = \frac{1}{n}\mathbf{y}^T \cdot \mathrm{d}\mathbf{x} + \frac{1}{n}(\mathrm{d}\mathbf{y})^T \cdot \mathbf{x} \tag{35}$$

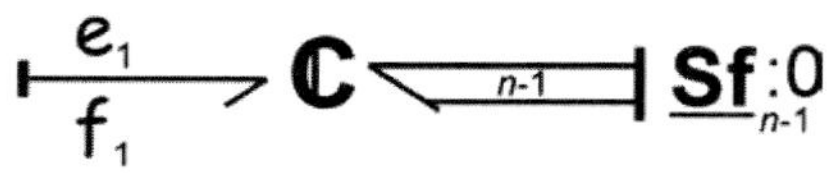

Figure 21. Bond graph of a 'one-port storage element'

Combination of Equations (34) and (35) gives:

$$(\mathrm{d}\mathbf{y})^T \cdot \mathbf{x} = (n-1)\,\mathbf{y}^T \cdot \mathrm{d}\mathbf{x} \tag{36}$$

and for $n = 1$:

$$(\mathrm{d}\mathbf{y})^T \cdot \mathbf{x} = 0 \tag{37}$$

for $n \neq 1$:

$$dF = \frac{1}{n-1}(\mathrm{d}\mathbf{y})^T \cdot \mathbf{x} \tag{38}$$

Homogeneous Energy Functions The **energy** of a system with k **state variables** q can be written $E(\mathbf{q}) = E(q_1, \ldots, q_k)$. If q_i is an **'extensive'** state variable, this means that

$$E(\alpha\mathbf{q}) = \alpha E(\mathbf{q}) \tag{39}$$

in other words $E(\alpha\mathbf{q}) = \alpha^n E(\mathbf{q})$ where $n = 1$. Hence $E(q)$ is **first order** homogeneous, so $e_i(\mathbf{q}) = \frac{\partial E}{\partial q_i}$ is **zero**th order $(n-1=0)$ homogeneous, which means that $e_i(q)$ is an **'intensive'** variable, i.e. $e_i(\alpha\mathbf{q}) = \alpha^0 e_i(\mathbf{q}) = e_i(\mathbf{q})$.

This means also that in case $n = 1$ and $k = 1$ $e(q)$ is constant, i.e. $\frac{\partial e}{\partial q} = \frac{\mathrm{d}e}{\mathrm{d}q} = 0$, which changes the behavior of this element into that of a source. This means that storage elements should in principle be multiports ($k > 1$). A one-port storage element is in principle an n-port storage element of which the flows of $n-1$ ports are kept zero. Accordingly, the corresponding $n-1$ states are constant and commonly not recognized as states. Such states are often considered parameters: if $E(q_1, q_2, ..., q_k)|_{dq_i=0,\forall i\neq 1} = E'(q_1)$, then $E'(q_1)$ is not necessarily first-order homogeneous in q_1 (cf. Figure 21).

For $n = 1$ and k independent extensities there are only $k-1$ independent intensities, because for $n = 1$ we find *Gibbs' fundamental relation*:

$$E(\mathbf{q}) = \mathbf{e}^T \cdot \mathbf{q} \tag{40}$$

As by definition holds that $dE = \mathbf{e}^T \cdot \mathbf{dq}$, this results in the *Gibbs-Duhem relation*:

$$(\mathbf{de})^T \cdot \mathbf{q} = 0 \tag{41}$$

which expresses that one of the intensities depends on the others.

Mathematical background: Legendre Transforms A *Legendre transform* $\pounds$ of a (homogeneous) function $F(x)$ with respect to x_i is by definition

$$\pounds\{F(\mathbf{x})\}_{x_i} = \pounds_{x_i} = F(\mathbf{x}) - y_i \cdot x_i \tag{42}$$

where $y_i = \frac{\partial F}{\partial x_i}$ and the *total Legendre transform* $\pounds$ of $F(x)$ is

$$\pounds\{F(\mathbf{x})\} = \pounds = F(\mathbf{x}) - \sum_{i=1}^{k} y_i x_i \tag{43}$$

Equation (33) shows that $\pounds = 0$ for $n = 1$. Now

$$d\pounds_{x_i} = dF - d(y_i x_i) = dF - y_i dx_i - x_i dy_i = \sum_{j \neq i} y_j dx_j - x_i dy_i \tag{44}$$

or

$$\pounds_{x_i} = \pounds_{x_i}(x_1, \ldots, x_{i-1}, y_i, x_{i+1}, \ldots, x_k) \tag{45}$$

which means that x_i is replaced by y_i as *independent* variable or 'coordinate'! Hence $\pounds = \pounds(\mathbf{y})$ and

$$d\pounds = -\sum_{i=1}^{k} x_i dy_i = -(\mathbf{dy})^T \cdot \mathbf{x} \tag{46}$$

Co-Energy Functions The complimentary energy or *co-energy* $E^*_{q_i}$ of $E(q)$ *with respect to* q_i is by definition a Legendre transform:

$$E^*_{q_i} = -\pounds\{E(\mathbf{q})\}_{q_i} = E^*_{q_i}(q_1, \ldots, q_{i-1}, e_i, q_{i+1}, \ldots, q_k) \tag{47}$$

Hence $E(\mathbf{q}) + E^*_{q_i}(\ldots, e_i, \ldots) = e_i \cdot q_i$. For the *total co-energy* $E^*(e)$ of $E(q)$ holds that $E^* = -\pounds$, hence

$$E(\mathbf{q}) + E^*(\mathbf{e}) = \mathbf{e}^T \cdot \mathbf{q} \tag{48}$$

which explains the terminology. For $n = 1$: $E^*(\mathbf{e}) = 0$, confirming the earlier conclusion that there are only $k - 1$ independent intensities e_i. For $n = 2$: $E(\mathbf{q}) = E^*(\mathbf{e}) = \frac{1}{2}\mathbf{e}^T \cdot \mathbf{q}$ (i.e. equal in value, but different in nature as is the case for linear constitutive relations that result in quadratic energy functions!) and for $n = 3$: $E(\mathbf{q}) = \frac{1}{3}\mathbf{e}^T \cdot \mathbf{q}$ while $E^*(\mathbf{e}) = \frac{2}{3}\mathbf{e}^T \cdot \mathbf{q}$. The following relations for co-energy functions can be useful:

$$dE^*_{q_i} = de_i \cdot q_i - \sum_{j \neq 1} e_j \cdot dq_j \tag{49}$$

$$dE^* = \sum_{i=1}^{k} de_i \cdot q_i = (\mathbf{de})^T \cdot \mathbf{q} = (n-1)\mathbf{e}^T \cdot \mathbf{dq} = (n-1)dE \tag{50}$$

$$E^* = (n-1)E = \left(\frac{n-1}{n}\right)\mathbf{e}^T \cdot \mathbf{q} = \left(1 - \frac{1}{n}\right)\mathbf{e}^T \cdot \mathbf{q} \tag{51}$$

Next several domain dependent manifestations of the Legendre transforms will be discussed.

Legendre transforms in thermodynamics In thermodynamic systems of 'simple' systems with internal energy U that is a function of the extensities entropy S, volume V, total mole number N and mole number N_i per species i , the conjugate intensities are respectively temperature $T = \frac{\partial U}{\partial S}$, pressure $p = \frac{\partial U}{-\partial V} = -\frac{\partial U}{\partial V}$ (energy increases in case of compression $-\partial V$), total material potential $\mu^{\mathrm{tot}} = \frac{\partial U}{\partial N}$ and chemical potential $\mu_i = \frac{\partial U}{\partial N_i}$. The Legendre transforms are the *free energy* F, the *enthalpy* H and the *Gibbs free enthalpy* G. The *free energy* F is a Legendre transform of the energy with respect to the entropy S:

$$\begin{aligned}
\pounds_S = F(T, V, \mathbf{N}, N) &= \\
= U - TS = -pV + \sum_{i=1}^{m-1} \mu_i N_i + \mu^{\mathrm{tot}} \cdot N &= \\
(= N \cdot f(T, v, \mathbf{c}))
\end{aligned} \tag{52}$$

where $\mathbf{N}^t = (N_1, \ldots, N_i, \ldots, N_{m-1})$, c is the vector of molar fractions $\frac{N_i}{N}$, $f = \frac{F}{N}$ the specific free energy and with variation

$$\mathrm{d}F = -S\mathrm{d}T - p\mathrm{d}V + \sum_{i=1}^{m-1} \mu_i \mathrm{d}N_i + \mu^{\mathrm{tot}} \cdot \mathrm{d}N \tag{53}$$

which means that the thermal domain does not influence the free energy at constant temperature. The *enthalpy* H is a Legendre transform of the energy with respect to the available volume V:

$$\pounds_V = H(S, p, \mathbf{N}, N) = U - (-pV) = U + pV (= N \cdot h(s, p, \mathbf{c})) \tag{54}$$

where $h = \frac{H}{N}$ the specific enthalpy and with variation

$$dH = TdS + Vdp + \sum_{i=1}^{m-1} \mu_i \cdot dN_i + \mu^{\mathrm{tot}} \cdot dN \tag{55}$$

which means that the mechanical domain does not influence the enthalpy at constant pressure. Finally, the *Gibbs free enthalpy* G is a Legendre transform of the energy with respect to both the entropy S and the available volume V:

$$\begin{aligned} \pounds_{S,V} &= G(T,p,\mathbf{N},N) = U - TS - (-pV) = \\ &= \mu^{tot} \cdot N + \sum_{i=1}^{m-1} \mu_i N_i = \\ &(= N \cdot g(T,p,\mathbf{c})) \end{aligned} \tag{56}$$

where $g = \frac{G}{N}$ the specific free enthalpy and with variation

$$dG = -SdT + Vdp + \sum_{i=1}^{m-1} \mu_i dN_i + \mu^{\mathrm{tot}} \cdot dN \tag{57}$$

which means that both the mechanical domain and the thermal domain do not influence the Gibbs free enthalpy at constant pressure and temperature. For one constituent ($m = 1$) it holds that $g = \mu^{\mathrm{tot}}(T,p)$, which means that the equilibrium of the material is determined by pressure and temperature. This explains why under common isothermal conditions pressure is considered the equilibrium-determining variable for fluid flow (gas or liquid), although they are not power-conjugate variables. This is reinforced by the fact that a linear relation between a pressure difference or gradient and a fluid flow is feasible, whereas this does not hold for the power-conjugate variables total material potential and fluid flow.

Legendre Transforms in Mechanics In mechanical systems with kinetic energy T and potential energy V, i.e. of which the total energy is the *Hamiltonian* H that is a function of two vectors of extensities $\mathbf{q}$, viz. the displacements vector $\mathbf{x}$ and the momenta vector $\mathbf{p}$:

$$E(\mathbf{q}) = H(\mathbf{x},\mathbf{p}) = T + V \tag{58}$$

while the conjugate intensities are the vector of the velocities $\mathbf{v} = \frac{\partial H}{\partial \mathbf{p}}$ and the vector of the forces $\mathbf{F} = \frac{\partial H}{\partial \mathbf{x}}$ respectively. The Legendre transforms of the *Hamiltonian* H are the *Lagrangian* L, the *co-Hamiltonian* and the *co-Lagrangian* or *Hertzian*. The *Lagrangian* L is a *negative* Legendre transform (co-energy) of the Hamiltonian with respect to the momenta $\mathbf{p}$:

$$H^*_{\mathbf{p}} = -\pounds_{\mathbf{p}} = \mathbf{v}^T \cdot \mathbf{p} - H = (T + T^*) - (T + V) = T^* - V = L(\mathbf{x},\mathbf{v}) \tag{59}$$

the *co-Hamiltonian* is a *negative* Legendre transform (co-energy) of the Hamiltonian with respect to both the displacements $\mathbf{x}$ and the momenta $\mathbf{p}$:

$$\begin{aligned} H^*_{\mathbf{x},\mathbf{p}} &= \mathbf{v}^T \cdot \mathbf{p} + \mathbf{F}^T \cdot \mathbf{q} - H = \\ &= (T + T^*) + (V + V^*) - (T + V) \\ &= T^* + V^* = H^*(\mathbf{F}, \mathbf{v}) \end{aligned} \tag{60}$$

and the *co-Lagrangian* or *Hertzian* is a *negative* Legendre transform (co-energy) of the Hamiltonian with respect to the displacements $\mathbf{x}$:

$$H^*_{\mathbf{x}} = \mathbf{F}^T \cdot \mathbf{q} - H = (V + V^*) - (T + V) = V^* - T = -L^*(\mathbf{F}, \mathbf{p}) \tag{61}$$

Note that adding the symplectic coupling between the kinetic and potential domains leads to the common Hamiltonian formulation of the equations of motion, where J is the symplectic matrix:

$$\frac{\mathrm{d}}{\mathrm{d}t}\begin{bmatrix} \mathbf{q} \\ \mathbf{p} \end{bmatrix} = \begin{bmatrix} 0 & +1 \\ -1 & 0 \end{bmatrix} \begin{bmatrix} \frac{\partial H}{\partial \mathbf{q}} \\ \frac{\partial H}{\partial \mathbf{p}} \end{bmatrix} + \begin{bmatrix} 0 \\ \mathbf{F}_{\text{external}} \end{bmatrix} = J \begin{bmatrix} \frac{\partial H}{\partial \mathbf{q}} \\ \frac{\partial H}{\partial \mathbf{p}} \end{bmatrix} + \begin{bmatrix} 0 \\ \mathbf{F}_{\text{external}} \end{bmatrix} \tag{62}$$

while after the Legendre transform into the Lagrangian the equations of motion are written:

$$\frac{\mathrm{d}}{\mathrm{d}t}\begin{bmatrix} \mathbf{q} \\ \frac{\partial L}{\partial \mathbf{v}} \end{bmatrix} = \begin{bmatrix} 0 & +1 \\ -1 & 0 \end{bmatrix} \begin{bmatrix} \frac{\partial L}{\partial \mathbf{q}} \\ \mathbf{v} \end{bmatrix} + \begin{bmatrix} 0 \\ \mathbf{F}_{\text{external}} \end{bmatrix} \tag{63}$$

or, by eliminating $\frac{\mathrm{d}\mathbf{q}}{\mathrm{d}t} = \mathbf{v}$:

$$\frac{\mathrm{d}}{\mathrm{d}t}\frac{\partial L}{\partial \mathbf{v}} + \frac{\partial L}{\partial \mathbf{q}} = \mathbf{F}_{\text{external}} \tag{64}$$

Note that in both common formulations velocity sources are excluded.

If one observes the common equation written for the dynamics of a mechanism based on the force balance $\tau^{\text{inertial}} + \tau^{\text{fictitious}} + \tau^{\text{conservative}} + \tau^{\text{dissipative}} = \tau^{\text{external}}$, where τ stands for generalized forces, i.e.

$$\mathbf{M}(\mathbf{q})\ddot{\mathbf{q}} + \frac{1}{2}\dot{\mathbf{q}}^t \frac{\partial \mathbf{M}(\mathbf{q})}{\partial \mathbf{q}}\dot{\mathbf{q}} + \mathbf{K}\mathbf{q} + \mathbf{B}\dot{\mathbf{q}} = \tau^{\text{external}} \tag{65}$$

where $\mathbf{q}$ stands for the generalized displacements, $\mathbf{M}$ for the mass matrix, $\mathbf{K}$ for the stiffness matrix and $\mathbf{B}$ for the viscous friction matrix, there are two surprising aspects on close inspection of the first two terms.

First of all, based on Newtons second law one would expect the first term in (33), if it represents the inertial forces, to be

$$\tau^{\text{inertial}}(\mathbf{q},\dot{\mathbf{q}}) = \frac{\mathrm{d}}{\mathrm{d}t}\left[\mathbf{M}(\mathbf{q})\dot{\mathbf{q}}\right] = \mathbf{M}(\mathbf{q})\ddot{\mathbf{q}} + \dot{\mathbf{q}}^t\frac{\partial \mathbf{M}^t(\mathbf{q})}{\partial \mathbf{q}}\dot{\mathbf{q}} = \\ = \mathbf{M}(\mathbf{q})\ddot{\mathbf{q}} + \dot{\mathbf{q}}^t\frac{\partial \mathbf{M}(\mathbf{q})}{\partial \mathbf{q}}\dot{\mathbf{q}} \tag{66}$$

using the symmetry of $\mathbf{M}(\mathbf{q})$ in the last step. In other words, it contains an additional term $\dot{\mathbf{q}}^t\frac{\partial \mathbf{M}(\mathbf{q})}{\partial \mathbf{q}}\dot{\mathbf{q}}$. Secondly, one would expect the fictitious forces to be minus the partial derivative of the kinetic co-energy with respect to the displacements:

$$\tau^{\text{fictitious}}(\mathbf{q},\dot{\mathbf{q}}) = -\frac{\partial E^*_{\text{kin}}(\mathbf{q},\dot{\mathbf{q}})}{\partial \mathbf{q}} = -\frac{\partial}{\partial \mathbf{q}}\left(\frac{1}{2}\dot{\mathbf{q}}^t\mathbf{M}(\mathbf{q})\dot{\mathbf{q}}\right) = \\ = -\frac{1}{2}\dot{\mathbf{q}}^t\frac{\partial \mathbf{M}(\mathbf{q})}{\partial \mathbf{q}}\dot{\mathbf{q}} \tag{67}$$

The conclusion is that the sum of the inertial and fictitious forces is indeed

$$\tau^{\text{inertial}}(\mathbf{q},\dot{\mathbf{q}}) + \tau^{\text{fictitious}}(\mathbf{q},\dot{\mathbf{q}}) = \mathbf{M}(\mathbf{q})\ddot{\mathbf{q}} + \dot{\mathbf{q}}^t\frac{\partial \mathbf{M}(\mathbf{q})}{\partial \mathbf{q}}\dot{\mathbf{q}} - \frac{1}{2}\dot{\mathbf{q}}^t\frac{\partial \mathbf{M}(\mathbf{q})}{\partial \mathbf{q}}\dot{\mathbf{q}} = \\ = \mathbf{M}(\mathbf{q})\ddot{\mathbf{q}} + \frac{1}{2}\dot{\mathbf{q}}^t\frac{\partial \mathbf{M}(\mathbf{q})}{\partial \mathbf{q}}\dot{\mathbf{q}} \tag{68}$$

such that Equation (65) as a whole is correct, but that the interpretation of the first term as the inertial forces and the second term as the fictitious forces is not correct. However, if one would forget to make a distinction between energy and co-energy, the minus sign would turn into a plus sign as " $\tau^{\text{fictitious}}$ " $= +\frac{\partial E_{\text{kin}}}{\partial \mathbf{q}}$ seems to be equal to $+\frac{1}{2}\dot{\mathbf{q}}^t\frac{\partial \mathbf{M}(\mathbf{q})}{\partial \mathbf{q}}\dot{\mathbf{q}}$ and the second term in Equation (65) would indeed seem to represent the fictitious forces. It then becomes hard to explain why the position dependent mass matrix does not generate an additional term due to the time derivative, unless one forgets that the common formulation of Newtons second law $\tau^{\text{inertial}} = \mathbf{M}\ddot{\mathbf{q}}$ only holds for constant mass matrices. In conclusion one could state that two misinterpretations compensate each other in a quite misleading manner, which has contributed to the common disregard of the concept of co-energy.

Legendre Transforms in Electrical Circuits In electrical circuits with capacitor charges $\mathbf{q}$ and conjugate voltages $\mathbf{u}$ and coil flux linkages λ and conjugate currents $\mathbf{i}$, the total stored energy is

$$E(\mathbf{q},\lambda) = E_C(\mathbf{q}) + E_L(\lambda) \tag{69}$$

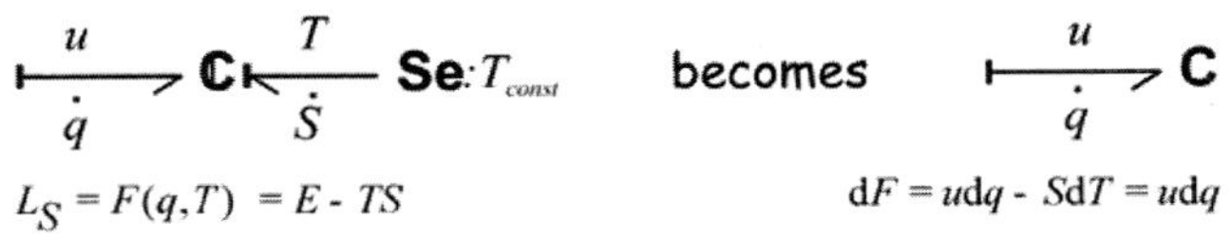

Figure 22. Use of free energy at constant temperature: no thermal port required

The total co-energy is

$$E^*(\mathbf{u}, \mathbf{i}) = \mathbf{u}^T \cdot \mathbf{q} + \mathbf{i}^T \cdot \mathbf{\Phi} - E = E \tag{70}$$

Note that the last equal sign only holds in case of circuit elements with linear constitutive relations. Nevertheless, this assumption is seldom made explicit when co-energy is used as the 'energy' of an electric circuit.

Legendre Transforms and Causality The above similarities between the Legendre transforms in these different domains are often not made explicit. For instance some properties of and processes with multiports that are quite common in thermodynamics are not common in mechanics and electrical circuit theory, although they may be used to enhance insight in similar ways.

In the domain independent approach as denoted in bond graphs one can conclude that if an effort is 'forced' on a port of a **C**-element ('derivative causality' or 'flow-out causality'), this means that the roles of e and q are interchanged in the set of independent variables, which means that the energy has to be *Legendre transformed* in order to maintain a generating function for the constitutive relations. Such a transformation is particularly useful when the *effort* e is *constant.* For example, an electrical capacitor in an isothermal environment with $T = T_{const}$ can be characterized by the free energy: $dF = udq - SdT = udq$ and the thermal port can be omitted as it does not result in changes of the free energy F (Figure 22):

$$P = u \cdot \dot{q} = \frac{dF}{dt} \tag{71}$$

Independent of the domain the function $e_i(q)$ that characterizes a storage port is called ***constitutive relation*** or *constitutive equation, constitutive law, state equation, characteristic equation,* etc. If $e_i(q)$ is **linear**, i.e. first order homogeneous, then $E(q)$ is second order homogeneous, i.e. $E(q)$ is **quadratic**. In this case, and only in this case: $E(\alpha\mathbf{q}) = \alpha^2 E(\mathbf{q})$, $E(\mathbf{q}) = \frac{1}{2}\mathbf{e}^T \cdot \mathbf{q} = E^*(\mathbf{e})$ and $dE^* = (\mathbf{de})^T \cdot \mathbf{q} = \mathbf{e}^T \cdot \mathbf{dq} = dE$, i.e. the values of energy and co-energy are equal.

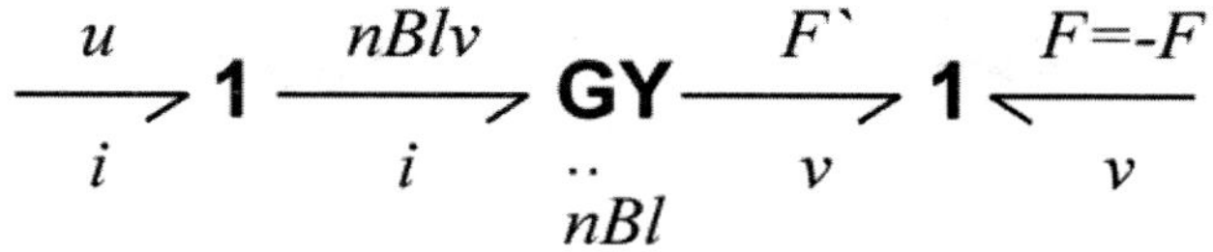

Figure 23. Basic transduction in a loudspeaker

Another property of constitutive relations of storage elements that is commonly used in thermodynamics is *Maxwell reciprocity* or *Maxwell symmetry.* From the principle of energy conservation can be derived that the mixed second derivatives of the energy should be equal $\frac{\partial^2 E}{\partial q_i \partial q_j} = \frac{\partial^2 E}{\partial q_j \partial q_i}$. This means that $\frac{\partial e_j}{\partial q_i} = \frac{\partial e_i}{\partial q_j}$, i.e. the *Jacobian* matrix of the constitutive relations is *symmetric.* Maxwell symmetry requires the 'true energy' form (integral causality) as Legendre transforms generally destroy the symmetry of the Jacobian .Yet another property that is commonly used in thermodynamics (le Chatelier-Braun principle) to check the *intrinsic stability* of a system that can be considered a multiport storage element is that the *Jacobian* of its constitutive relations and all its sub-blocks are also *positive-definite*: $\det\left(\frac{\partial \mathbf{e}}{\partial \mathbf{q}}\right) > 0$ and that the *diagonal elements* of the *Jacobian* are *positive*: $\frac{\partial e_i}{\partial q_i} > 0\ \forall i$. Note that similar properties were found for the multiport $\mathbf{R(S)}$ earlier.

Loudspeaker example This example will be treated in two ways: first a two-port storage will be constructed by a bottom-up approach, starting from the dominant behavior and next the involved two-port storage will be taken as a starting point in a top-down approach.

The *dominant* behavior of a loudspeaker, in particular the electromechanical transduction based on the Lorentz force, $F = n\,(i \times B)\,l$ and on $\frac{\mathrm{d}\lambda}{\mathrm{d}t} = u = n\,(v \times B)\,l$, can in the given configuration (see Figure 23) be approximated by a gyrator, $F = (nBl)\,i$ and $u = (nBl)\,v$.

Making the relation between electric and magnetic domain explicit by means of an **SGY** shows that the *storage* of magnetic energy in the voice coil can be represented by a magnetic **C** (Figure 24) for which can be written

$$\frac{d\lambda}{dt} = \frac{dLi}{dt} + nBlv = \frac{dLi}{dt} + nBl\frac{dx}{dt} \tag{72}$$

and after integration:

$$\lambda = Li + nBlx \tag{73}$$

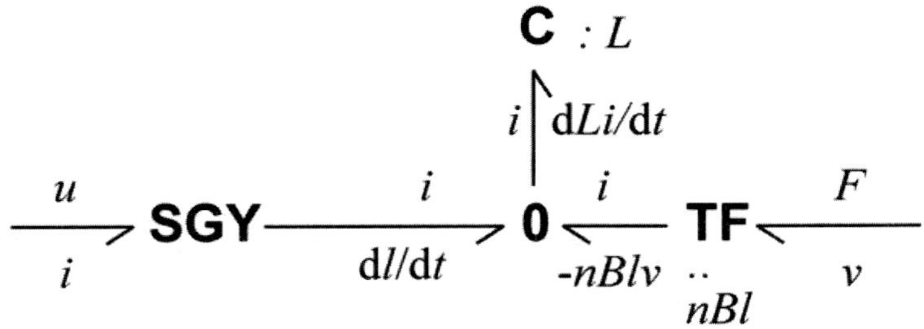

Figure 24. Addition of magnetic storage with an explicit magnetic domain

If this constitutive relation is put in preferred integral causality

$$i = \frac{1}{L}\lambda - \frac{nBl}{L}x \tag{74}$$

and combined with the relation of the mechanical port

$$F = -nBli = \frac{-nBl}{L}\lambda + \frac{(nBl)^2}{L}x$$

into a relation of a two-port **C** in matrix form

$$\begin{bmatrix} i \\ F \end{bmatrix} = \begin{bmatrix} \frac{1}{L} & -\frac{nBl}{L} \\ -\frac{nBl}{L} & \frac{(nBl)^2}{L} \end{bmatrix} \begin{bmatrix} \lambda \\ x \end{bmatrix} \tag{75}$$

it can be concluded that this two-port satisfies Maxwell symmetry, but that it is singular ($\frac{1}{L}\frac{(nBl)^2}{L} - \left(-\frac{nBl}{L}\right)^2 = 0$), such that a positive mechanical spring constant K representing the connection to the frame of the moving voice coil, has to be added to make it intrinsically stable: $\frac{1}{L}\left(\frac{(nBl)^2}{L} + K\right) - \left(-\frac{nBl}{L}\right)^2 > 0$ (Figure 25).

The junction structure with the transformer and the two **C**'s coincides with the so-called congruent canonical decomposition of a linear two-port **C** of which the energy can be used as a generating function of the constitutive relations (Breedveld, 1984b).

When starting again with this two-port **C**, but without a mechanical

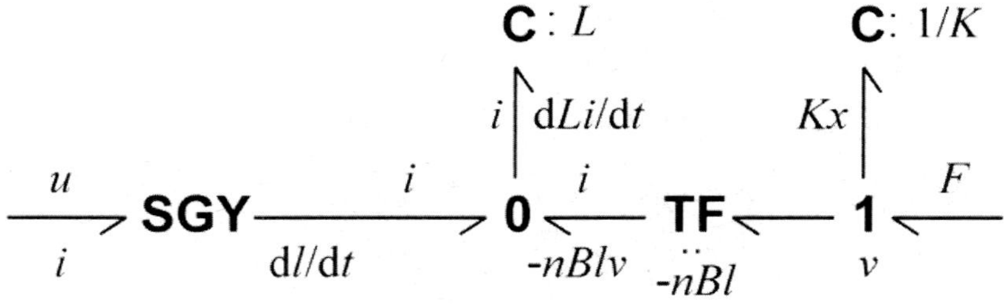

Figure 25. Addition of elastic cone suspension resulting in the decomposition of a two-port storage element

spring, the energy of this two-port should be written in terms of λ and x:

$$
\begin{aligned}
dE(\lambda, x) &= i\mathrm{d}\lambda + F\mathrm{d}x = \\
&= \left(\frac{\lambda}{L} - \frac{(nBl)\,x}{L}\right)\mathrm{d}\lambda + \left(-\frac{(nBl)\,\lambda}{L} + \frac{(nBl)^2\,x}{L}\right)\mathrm{d}x = \\
&= \mathrm{d}\frac{(\lambda - (nBl)\,x)^2}{2L}
\end{aligned}
\tag{76}
$$

such that $E(\lambda, x) = \frac{(\lambda-(nBl)x)^2}{2L}$, but is commonly mistaken by what is actually the co-energy in terms of i and x: $E^*(i, x) = \frac{Li^2}{2} + nBlix$. If the latter is the case and the force is incorrectly derived by taking the partial derivative of this co-energy as if it were an energy $\frac{\partial E^*(i,x)}{\partial x} = \frac{i^2}{2}\frac{\mathrm{d}L}{\mathrm{d}x} + nBli$, a sign error is obtained, as the force should be

$$
\begin{aligned}
F &= \frac{\partial E(\lambda, x)}{\partial x} = \frac{\partial}{\partial x}\frac{(\lambda - (nBl)\,x)^2}{2L} = \\
&= -(nBl)\left(\frac{\lambda}{L} - \frac{(nBl)\,x}{L}\right) - \frac{1}{2}\left(\frac{\lambda}{L} - \frac{(nBl)\,x}{L}\right)^2\frac{\mathrm{d}L}{\mathrm{d}x} = \\
&= -(nBl)\,i - \frac{1}{2}i^2\frac{\mathrm{d}L}{\mathrm{d}x}
\end{aligned}
\tag{77}
$$

Note that the mere difference in the form of a minus sign between these results only occurs for the particular case in which the current i depends linearly on the flux linkage λ. In case of a nonlinear relation, the error is larger than 'just' a sign error that is commonly not noticed as it is compensated by an implicit change in orientation: in contrast with the global convention, the mechanical port is then taken positively outbound.

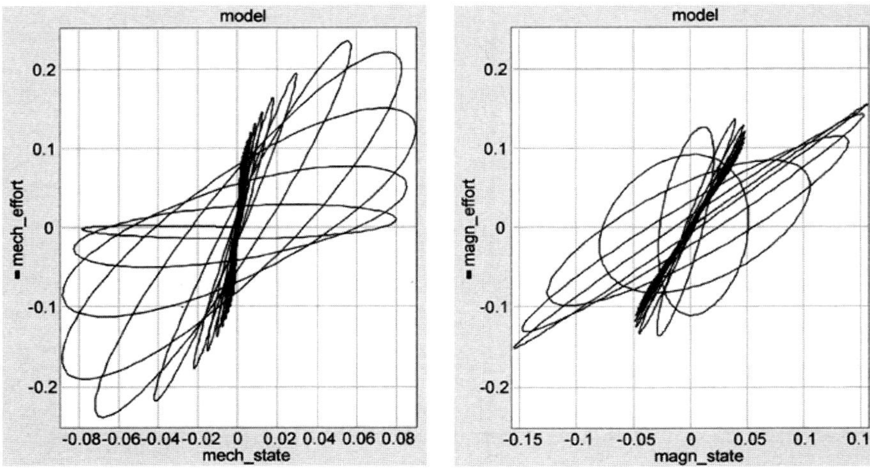

Figure 26. Simulation of the mechanical and magnetic cycle with frequency sweep input

If a spring is added to satisfy intrinsic stability the energy increases with the potential energy $E(x) = \frac{Kx^2}{2}$ and similar constitutive relations are found from the energy function:

$$\begin{bmatrix} i \\ F \end{bmatrix} = \begin{bmatrix} \frac{\partial i}{\partial \lambda} & \frac{\partial i}{\partial x} \\ \frac{\partial F}{\partial \lambda} & \frac{\partial F}{\partial x} \end{bmatrix} \begin{bmatrix} \lambda \\ x \end{bmatrix} = \begin{bmatrix} \frac{\partial^2 E}{\partial \lambda^2} & \frac{\partial^2 E}{\partial x \partial \lambda} \\ \frac{\partial^2 E}{\partial \lambda \partial x} & \frac{\partial^2 E}{\partial x^2} \end{bmatrix} \begin{bmatrix} \lambda \\ x \end{bmatrix} =$$
$$= \begin{bmatrix} \frac{1}{L} & -\frac{nBl}{L} \\ -\frac{nBl}{L} & \frac{(nBl)^2}{L} + K \end{bmatrix} \begin{bmatrix} \lambda \\ x \end{bmatrix} \qquad (78)$$

which is intrinsically stable as long as $\frac{(nBl)^2}{L^2} + \frac{K}{L} - \frac{(nBl)^2}{L^2} = \frac{K}{L} > 0$. Figure 26 shows the simulation results of a loudspeaker model containing this two-port, with a sinusoidal frequency sweep input at the electric port. Both ports indeed perform a cycle process in which the magnetic port cycles clockwise and the mechanical port counter clockwise (large loops, i.e. more energy transferred per cycle, represent a simulation at the resonance frequency of the speaker), which demonstrates that indeed per cycle a net amount of magnetic (electric) energy is transduced into mechanical energy. The areas of the loops are equal in the conservative case.

Obviously, if the co-energy is explicitly identified as being different from the energy, the derivations lead to the correct results too:

$$E^*(i,x) = -\left(E(\lambda,x) - i\lambda\right) = i\lambda - E(\lambda,x) = \frac{Li^2}{2} + nBlix - \frac{Kx^2}{2} \quad (79)$$

$$dE^*(i,x) = \lambda di - Fdx = Lidi + nBlidx - Kxdx \quad (80)$$

hence

$$E^*(i,x) = \frac{Li^2}{2} + nBlix - \frac{Kx^2}{2} \text{ and } \lambda = \frac{\partial E^*(i,x)}{\partial i} = Li + nBlx \quad (81)$$

so

$$u = \frac{d\lambda}{dt} = \frac{d}{dt}\frac{\partial E^*(i,x)}{\partial i} = L\frac{di}{dt} + nBlv = u_{\text{self ind}} + u_{\text{Lorentz}} \quad (82)$$

and

$$F = \frac{-\partial E^*(i,x)}{\partial x} = -nBli + Kx \quad (83)$$

or in matrix form:

$$\begin{bmatrix} \lambda \\ F \end{bmatrix} = \begin{bmatrix} \frac{\partial \lambda}{\partial i} & \frac{\partial \lambda}{\partial x} \\ \frac{\partial F}{\partial i} & \frac{\partial F}{\partial x} \end{bmatrix} = \begin{bmatrix} \frac{\partial^2 E^*}{\partial i^2} & \frac{\partial^2 E^*}{\partial i \partial x} \\ \frac{-\partial^2 E^*}{\partial x \partial i} & \frac{-\partial^2 E^*}{\partial x^2} \end{bmatrix} = \begin{bmatrix} L & nBl \\ -nBl & K \end{bmatrix} \begin{bmatrix} i \\ x \end{bmatrix} \quad (84)$$

i.e. no symmetry of the Jacobian if energy and co-energy are confused.

This example demonstrates that the concept of multiport storage, in particular two-port storage, may not only lead to conceptual models of transduction phenomena, but also that reversible transduction requires a cycle process in principle as it is impossible to fully exclude storage. This means that a transducer cannot continuously transduce a DC input. By changing its configuration into that of an electric motor, the configuration that allows rotation takes care of the cycle of the conductors in the magnetic field.

Hence, pure 'DC'-type transduction can only be achieved by some form of 'carrier' that performs a cycle: gears carrying teeth (with elastic deformation!) during contact, a cooling fluid, a rolling wheel that cycles its interaction point, etc., which is a useful insight during conceptual design of transducers.

5 Conclusion

In this chapter the basics of the port-based approach were introduced as well as their natural notation, viz. bond graphs. The main advantages of using bond graphs are

1. the *domain-independence* of the elementary behaviors and their graphical notation that allow quick analysis of dynamic interaction across domain boundaries;
2. that *not* all ports have an *a-priori fixed* causality, thus allowing flexible re-use of submodels;
3. the *combination* of physical and computational structure *in one notation*, thus allowing

    a) *direct physical interpretation* of required changes in model structure, e.g. for controller design
    b) *direct feedback* on modeling decisions
    c) *direct graphical input* for simulation software.

All these features contribute to rapid insight and the ability of efficient iteration during the modeling process.

It was also demonstrated that various domains of physics that use some form of energy-based model formulation technique are all strongly related, even though terminology and a loss of conceptual distinction between conserved energy and its non-conserved Legendre transforms commonly obstructs this insight. Last, but not least, one should keep in mind that the conceptual framework is justified under all circumstances where energy conservation and positive entropy production are valid concepts and even in a case where a globally conserved quantity other than energy can be defined. Obeying the grammar rules of the port-based language automatically generates half of the structure relations, reduces the risk of sign errors and often increases insight.

Bibliography

Uri M. Ascher and Linda R. Petzold. *Computer Methods for Ordinary Differential Equations and Differential–Algebraic Equations.* SIAM, Philadelphia, 1998.

Albert M. Bos. *Modelling Multibody Systems in Terms of Multibond Graphs.* PhD thesis, Electrical Engineering, University of Twente, Enschede, Netherlands, 1986.

Albert M. Bos and Martin J.L. Tiernego. Formula manipulation in the multibond graph modelling and simulation of large mechanical systems. *J. Franklin Inst.*, 319(1/2):51–65, 1985.

Peter C. Breedveld. Thermodynamic bond graphs and the problem of thermal inertance. *J. Franklin Inst.*, 314(1):15–40, 1982.

Peter C. Breedveld. Physical systems theory in terms of bond graphs. ISBN 90-9000599-4, 1984a.

Peter C. Breedveld. Decomposition of multiport elements in a revised multibond graph notation. *J. Franklin Inst.*, 318(4):253–273, 1984b.

Peter C. Breedveld. Multibond graph elements in physical systems theory. *J. Franklin Inst.*, 319(1/2):1–36, 1985.

Peter C. Breedveld. Insight in rigid body motion stability via an alternative for the Eulerian junction structure. In *Proceedings 1999 International Conference on Bond Graph Modeling and Simulation (ICBGM'99)*, Western Multi Conference Simulation Series Vol. 31, No. 1, pages 269–274, San Francisco, USA, 1999.

Herbert B. Callen. *Thermodynamics*. John Wiley, New York, 1960.

Goran Golo. *Interconnection Structures in Port-based Modelling: Tools for Analysis and Simulation*. PhD thesis, University of Twente, Twente University Press, 2002.

Goran Golo, Peter C. Breedveld, Bernhard M. Maschke, and Arjan J. van der Schaft. Input–output representations of Dirac structures and junction structures in bond graphs. In *Proc. 14th Int. Symp. on Math. Theory of Networks and Systems (MTNS 2000), Perpignan, France, June 19–23*, 2000.

William H. Hayt Jr., Jack E. Kemmerly, and Steven M. Durbin. *Engineering Circuit Analysis*. McGraw Hill, New York, 6th edition, 2002.

Neville J. Hogan and Ernest D. Fasse. Conservation principles and bond graph junction structures. In *Proc. ASME 1988 WAM*, DSC-Vol. 8, pages 9–14, 1988.

Dean C. Karnopp and Ronald C. Rosenberg. *Analysis and Simulation of Multi Port Systems - The Bond Graph Approach to Physical System Dynamics*. MIT press, Cambridge, USA, 1968.

Norbert E. Ligterink, Peter C. Breedveld, and Arjan J. van der Schaft. Physical model reduction of interacting, continuous systems. In *Proceedings 17th International Symposium on Mathematical Theory of Networks and Systems, Kyoto, Japan, July 24-28*, 2006.

Bernhard M. Maschke and Arjan J. van der Schaft. Canonical interdomain coupling in distributed parameter systems: an extension of the symplectic gyrator. In *Proc. Int. Mechanical Engineering Congress and Exposition, New York, USA, November 11-16*, 2001.

Bernhard M. Maschke, Arjan J. van der Schaft, and Peter C. Breedveld. An intrinsic Hamiltonian formulation of the dynamics of LC–circuits. *Trans. IEEE on Circuits and Systems, I: Fundamental Theory and Applications*, 42(2):73–82, 1995.

Henry M. Paynter. *Analysis and Design of Engineering Systems*. MIT Press, Cambridge (Mass), 1961.

Henry M. Paynter and Ilene J. Busch-Vishniac. Wave-scattering approaches to conservation and causality. *J. Franklin Inst.*, 325(3):295–313, 1988.

Stefano Stramigioli. *From Differentiable Manifolds to Interactive Robot Control.* PhD thesis, Delft University of Technology, Netherlands, 1998.

Martin J.L. Tiernego and Jan J. van Dixhoorn. Three-axis platform simulation: Bond graph and Lagrangian approach. *J. Franklin Inst.*, 308(3): 185–204, 1979.

Johannes van Dijk and Peter C. Breedveld. Simulation of system models containing zero-order causal paths - part I: Classification of zero-order causal paths and part II: Numerical implications of class-1 zero-order causal paths. *J. Franklin Inst.*, 328(5/6):959–979 and 981–1004, 1991.

Harold A. Wheeler and D. Dettinger. Measuring the efficiency of a superheterodyne converter by the input impedance circle diagram. Wheeler Monograph 9, 1949.

Numerical methods for simulation in applied dynamics

Martin Arnold
Martin Luther University Halle–Wittenberg, Institute of Mathematics
D - 06099 Halle (Saale), Germany.

Abstract Applied dynamics may be considered as integration platform for simulation in various fields of engineering like vehicle system dynamics, dynamics of machines and mechanisms and robotics. In the present contribution we discuss classical numerical simulation techniques of nonlinear system dynamics, their use in multibody system simulation and extensions to typical problems of applied dynamics like continuous and discrete controllers and multidisciplinary applications. A frequently used alternative approach to the analysis of multidisciplinary problems is based on the coupling of two or more monodisciplinary simulation packages. Typical numerical problems of such co-simulation techniques will be considered and illustrated by numerical tests.

1 Introduction

Model based simulation in applied dynamics relies on advanced methods for model setup, robust and efficient numerical solution techniques and powerful simulation tools for industrial applications.

In many applications, the mechanical components are described by rigid or flexible multibody system models that interact with electrical, hydraulic and other system components (Kortüm and Schiehlen, 1985; Schiehlen and Eberhard, 2008), see also (Kortüm et al., 2001) for a summary from the viewpoint of vehicle system dynamics and (Rulka, 1998) for a comprehensive overview on methods and algorithms that are tailored to industrial simulation packages. A first systematic treatment of numerical problems and numerical methods in multibody dynamics with focus on applications in vehicle system dynamics is the monograph of Eich–Soellner and Führer (Eich-Soellner and Führer, 1998), see also (Simeon et al., 1991).

In its simplest form the equations of motion of a multibody system are given by a non-linear second order system of ordinary differential equations

(ODEs) of moderate dimension that may be solved numerically by standard methods.

The structure of the model equations gets more complicated if the multibody system has kinematically *closed loops* (Schiehlen, 1997; Shabana, 2001). Multibody system models of wheel suspensions are a typical example in road vehicle simulation. The complex kinematics may be modelled by *constraints*, the resulting model equations form a second order differential-algebraic equation (DAE).

Robust and efficient DAE time integration methods are the backbone of state-of-the-art multibody system simulation software. Another important issue in simulation is the handling of *discrete* controllers that are beyond the classical time continuous world of mechanics.

Advanced multibody system simulation packages have been open to multidisciplinary applications for a long time. The classical approach to the integration of hydraulic and electrical system components in multibody models are special force elements with time continuous or time discrete inner state variables (Kortüm et al., 2001).

Alternatively, novel modelling and simulation techniques for heterogeneous technical systems may be used in more complex multidisciplinary applications. Here, the multidisciplinary aspect is considered by general port-based modelling techniques (Breedveld, 2008), by universal modelling languages like Modelica (Tiller, 2001) or by the coupling of several monodisciplinary simulation packages in a co-simulation environment (Veitl et al., 1999; Kübler and Schiehlen, 2000).

In this contribution we give an overview on classical and more advanced simulation algorithms in applied dynamics that are closely related to multibody numerics and multibody system simulation packages. Throughout the paper the analysis of distributed physical phenomena like the elastic deformation of car components or the temperature field in disk brakes is restricted to low-dimensional modal approximations.

The material is organized as follows: Basic simulation algorithms in multibody system analysis are summarized in Section 2. The topic of Section 3 are ODE time integration methods for technical simulation. In Section 4 we discuss the general structure of model equations including closed-loop systems, inner state variables of force elements and time discrete subsystems. Special DAE time integration methods for the robust and efficient numerical solution of these model equations are presented in Section 5. Finally, in Section 6 we focus on algorithms and tools for multidisciplinary applications.

2 Basic Simulation Algorithms

If a mechanical multibody system is described by a minimum set of generalized coordinates $\boldsymbol{q}(t) \in \mathbb{R}^{n_q}$ the equations of motion form a second order ODE

$$\boldsymbol{M}(\boldsymbol{q})\ddot{\boldsymbol{q}}(t) = \boldsymbol{f}(\boldsymbol{q}, \dot{\boldsymbol{q}}, t) \tag{1}$$

with the symmetric, positive definite mass matrix $\boldsymbol{M}(\boldsymbol{q})$ containing mass and inertia properties of all bodies and the vector of applied and gyroscopic forces $\boldsymbol{f}(\boldsymbol{q}, \dot{\boldsymbol{q}}, t)$ (Shabana, 2001; Schiehlen and Eberhard, 2008).

Despite its implicit structure Eq. (1) can be solved numerically with a complexity that grows only linearly with the number N of bodies in the system. *Explicit $\mathcal{O}(N)$–formalisms* exploit the topology of the multibody system to evaluate the right-hand side $\boldsymbol{M}^{-1}(\boldsymbol{q})\boldsymbol{f}(\boldsymbol{q}, \dot{\boldsymbol{q}}, t)$ for given $t, \boldsymbol{q}, \dot{\boldsymbol{q}}$ efficiently, see, e.g., (Brandl et al., 1988; Roberson and Schwertassek, 1988). In *residual formalisms*, the residual $\boldsymbol{r}(\boldsymbol{q}, \dot{\boldsymbol{q}}, \boldsymbol{a}, t) := \boldsymbol{M}(\boldsymbol{q})\boldsymbol{a} - \boldsymbol{f}(\boldsymbol{q}, \dot{\boldsymbol{q}}, t)$ is evaluated for given $t, \boldsymbol{q}, \dot{\boldsymbol{q}}, \boldsymbol{a}$ by an algorithm with $\mathcal{O}(N)$ complexity, see, e.g., (Eichberger, 1993).

In contrast to structural dynamics the classical time integration methods of multibody dynamics are not tailored to the second order structure of (1) since additional first order equations have to be considered frequently in practical applications, see Section 4. With the velocity vector $\boldsymbol{v} := \dot{\boldsymbol{q}}$ the equations of motion get the state space form

$$\dot{\boldsymbol{x}}(t) = \boldsymbol{\varphi}(\boldsymbol{x}, t) \tag{2}$$

with

$$\boldsymbol{x}(t) := \begin{pmatrix} \boldsymbol{q}(t) \\ \boldsymbol{v}(t) \end{pmatrix}, \quad \boldsymbol{\varphi}(\boldsymbol{x}, t) := \begin{pmatrix} \boldsymbol{v} \\ \boldsymbol{M}^{-1}\boldsymbol{f}(\boldsymbol{q}, \boldsymbol{v}, t) \end{pmatrix}.$$

Static analysis The computation of static equilibria is often the first step in the analysis of a multibody system model. They define, e.g., the working point for the linearization of (1) and provide initial values for the dynamical simulation. An equilibrium position $\boldsymbol{q}^*$ is determined by the equilibrium conditions $\dot{\boldsymbol{q}} = \ddot{\boldsymbol{q}} = \boldsymbol{0}$ at $t = t_0$:

$$\boldsymbol{0} = \boldsymbol{f}(\boldsymbol{q}^*, \boldsymbol{0}, t_0)\,. \tag{3}$$

Eqs. (3) form a system of n_q non-linear equations for the n_q unknowns $\boldsymbol{q}^*$ that is solved by Newton's method (Kelley, 2003) being available as free software in a number of implementations that meet the demands on robustness and efficiency in engineering applications, see, e.g., the results of the MINPACK project (Moré et al., 1984) at `http://www.netlib.org/minpack/`.

Alternatively, the equilibrium position $\boldsymbol{q}^*$ of a stable system may also be obtained as limit $\boldsymbol{q}^* = \lim_{t\to\infty} \boldsymbol{q}(t)$ by a long-term time integration of the equations of motion (1) without external excitations. The time integration to the equilibrium position needs more computing time than Newton's method but in critical applications it is often the more robust approach.

Problems with the same mathematical structure as the equilibrium equations (3) result from the analysis of steady states. If a car runs with constant speed v_0 in a constant radius path, its position and velocity along this path are given by $s(t) = s_0 + v_0(t - t_0)$ and by v_0, respectively, and a force acts in normal direction to keep the radius constant. The remaining $n_q - 1$ components $\bar{\boldsymbol{f}}(\boldsymbol{q}, \dot{\boldsymbol{q}}, t)$ of the force vector $\boldsymbol{f}$ in (1) have to vanish and give the first $n_q - 1$ equations of the n_q steady state conditions

$$\mathbf{0} = \bar{\boldsymbol{f}}(\boldsymbol{q}^*, \boldsymbol{v}^*, t_0)\,, \quad s_0 = s(t_0) \tag{4}$$

that have to be solved w.r.t. $\boldsymbol{q}^* \in \mathbb{R}^{n_q}$. All components of the steady state velocity vector $\boldsymbol{v}^*$ are equal to zero except the one that is tangential to the constant radius path ($v_0^* = v_0$). For steady state problems (4), Newton's method (or, alternatively, the long-term time integration of a stable system into its steady state) is again the method-of-choice.

There are some other static analysis methods that are closely related to the equilibrium problem (3): the computation of nominal forces in vehicle dynamics (Eich-Soellner and Führer, 1998) and the trimming of aircraft models in aeronautics (Krüger and Spieck, 2004). In both applications the actual values of some parameters $\boldsymbol{\theta} \in \mathbb{R}^{n_\theta}$ in the force elements are left undefined in the model setup: $\boldsymbol{f} = \boldsymbol{f}(\boldsymbol{q}, \dot{\boldsymbol{q}}, t; \boldsymbol{\theta})$. Later, these parameters are adjusted to meet equilibrium conditions

$$\mathbf{0} = \boldsymbol{f}(\boldsymbol{q}_0, \mathbf{0}, t_0; \boldsymbol{\theta}^*) \tag{5}$$

for a pre-defined position $\boldsymbol{q}_0$ of the multibody system. The actual values $\boldsymbol{\theta}^*$ of nominal forces or trim parameters are important criteria for the evaluation of a vehicle or aircraft design.

Eq. (5) has the same mathematical structure as the classical equilibrium problem (3) and could in principle be solved again by Newton's method. However, in practical applications the system (5) is often underdetermined. Following the approach of Levenberg and Marquardt, (5) may be substituted by

$$\|\boldsymbol{f}(\boldsymbol{q}_0, \mathbf{0}, t_0; \boldsymbol{\theta}^*)\|_2 + \alpha \|\boldsymbol{\theta}^* - \boldsymbol{\theta}_0\|_2 \to \min$$

to guarantee uniqueness of $\boldsymbol{\theta}^*$. Here, $\alpha > 0$ denotes a small scalar regularization parameter and $\boldsymbol{\theta}_0$ is some initial guess for the unknown system parameters (Kelley, 1999).

Linearization The linearization of the equations of motion (1) is the key to the linear stability analysis near an equilibrium and to other methods of linear system analysis (Kortüm and Lugner, 1994). Writing the right hand side of (1) as $\boldsymbol{f}(\boldsymbol{q}, \dot{\boldsymbol{q}}, \boldsymbol{u}(t))$ with system inputs $\boldsymbol{u}(t) \in \mathbb{R}^{n_u}$ the linearization at an equilibrium $\boldsymbol{q} = \boldsymbol{q}^*$, $\dot{\boldsymbol{q}} = \boldsymbol{0}$, $\boldsymbol{u} = \boldsymbol{0}$ yields

$$\boldsymbol{M}\ddot{\bar{\boldsymbol{q}}} = -\boldsymbol{D}\dot{\bar{\boldsymbol{q}}} - \boldsymbol{K}\bar{\boldsymbol{q}} + \boldsymbol{B}\boldsymbol{u}(t) \tag{6}$$

with $\boldsymbol{M} = \boldsymbol{M}(\boldsymbol{q}^*)$,

$$\boldsymbol{D} = -\frac{\partial \boldsymbol{f}}{\partial \dot{\boldsymbol{q}}}(\boldsymbol{q}^*, \boldsymbol{0}, \boldsymbol{0}), \quad \boldsymbol{K} = -\frac{\partial \boldsymbol{f}}{\partial \boldsymbol{q}}(\boldsymbol{q}^*, \boldsymbol{0}, \boldsymbol{0}), \quad \boldsymbol{B} = \frac{\partial \boldsymbol{f}}{\partial \boldsymbol{u}}(\boldsymbol{q}^*, \boldsymbol{0}, \boldsymbol{0}).$$

The Jacobians $\partial \boldsymbol{f}/\partial \boldsymbol{q}$, $\partial \boldsymbol{f}/\partial \dot{\boldsymbol{q}}$ and $\partial \boldsymbol{f}/\partial \boldsymbol{u}$ are approximated by classical finite differences (Kelley, 2003). We obtain for the i-th column of $\partial \boldsymbol{f}/\partial \boldsymbol{q}$

$$\left(\frac{\partial \boldsymbol{f}}{\partial \boldsymbol{q}}(\boldsymbol{q}^*, \boldsymbol{0}, \boldsymbol{0})\right)_{\cdot,i} \approx \frac{\boldsymbol{f}(\boldsymbol{q}^* + \Delta \cdot \boldsymbol{e}_i, \boldsymbol{0}, \boldsymbol{0}) - \boldsymbol{f}(\boldsymbol{q}^*, \boldsymbol{0}, \boldsymbol{0})}{\Delta} \tag{7}$$

with the i-th unit vector $\boldsymbol{e}_i$ and a small scalar parameter Δ that satisfies $0 < |\Delta| \ll 1$.

The numerical effort for the linearization (6) is dominated by the difference approximations (7) with a total of $2n_q + n_u + 1$ evaluations of $\boldsymbol{f}$ that require $2n_q + n_u + 1$ evaluations of each force element in the multibody system model. Substantial savings are achieved by the simultaneous difference approximation of several columns of $\partial \boldsymbol{f}/\partial \boldsymbol{q}$, $\partial \boldsymbol{f}/\partial \dot{\boldsymbol{q}}$ and $\partial \boldsymbol{f}/\partial \boldsymbol{u}$ that requires some graph theoretical preparations based on the topology of the multibody system (Coleman et al., 1984; Arnold et al., 2006).

3 Time Integration: Basics

The state space form (2) of the equations of motion may be integrated by any standard ODE solver, see, e.g., (Ascher and Petzold, 1998) for a comprehensive introduction to this subject. Starting from initial values $\boldsymbol{x}_0 = \boldsymbol{x}(t_0)$ the numerical solution $\boldsymbol{x}_{n+1} \approx \boldsymbol{x}(t_{n+1})$ is obtained in time steps $t_n \to t_{n+1}$ of stepsize $h_n := t_{n+1} - t_n$, $(n = 0, 1, \ldots)$.

Explicit Runge–Kutta methods The most simple ODE time integration method goes back to Euler who proposed to substitute at $t = t_n$ the time derivative in the left hand side of (2) by a difference approximation:

$$\frac{\boldsymbol{x}_{n+1} - \boldsymbol{x}_n}{t_{n+1} - t_n} = \boldsymbol{\varphi}(\boldsymbol{x}_n, t_n)\,. \tag{8}$$

With (8), the numerical solution $\boldsymbol{x}_{n+1} \approx \boldsymbol{x}(t_{n+1}) = \boldsymbol{x}(t_n + h_n)$ is explicitly defined as function of $\boldsymbol{x}_n \approx \boldsymbol{x}(t_n)$:

$$\boldsymbol{x}_{n+1} = \boldsymbol{x}_n + h_n \boldsymbol{\varphi}(\boldsymbol{x}_n, t_n) . \tag{9}$$

This *explicit Euler method* is robust and easy to implement. Its practical use is, however, restricted to applications with low accuracy requirements like, e.g., real-time applications. On a finite time interval $[t_0, t_e]$, the first order difference approximation of $\dot{\boldsymbol{x}}(t_n)$ in each time step, see (2) and (8), results in a *global error* $\epsilon := \max_{0 \le n \le N} \|\boldsymbol{x}_n - \boldsymbol{x}(t_n)\|$ that is proportional to the maximum time stepsize $h := \max_{0 \le n \le N} h_n$. Here, index "N" denotes the last time step $t_{N-1} \to t_N$ in $[t_0, t_e]$, i.e., $t_N \le t_e < t_{N+1}$.

As a practical consequence, the maximum time stepsize h of the explicit Euler method has to be reduced by a factor of 10 to gain only one additional digit of accuracy. With ten times more computing time, the global error is only reduced by a factor of 10. To avoid this very large numerical effort, additional information about the ODE (2) has to be used in each time step $t_n \to t_{n+1}$. Higher order explicit Runge–Kutta methods get this information from additional function evaluations of the right hand side of (2) at some intermediate nodes $t_n + c_i h_n \in [t_n, t_{n+1}]$, $(i = 1, \ldots, s)$. As an alternative, linear multistep methods re-use information $(\boldsymbol{x}_{n-1}, \boldsymbol{\varphi}(\boldsymbol{x}_{n-1}, t_{n-1}))$, ... , $(\boldsymbol{x}_{n-(k-1)}, \boldsymbol{\varphi}(\boldsymbol{x}_{n-(k-1)}, t_{n-(k-1)}))$ from $k - 1$ previous time steps to improve their accuracy.

In an *explicit Runge–Kutta method*, the right hand side of (8) is substituted by a weighted linear combination of function values at $t_n + c_i h_n$, $(i = 1, \ldots, s)$:

$$\frac{\boldsymbol{x}_{n+1} - \boldsymbol{x}_n}{h_n} = \sum_{i=1}^{s} b_i \, \boldsymbol{\varphi}(\boldsymbol{X}_{ni}, t_n + c_i h_n) . \tag{10a}$$

The *stage values* $\boldsymbol{X}_{ni} \approx \boldsymbol{x}(t_n + c_i h_n)$, $(i = 1, \ldots, s)$ are defined in a similar way by

$$\frac{\boldsymbol{X}_{ni} - \boldsymbol{x}_n}{h_n} = \sum_{j=1}^{i-1} a_{ij} \, \boldsymbol{\varphi}(\boldsymbol{X}_{nj}, t_n + c_j h_n) , \quad (i = 1, \ldots, s). \tag{10b}$$

They are obtained recursively from $\boldsymbol{X}_{n1} = \boldsymbol{x}_n$ and the explicit expressions

$$\boldsymbol{X}_{ni} = \boldsymbol{x}_n + h_n \sum_{j=1}^{i-1} a_{ij} \, \boldsymbol{\varphi}(\boldsymbol{X}_{nj}, t_n + c_j h_n) , \quad (i = 2, \ldots, s).$$

Because of the s function evaluations $\boldsymbol{\varphi}(\boldsymbol{X}_{ni}, t_n + c_i h_n)$ the numerical effort per time step is s times larger than in the explicit Euler method (8).

This potential drawback may be compensated by a higher order of convergence that is achieved by an appropriate definition of nodes c_i, weights b_i and Runge–Kutta parameters a_{ij}.

One of the most often used higher order explicit Runge–Kutta methods is the 5th order method of Dormand and Prince (Dormand and Prince, 1980), a method with $s = 6$ stages that has (asymptotically for $h \to 0$) a global error $\epsilon = Ch^5$ with some constant $C > 0$. A stepsize reduction by a factor of 4 will result in an error reduction by a factor of $4^5 \approx 1{,}000 = 10^3$. Three more digits of accuracy may be obtained with a computing time that is increased only by a factor of 4.

DOPRI5 is a free FORTRAN implementation of this method, see (Hairer et al., 1993). The 5th order method of Dormand and Prince is also available as default ODE solver ode45 in MATLAB[1], see (Moler, 2004). In both implementations, the time stepsize h_n is adapted automatically to meet user defined error tolerances, see below.

Multistep methods *Linear multistep methods of Adams type* are an alternative to higher order explicit Runge–Kutta methods. They substitute the right hand side of (8) by a weighted linear combination of function values at $t_n, t_{n-1}, \ldots, t_{n-(k_n-1)}$, (*explicit* Adams methods or *Adams–Bashforth* methods) or $t_{n+1}, t_n, t_{n-1}, \ldots, t_{n-(k_n-1)}$, (*implicit* Adams methods or *Adams–Moulton* methods), see (Ascher and Petzold, 1998):

$$\frac{\boldsymbol{x}_{n+1} - \boldsymbol{x}_n}{h_n} = \sum_{j=0}^{k_n} \beta_{n,j} \boldsymbol{\varphi}(\boldsymbol{x}_{n+1-j}, t_{n+1-j}) \tag{11}$$

with $\beta_{n,0} = 0$ for the explicit methods. The weights $\beta_{n,j}$ are chosen to guarantee convergence with order k_n for the explicit methods and with order $k_n + 1$ for the implicit methods.

Since the function values $\boldsymbol{\varphi}(\boldsymbol{x}_{n+1-j}, t_{n+1-j})$ with $j \geq 2$ are already available from previous time steps, they do not have to be re-evaluated in time step $t_n \to t_{n+1}$. The numerical effort per time step is substantially smaller than in higher order explicit Runge–Kutta methods. On the other hand, the re-use of information from previous time steps may cause problems if there are discontinuities in the right hand side $\boldsymbol{\varphi}$.

In a practical implementation, the explicit Adams method is just used to get an initial guess (*predictor*) for the iterative computation of the numerical solution $\boldsymbol{x}_{n+1}$ in the corresponding implicit Adams method (*corrector*). For practical use, this predictor-corrector method is combined with sophisticated stepsize and order control algorithms to adapt the stepsize h_n and

[1]MATLAB is a trademark of The MathWorks, Inc., `http://www.mathworks.com/`.

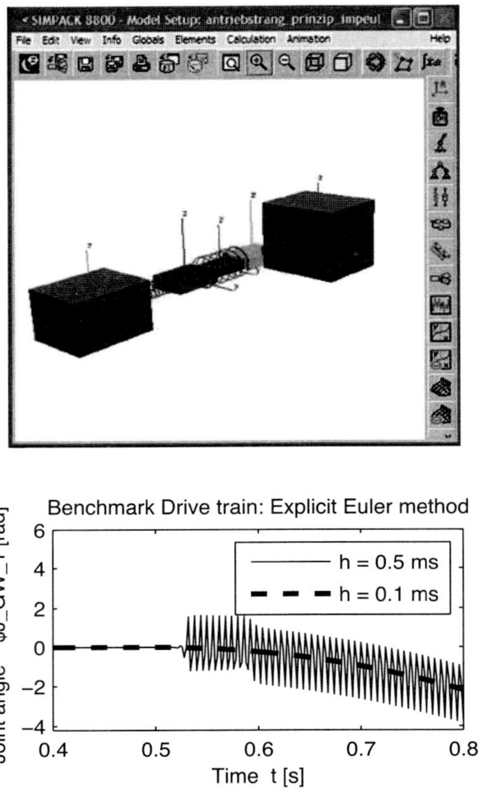

Figure 1. Numerical instability of explicit Euler method (8) applied to a stiff drive train model.

order $k_n + 1$ to the solution behaviour and to user defined error bounds, see Hairer et al. (1993). A free FORTRAN implementation is available in the LSODE package as `http://www.netlib.org/odepack/opkdmain.f`, the corresponding MATLAB solver is ode113.

Stepsize restrictions for stiff problems In many practical applications, explicit Runge–Kutta methods and predictor-corrector methods of Adams type are the most efficient ODE time integration methods. They fail, however, for a problem class that is illustrated by the simplified drive train model in Fig. 1. The multibody system model consists of 7 bodies that are connected by 5 torsional springs. The simulation starts at $t_0 = 0\,\mathrm{s}$ from an equilibrium position. At $t = 0.5\,\mathrm{s}$, a constant driving torque is switched on. As a typical simulation result, Fig. 1 shows one of the joint angles.

From practical experience, one would expect that even a first order method like (8) should reproduce this rather simple dynamical behaviour with time stepsizes h_n in the range of $1.0\,\mathrm{ms} \ldots 10.0\,\mathrm{ms}$. Numerical tests show, however, that the explicit Euler method (8) with fixed time stepsize $h_n = h$ is unstable for all stepsizes $h > 0.6\,\mathrm{ms}$. For smaller stepsizes, the simulation results are affected by unphysical oscillations, see the solid line in Fig. 1. The dashed line shows the physical correct solution that is obtained only for the very small time stepsize $h = 0.1\,\mathrm{ms}$.

Similar stability problems are observed, if the drive train model is integrated by higher order explicit Runge–Kutta methods or predictor-corrector methods of Adams type. In a fixed stepsize implementation, all these methods get unstable unless a stability condition $h \leq h^*$ with a method dependent small upper bound $h^* \approx 0.5\,\mathrm{ms}$ is satisfied. In a variable stepsize implementation, the stepsize control enforces $h_n \leq h^*$ automatically resulting in a very large number of time steps even for higher order methods.

It is well known that the instability of explicit time integration methods for applications like the drive train model in Fig. 1 is caused by stiff (torsional) springs acting on bodies with small masses. This instability phenomenon was studied in great detail for problems like Dahlquist's test equation

$$\dot{x} = \lambda x \quad \text{with} \quad \lambda \in \mathbb{C}, \;\; \mathrm{Re}\,\lambda < 0,$$

see (Ascher and Petzold, 1998), or damped oscillators with harmonic excitations:

$$m\ddot{q} + d\dot{q} + kq = \sin \omega t. \tag{12}$$

If the damping constant d satisfies $d^2 \leq 4km$, the explicit Euler method with fixed stepsize $h_n = h$ is unstable unless

$$h \leq h^* := \sqrt{m/k}.$$

The maximum time stepsize h is strongly restricted for stability reasons if the ratio of mass m and stiffness constant k is small.

Since these stability problems were observed first for mechanical systems with stiff springs, an ODE (2) is called *stiff* if it may be solved with moderate numerical effort by implicit methods like BDF (see below) but causes stability problems in the time integration by explicit Runge–Kutta methods and predictor-corrector methods of Adams type (*nonstiff integrators*), see (Hairer and Wanner, 1996).

For linear ODEs with time invariant Jacobian $\boldsymbol{A} := (\partial\boldsymbol{\varphi}/\partial\boldsymbol{x})(\boldsymbol{x},t)$, stiffness may often be characterized by the eigenvalues $\lambda_1, \ldots, \lambda_{n_x}$ of $\boldsymbol{A}$: The system is stiff if

$$\operatorname{Re}\lambda_i \le 0,\ (\ i = 1, \ldots, n_x),\ \text{ and } \ \min_i |\lambda_i| \ll \max_i |\lambda_i|. \tag{13}$$

For stability reasons, the stepsize of nonstiff integrators is restricted by

$$h_n \le h^* \ \text{ with } \ h^* \cdot \max_i |\lambda_i| \le C_s$$

with a method dependent constant C_s that is typically of size $C_s \approx 2.0$, see (Ascher and Petzold, 1998; Hairer and Wanner, 1996).

In applied dynamics, stiffness of the model equations is caused by components that are designed to bring an externally excited technical system rapidly back to some nominal state. Hydraulic components and very stiff or strongly damped spring-damper elements are well known sources of stiffness in the model equations that often result in stepsize bounds $h^* \le 0.01$ ms for nonstiff integrators.

Backward differentiation formulae Because of its close connection to the explicit Euler method (8), the most simple *stiff integrator*

$$\frac{\boldsymbol{x}_{n+1} - \boldsymbol{x}_n}{h_n} = \boldsymbol{\varphi}(\boldsymbol{x}_{n+1}, t_{n+1}) \tag{14}$$

is called *implicit* or *backward Euler method.* In contrast to Eqs. (8) and (9), Eq. (14) cannot be transformed to an explicit expression for $\boldsymbol{x}_{n+1}$ but defines $\boldsymbol{x}_{n+1} \approx \boldsymbol{x}(t_{n+1})$ only implicitly as solution of a system of n_x nonlinear equations. On the other hand, there is no stepsize restriction for the implicit method (14) in the application to linear stiff problems like (12) or $\dot{\boldsymbol{x}} = \boldsymbol{A}\boldsymbol{x}$ with (13) such that in the application to stiff ODEs the larger numerical effort per time step is more than compensated by time stepsizes h_n that are often two or three magnitudes larger than the stability bounds h^* of nonstiff integrators (Hairer and Wanner, 1996).

Despite its very good stability properties, the practical use of the backward Euler method (14) is restricted to applications with low accuracy requirements since its order of convergence is only $p = 1$. The most often used time integration methods for stiff technical systems are the backward differentiation formulae (BDF, also: Gear's method):

$$\frac{1}{h_n} \sum_{j=0}^{k_n} \alpha_{n,j} \boldsymbol{x}_{n+1-j} = \boldsymbol{\varphi}(\boldsymbol{x}_{n+1}, t_{n+1}) \tag{15}$$

that generalize the backward Euler method ($k_n = 1$) to higher order methods with $k_n > 1$. The parameter k_n defines the order of the method, $1 \le k_n \le 6$, and $\alpha_{n,j}$ are the BDF coefficients that are determined by k_n and by the stepsizes h_n, h_{n-1}, ..., see (Ascher and Petzold, 1998).

For given state vectors $\boldsymbol{x}_{n+1-j} \approx \boldsymbol{x}(t_{n+1-j})$, $(j = 1, \ldots, k)$ the BDF equations (15) define $\boldsymbol{x}_{n+1}$ implicitly as solution of the system of n_x nonlinear equations

$$\mathbf{0} = \boldsymbol{\Phi}(\boldsymbol{x}) := \frac{\alpha_{n,0}}{h_n}\boldsymbol{x} - \boldsymbol{\varphi}(\boldsymbol{x}, t_{n+1}) + \frac{1}{h_n}\sum_{j=1}^{k_n} \alpha_{n,j}\boldsymbol{x}_{n+1-j} \,. \tag{16}$$

The application of Newton's method to (16) yields

$$\boldsymbol{x}_{n+1}^{(l+1)} := \boldsymbol{x}_{n+1}^{(l)} - \boldsymbol{J}^{-1} \cdot \boldsymbol{\Phi}(\boldsymbol{x}_{n+1}^{(l)}), \ \ (l \ge 0) \tag{17}$$

with an initial guess $\boldsymbol{x}_{n+1}^{(0)}$ that is obtained by polynomial extrapolation of $\boldsymbol{x}_n$, $\boldsymbol{x}_{n-1}$, ... In (17), matrix $\boldsymbol{J}$ approximates the Jacobian of (16):

$$\boldsymbol{J} \approx \frac{\partial \boldsymbol{\Phi}}{\partial \boldsymbol{x}}(\boldsymbol{x}_{n+1}^{(l)}) = \frac{\alpha_{n,0}}{h_n}\boldsymbol{I}_{n_x} - \frac{\partial \boldsymbol{\varphi}}{\partial \boldsymbol{x}}(\boldsymbol{x}_{n+1}^{(l)}, t_{n+1}) \,. \tag{18}$$

Typically, the approximation of $\partial\boldsymbol{\varphi}/\partial\boldsymbol{x}$ by finite differences according to (7) dominates the overall computing time in the dynamical simulation because of the large number of function evaluations of $\boldsymbol{\varphi}$ that require the evaluation of all force elements in the multibody system and the computation of $[\boldsymbol{M}^{-1}\boldsymbol{f}](\boldsymbol{q}, \boldsymbol{v}, t)$ by the $\mathcal{O}(N)$–formalism, see (2). Therefore these time consuming re-evaluations of the Jacobian $\boldsymbol{J}$ have to be avoided as far as possible (Arnold et al., 2006; Ascher and Petzold, 1998).

In practical computations stepsize h_n and order k_n are again adapted automatically to meet the user defined error tolerances. The combination of stepsize and order control with the algorithms for Newton's method (17) and for the Jacobian re-evaluations results in a fairly complicated structure of state-of-the-art BDF solvers. Free FORTRAN codes are LSODE at `http://www.netlib.org/odepack/opkdmain.f` and DASSL at `http://www.netlib.org/ode/`. The MATLAB BDF solver is ode15s, see (Moler, 2004).

From the viewpoint of multibody dynamics the solver DASSL (Brenan et al., 1996) is especially interesting because of its interface in residual form

$$\mathbf{0} = \boldsymbol{F}(\boldsymbol{x}, \dot{\boldsymbol{x}}, t) \tag{19}$$

that is a generalization of (2) with $\boldsymbol{F}(\boldsymbol{x},\dot{\boldsymbol{x}},t)=\dot{\boldsymbol{x}}-\boldsymbol{\varphi}(\boldsymbol{x},t)$ and optionally allows also the use of very efficient residual formalisms (Eichberger, 1993):

$$\boldsymbol{x}(t):=\begin{pmatrix}\boldsymbol{q}(t)\\ \boldsymbol{v}(t)\end{pmatrix},\quad \boldsymbol{F}(\boldsymbol{x},\dot{\boldsymbol{x}},t):=\begin{pmatrix}\dot{\boldsymbol{q}}-\boldsymbol{v}\\ \boldsymbol{M}(\boldsymbol{q})\dot{\boldsymbol{v}}-\boldsymbol{f}(\boldsymbol{q},\boldsymbol{v},t)\end{pmatrix}.$$

Furthermore, DASSL may also be used in the time integration of closed loop systems, see Section 5 below.

BDF for (19) have the form

$$\mathbf{0}=\boldsymbol{F}(\boldsymbol{x}_{n+1},\frac{1}{h_n}\sum_{j=0}^{k_n}\alpha_{n,j}\boldsymbol{x}_{n+1-j},t_{n+1}) \tag{20}$$

with the Jacobian

$$\boldsymbol{J}\approx\frac{\alpha_{n,0}}{h_n}\frac{\partial\boldsymbol{F}}{\partial\dot{\boldsymbol{x}}}(\boldsymbol{x},\dot{\boldsymbol{x}},t)+\frac{\partial\boldsymbol{F}}{\partial\boldsymbol{x}}(\boldsymbol{x},\dot{\boldsymbol{x}},t) \tag{21}$$

instead of (18). To keep the efficiency of classical BDF solvers in the application to multibody systems DASSL's standard algorithm for Jacobian re-evaluation has to be modified to exploit the special block structure of $\partial\boldsymbol{F}/\partial\dot{\boldsymbol{x}}$, see (Rulka, 1998; Arnold et al., 2006):

Explicit formalism: $$\frac{\partial\boldsymbol{F}}{\partial\dot{\boldsymbol{x}}}=\begin{pmatrix}\boldsymbol{I}_{n_q}&\mathbf{0}\\ \mathbf{0}&\boldsymbol{I}_{n_q}\end{pmatrix},$$

Residual formalism: $$\frac{\partial\boldsymbol{F}}{\partial\dot{\boldsymbol{x}}}=\begin{pmatrix}\boldsymbol{I}_{n_q}&\mathbf{0}\\ \mathbf{0}&\boldsymbol{M}(\boldsymbol{q})\end{pmatrix}.$$

Stepsize control: Basics Stepsize control algorithms in ODE time integration are designed to get with minimal numerical effort a numerical solution $\boldsymbol{x}_0,\boldsymbol{x}_1,\ldots,\boldsymbol{x}_N$ that approximates the solution of (2) up to errors being bounded by user defined tolerances ATOL and RTOL for absolute and relative errors, respectively. Under mild regularity assumptions, the global error $\epsilon=\max_{0\le n\le N}\|\boldsymbol{x}_n-\boldsymbol{x}(t_n)\|$ in ODE time integration is proportional to the maximum *local* error, i.e., to the maximum error in one single time step $t_n\to t_{n+1}$. Therefore, all common stepsize control algorithms proceed stepwise and adjust the stepsize h_n to keep the local error of the current time step below the user defined error bounds (Ascher and Petzold, 1998; Hairer et al., 1993).

Most integrators support a componentwise definition ATOL_i, RTOL_i of error bounds. At the end of each time step $t_n\to t_{n+1}$ an error indicator

$$\mathrm{E}_n:=\Big(\frac{1}{n_x}\sum_{i=1}^{n_x}(\frac{\mathrm{err}_{n,i}}{\mathrm{ATOL}_i+\mathrm{RTOL}_i\cdot|x_{n,i}|})^2\Big)^{1/2}$$

is evaluated that compares an error estimate $\mathbf{err} = (\mathrm{err}_{n,i})_{i=1}^{n_x}$ with the user defined error bounds (Hairer et al., 1993).

The error test is passed if $\mathrm{E}_n \leq 1.0$. In that case, time step $t_n \to t_{n+1}$ with the current time stepsize h_n is accepted and the time integration may be continued with time step $t_{n+1} \to t_{n+2}$. A comparison of the error indicator E_n with its critical value 1.0 and an asymptotic error analysis for $h_n \to 0$ show that the optimal time stepsize of a p-th order method would have been

$$h_n^{\mathrm{opt}} := \frac{h_n}{\mathrm{E}_n^{1/(p+1)}} . \tag{22}$$

Therefore, the stepsize h_{n+1} for the new time step $t_{n+1} \to t_{n+2}$ is chosen as $h_{n+1} := \alpha_{\mathrm{s}} h_n^{\mathrm{opt}}$ with some safety factor $\alpha_{\mathrm{s}} \in [0.8, 0.9]$.

If the error indicator E_n exceeds its critical value 1.0, stepsize h_n was too large to meet the user defined accuracy requirements. The current time step $t_n \to t_{n+1}$ is rejected and has to be repeated with a smaller stepsize $h_n^{\mathrm{new}} < h_n$. Based on the same asymptotic analysis as before, h_n^{new} is defined as $h_n^{\mathrm{new}} := \alpha_{\mathrm{s}} h_n^{\mathrm{opt}}$, see (22).

Further details of stepsize control algorithms and their combination with order control in multistep methods are discussed in textbooks on ODE and DAE time integration, see, e.g., (Ascher and Petzold, 1998; Brenan et al., 1996; Hairer et al., 1993).

Practical aspects of stepsize and order control The use of variable time stepsizes h_n that are automatically adapted to the changing solution behaviour during time integration has been state-of-the-art in technical simulation for more than a decade, see also the detailed discussion in Section 4.1.5 of (Eich-Soellner and Führer, 1998). The theoretical justification of standard stepsize and order control algorithms relies, however, on smoothness assumptions on the model equations that are often violated in complex models from applied dynamics. Typical examples are spring-damper elements with non-differentiable force characteristics, stick-slip phenomena, spline approximations of system inputs $\boldsymbol{u}(t)$ and lookup tables, see (Eich-Soellner and Führer, 1998; Pfeiffer, 2008; Weidemann et al., 2006).

In large scale applications, the optimal use of variable stepsize variable order solvers requires expert knowledge based on previous practical experience and some extra effort for tuning the tolerance settings. Therefore, unexperienced users of industrial simulation packages tend to prefer robust low order fixed stepsize integrators despite their well known drawbacks w.r.t. reliability and efficiency. Recently, this problem was studied in detail for a benchmark problem from railway dynamics (Arnold et al., 2007b; Weidemann et al., 2006), see also Fig. 2.

Figure 2. Multibody system models in railway dynamics. Benchmark problem "passenger car" (Iwnicki, 1999; Weidemann et al., 2006).

The multibody system model of a passenger car in the industrial multibody simulation package SIMPACK (`http://www.simpack.com/`) is very similar to the so-called Manchester Benchmark vehicle (Iwnicki, 1999). It consists of 17 bodies with altogether $n_q = 53$ joint coordinates. The force elements have $n_c = 18$ inner state variables $\boldsymbol{c}(t)$, see Section 4 below, resulting in model equations (2) with $n_x = 2n_q + n_c = 124$ first order differential equations. One of the most crucial model components in railway dynamics is the wheel-rail contact model. Here, an elastic contact model with stiffness constant $c = 5 \cdot 10^8\,\mathrm{N/m}$ was combined with the creep force calculation by FASTSIM (Kalker, 1990).

The passenger car starts with a speed of 200 km/h in a straight track with medium track irregularities. At $x = 100\,\mathrm{m}$, after $t = 1.8\,\mathrm{s}$, the car enters a curve of radius $r = 1750\,\mathrm{m}$. For a derailment analysis, a strong track irregularity was modelled by a lateral track shift of $\Delta_y = 60.0\,\mathrm{mm}$ for $x \in [335\,\mathrm{m}, 350\,\mathrm{m}]$ corresponding to $t \in [6.0\,\mathrm{s}, 6.3\,\mathrm{s}]$ for the wheelsets of the front bogie.

For the leading wheelset, the upper plots of Fig. 3 show the reference solution for lateral displacement y and derailment coefficient Y/Q (outer wheel, filtered data). This reference is computed by SIMPACK's DASSL based integrator SODASRT using the very tight tolerances ATOL $= 10^{-5}$, RTOL $= 10^{-6}$. The strong changes in the track definition at $x = 100\,\mathrm{m}$, $x = 335\,\mathrm{m}$ and $x = 350\,\mathrm{m}$ affect the leading wheelset at $t = 1.8\,\mathrm{s}, \ldots$ (indicated by gray marks). Due to the sudden lateral shifts of the track at $t = 6.0\,\mathrm{s}$ and $t = 6.3\,\mathrm{s}$, the simulation results for Y/Q reach twice the limit value of -0.8, the vehicle is on the verge of derailing.

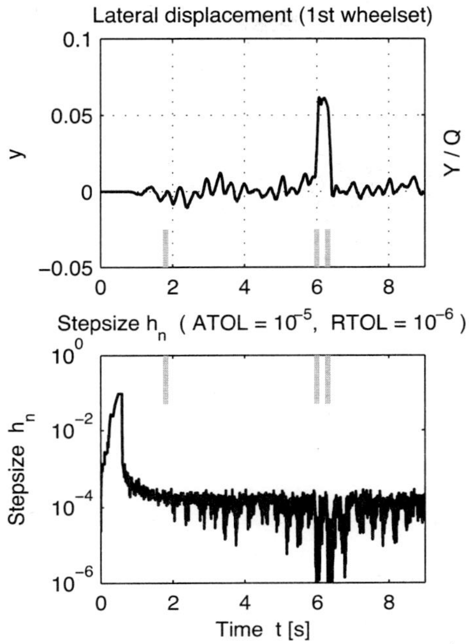

Figure 3. Variable stepsize time integration of benchmark problem: BDF solver SODASRT with ATOL $= 10^{-5}$, RTOL $= 10^{-6}$, see (Weidemann et al., 2006).

The lower plots of Fig. 3 show in logarithmic scale the time stepsize h_n and the increasing numerical effort during time integration that is roughly proportional to the number of function evaluations of the right hand side φ in (2). Starting from $h_0 = 0.02\,\text{ms}$, the time stepsize h_n increases up to $h_n = 50.0\,\text{ms}$. Later, the growing influence of track irregularities enforces a stepsize reduction to $h_n \approx 0.1\,\text{ms}$. A large part of the numerical effort is necessary to guarantee the requested accuracy of simulation results whenever the wheelsets of the front or back bogie pass the lateral track shift. At $t = 6.0\,\text{s}$, $t = 6.3\,\text{s}$ and $t = 6.6\,\text{s}$, the stepsize reaches its minimum value $h_n < 0.001\,\text{ms}$.

Tight tolerances make variable stepsize variable order integrators robust and yield very precise simulation results. With more than 300,000 function evaluations in a time interval of 9.0 s, the reference solution (ATOL $= 10^{-5}$,

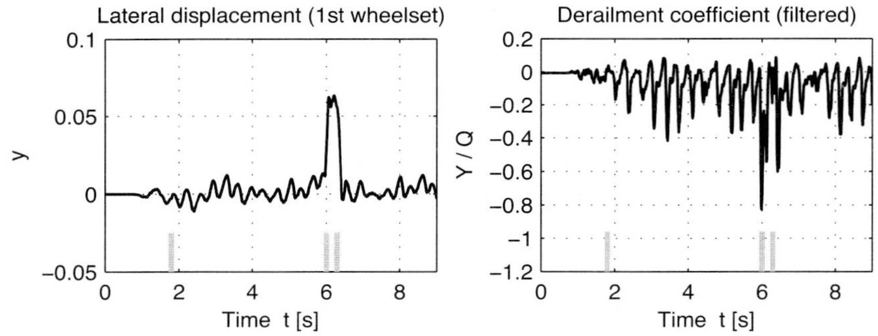

Figure 4. Fixed stepsize time integration of benchmark problem: Linearly implicit solver FIXINT with $h = 0.1$ ms, see (Weidemann et al., 2006).

RTOL $= 10^{-6}$) is, however, not competitive with low order fixed stepsize time integration methods like the linearly implicit Euler method (Rill, 1994; Arnold et al., 2007a).

In a case study, SIMPACK's linearly implicit FIXINT solver was applied with different fixed time stepsizes $h_n = h$, see (Weidemann et al., 2006). For $h = 1.0$ ms, the simulation fails, but the results for $h = 0.1$ ms, see Fig. 4, look promising since the qualitative solution behaviour is well reproduced, critical output values like the peaks of the derailment coefficient Y/Q at $t = 6.0$ s and $t = 6.3$ s are computed sufficiently accurate and the numerical effort is reduced to only 90,125 function evaluations. The quantitative comparison with the reference solution shows, however, rather large errors, see Figs. 3 and 4. The maximum absolute errors in the lateral displacement are $8.1 \cdot 10^{-3}$ (position) and $2.3 \cdot 10^{-1}$ (velocity), the maximum error in the output value Y/Q is $3.2 \cdot 10^{-1}$, see (Arnold et al., 2007b).

There is a trade-off between accuracy and numerical effort. For a fixed stepsize method, the optimal solver configuration is found by variation of one single parameter (stepsize h). In theory, the optimal performance of a variable stepsize variable order solver with error control should be achieved defining appropriate bounds ($\text{ATOL}_i, \text{RTOL}_i$), ($i = 1, \ldots, n_x$) for the absolute and relative errors.

In applied problems (2) with $n_x \gg 10$ state variables, the individual definition of $2n_x$ error bounds ATOL_i, RTOL_i would be very time consuming. Furthermore, the error tolerances for the components of state vector $\boldsymbol{x}(t)$ affect only indirectly the accuracy of practically important output values like

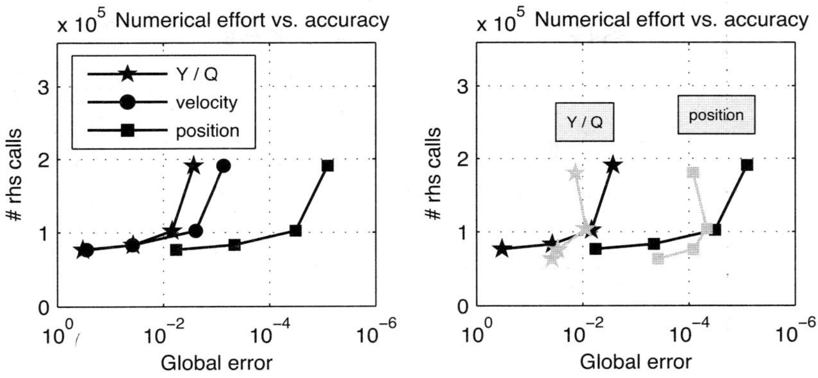

Figure 5. Numerical effort and accuracy of BDF solver SODASRT for various tolerance settings, see (23). Left plot: errors in y, $\dot{y}$ and Y/Q. Right plot: errors in y and Y/Q (gray color: SODASRT with order $k_n \leq 2$).

the derailment coefficient Y/Q in Fig. 3. For both reasons, it is common practice in large scale applications to use identical error bounds for all n_x solution components:

$$\text{ATOL}_i := \text{ATOL}, \quad \text{RTOL}_i := \text{RTOL}, \; (\, i = 1, \dots, n_x\,). \tag{23a}$$

The left diagram of Fig. 5 compares numerical effort and accuracy of simulation results for

$$\text{ATOL} = \text{TOL}, \quad \text{RTOL} = 0.1 * \text{TOL} \tag{23b}$$

and

$$\text{TOL} = 2.0 \cdot 10^{-4}, \quad \text{TOL} = 1.0 \cdot 10^{-4}, \quad \text{TOL} = 5.0 \cdot 10^{-5}, \quad \text{TOL} = 2.0 \cdot 10^{-5}. \tag{23c}$$

With $\text{TOL} = 5.0 \cdot 10^{-5}$ the absolute errors are kept below 10^{-4} (position coordinates) and 10^{-2} (velocity coordinates and output value Y/Q), respectively. Compared with the reference solution, the numerical effort is reduced by a factor of three (102,285 vs. 309,073 function evaluations of φ). For $\text{TOL} = 1.0 \cdot 10^{-4}$, the errors are still acceptable and the numerical effort is decreased by another 20% (only 83,307 function evaluations).

Identical tolerance settings ATOL, RTOL for all n_x solution components simplify the handling of error controlled solvers substantially but do not consider characteristic differences between the components x_i of the state

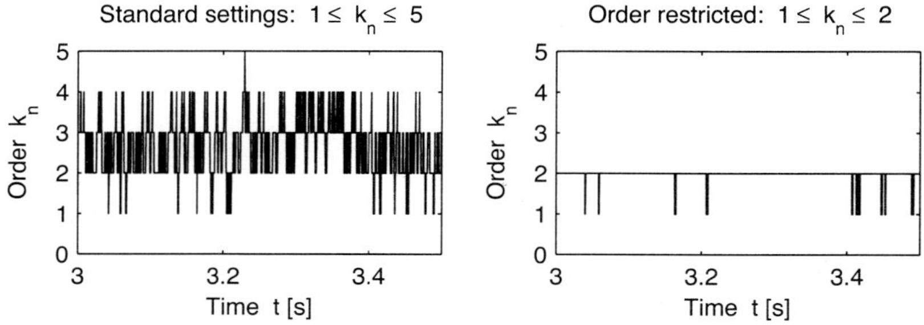

Figure 6. Order control of BDF solver SODASRT with $\text{ATOL} = 10^{-4}$, $\text{RTOL} = 10^{-5}$. Left plot: standard settings. Right plot: order restricted to $k_n \leq 2$.

vector $\boldsymbol{x}(t)$ in (2). The position coordinates $\boldsymbol{q}(t)$ denote often the deviation from some nominal value and are typically in the range of 0.1 m. The corresponding velocity coordinates $\boldsymbol{v}(t) = \dot{\boldsymbol{q}}(t)$ are substantially larger and have absolute values in the range of 1.0 m/s. Therefore, the error bounds for velocity coordinates ($\text{ATOL}_{\dot{\boldsymbol{q}}}$) should be larger than the ones for position coordinates ($\text{ATOL}_{\boldsymbol{q}}$). With $\text{ATOL}_{\dot{\boldsymbol{q}}} = 10 \cdot \text{ATOL}_{\boldsymbol{q}}$, a speed-up of 30% was achieved (Arnold et al., 2007b). The optimal definition of different error bounds ATOL_i, RTOL_i, $(\, i = 1, \ldots, n_x\,)$ for each individual solution component x_i requires a more detailed error analysis but may further decrease the numerical effort.

In the right plot of Fig. 5 and in Fig. 6 the interaction of stepsize and order control was studied. By default, the BDF solver SODASRT starts at $t = 0$ with order $k_0 = 1$ and selects in each time step the order k_n, $1 \leq k_n \leq 5$, such that the local error is kept below the error bounds ATOL_i, RTOL_i and the time stepsize h_n is maximized (Brenan et al., 1996). The combined order and stepsize control algorithm relies on smoothness assumptions on the right hand side $\boldsymbol{\varphi}$ that are not satisfied in the present benchmark problem (stick-slip effects, track irregularities, lookup tables and spline approximations in some force elements), see (Weidemann et al., 2006).

Practical consequences are illustrated by the left plot of Fig. 6 that shows the order k_n vs. time t for $t \in [3.0\,\text{s}, 3.5\,\text{s}]$. The order control fails to find an optimal order but switches frequently between $k_n = 2$, $k_n = 3$ and $k_n = 4$. If the order is restricted to $k_n \leq 2$, then stepsize and order get much smoother, see the right plot of Fig. 6. In the application to stiff systems with nearly undamped high-frequency oscillations, BDF with $k_n \leq 2$ are especially attractive because of their optimal stability properties

(*A-stability*), see (Ascher and Petzold, 1998; Brenan et al., 1996).

In the right plot of Fig. 5, the solver performance with standard settings (black) is compared to the performance with order restriction $k_n \leq 2$ (gray). For coarse tolerances ($\text{TOL} = 2.0 \cdot 10^{-4}$, $\text{TOL} = 1.0 \cdot 10^{-4}$), see (23), the simulation with $k_n \leq 2$ gets faster by about 20% and the accuracy of the simulation results is *improved.* For moderate tolerances ($\text{TOL} = 5.0 \cdot 10^{-5}$), there are no essential differences between the results with default settings ($1 \leq k_n \leq 5$) and with restricted order ($1 \leq k_n \leq 2$). For very fine tolerances, the low order methods are not competitive and result in increasing errors.

The test results for the benchmark from railway dynamics illustrate some typical problems and strategies in the practical application of error controlled solvers with stepsize and order control. With appropriate tolerance settings, the numerical effort may often be decreased substantially without deteriorating the accuracy of simulation results. For low accuracy requirements and systems with non-smooth model components, the performance of BDF solvers may be improved restricting the order to $k_n \leq 2$.

4 General Structure of Model Equations

The numerical methods that are discussed in Sections 2 and 3 may be extended from the equations of motion (1) to more complex model equations in applied dynamics.

Flexible multibody systems The elastic deformation of flexible bodies is not covered by the classical equations of motion (1) of rigid multibody systems. Extensions to flexible multibody systems are discussed in detail in the contribution of O. Brüls et al. to this proceedings volume (Brüls et al., 2008).

In the floating frame of reference formulation (Shabana, 1998; Wallrapp, 1991; Schwertassek and Wallrapp, 1999; Schiehlen and Eberhard, 2008) the configuration $\vec{u}^{(i)}(\vec{x}, t)$ of a flexible body $(.)^{(i)}$ in $\mathbb{R}^3$ is represented by the gross motion of a body fixed reference frame with coordinates $\boldsymbol{q}_{\mathrm{r}}^{(i)}(t)$ and the (small) elastic deformation $\vec{w}^{(i)}(\vec{x}, t)$ w.r.t. this reference configuration:

$$\vec{u}^{(i)}(\vec{x}, t) = \vec{r}(\boldsymbol{q}_{\mathrm{r}}^{(i)}(t)) + \vec{w}^{(i)}(\vec{x}, t) \,. \tag{24}$$

Following a Rayleigh–Ritz approach a low-dimensional modal approximation of $\vec{w}^{(i)}(\vec{x}, t)$ is used:

$$\vec{w}^{(i)}(\vec{x}, t) \approx \sum_{j=1}^{\sigma_i} p_j^{(i)}(t)\, \vec{w}_j^{(i)}(\vec{x}) \,. \tag{25}$$

In general, the σ_i modes $\vec{w}_j^{(i)}$ are selected based on a finite element analysis of the flexible body $(.)^{(i)}$. Typical mode functions $\vec{w}_j^{(i)}$ are eigenmodes in the frequency range up to 50 ... 100 Hz, static modes or frequency response modes (Schwertassek and Wallrapp, 1999; Dietz, 1999).

With (24) and (25) the equations of motion for flexible multibody systems get the basic structure (1) with a coordinate vector $\boldsymbol{q}(t)$ that contains the generalized coordinates of rigid bodies and the elastic coordinates $\boldsymbol{q}_{\mathrm{r}}^{(i)}(t)$ and $(\, p_1^{(i)}(t), \ldots, p_{\sigma_i}^{(i)}(t)\,)$ of flexible bodies (Schiehlen and Eberhard, 2008). $\boldsymbol{M}(\boldsymbol{q})$ and $\boldsymbol{f}(\boldsymbol{q}, \dot{\boldsymbol{q}}, t)$ are extended by the modal mass, damping and stiffness matrices and by the coupling terms between gross motion and elastic deformation $\vec{w}^{(i)}$, see (Wallrapp, 1991; Simeon, 2000). In industrial applications these data are generated automatically using standardized interfaces to finite element tools (Wallrapp, 1994).

Today, the selection of an appropriate set of modes $\{\, \vec{w}_1^{(i)}, \ldots, \vec{w}_{\sigma_i}^{(i)} \,\}$ still relies much on engineering intuition. Algorithms for (semi-)automatic mode selection are a topic of active research, see (Dietz, 1999) for a criterion that is based on a linear error analysis, (Simeon, 2000) for a more general approach that could be used in an adaptive mode selection strategy and (Heckmann et al., 2005) for the use of *thermal response* modes in thermoelastic applications.

Recently, the floating frame of reference formulation has been extended to a modal multifield approach that considers elastic deformation as well as electrostatic fields in piezo-elements and spatial temperature distributions in thermoelastic problems like the thermal deformation of disk brakes (Heckmann et al., 2005).

Inner state variables Frequently, force elements in multibody system models describe engineering systems that have their own, internal dynamics (Kortüm and Lugner, 1994; Eich-Soellner and Führer, 1998). Typical examples of such *dynamical* force elements include hydraulic components, control devices and advanced tyre models. From the mathematical viewpoint these force elements are characterized by *inner* state variables. Elements like hydraulic components have time continuous state variables $\boldsymbol{c}(t)$ with state equations

$$\dot{\boldsymbol{c}}(t) = \boldsymbol{d}(\boldsymbol{c}, \boldsymbol{q}, \dot{\boldsymbol{q}}, t)\,. \tag{26}$$

The internal dynamics of discrete controllers is characterized by time discrete state variables $\boldsymbol{r}$ with $\boldsymbol{r} = \boldsymbol{r}_j$ in the sampling interval $[T_j, T_{j+1})$ and

$$\boldsymbol{r}_{j+1} = \boldsymbol{k}(\boldsymbol{r}_j, \boldsymbol{r}_{j-1}, \ldots, \boldsymbol{q}, \dot{\boldsymbol{q}}, T_{j+1})\,. \tag{27}$$

In most applications the sampling points are equally spaced: $T_j = j \cdot \Delta_t$, typical sampling rates are in the range of $\Delta_t = 1\,\mathrm{ms} \ \ldots\ 10\,\mathrm{ms}$.

In the multibody system, these force elements contribute to the force vector $\boldsymbol{f}$ in (1):

$$\boldsymbol{f} = \boldsymbol{f}(\boldsymbol{q}(t), \dot{\boldsymbol{q}}(t), \boldsymbol{c}(t), \boldsymbol{r}_j, t) \quad \text{if} \quad t \in [T_j, T_{j+1})\,. \tag{28}$$

With (27) and (28) the equations of motion are not longer continuous in time but form a hybrid system with continuous and discrete components. In the dynamical simulation the variable order, variable stepsize integrators of Section 3 may be used only within a single sampling interval $[T_j, T_{j+1}]$. At $t = T_{j+1}$ the integration has to be stopped, the discrete variables $\boldsymbol{r}$ are updated according to (27) and the time integration may be continued for $t \in [T_{j+1}, T_{j+2}]$.

The frequent re-initializations of the ODE solvers have a very negative effect on their performance that is compensated in part by an adapted algorithm for Jacobian re-evaluations (Rulka, 1998; Arnold et al., 2006). Alternatively, special Runge–Kutta type methods are proposed for a more efficient re-initialization after the sampling point T_{j+1}, see (von Schwerin, 1999).

Time events The discontinuous changes (27) of state variables are special *time events* (Ascher and Petzold, 1998; Pfeiffer, 2008). The term "time event" refers to a discontinuity ("event") in state variables or model equations at an isolated time $t = T^*$ during simulation. For model equations (1) with $\boldsymbol{f} = \boldsymbol{f}(\boldsymbol{q}, \dot{\boldsymbol{q}}, \boldsymbol{c}, \boldsymbol{r}_j, \boldsymbol{u}(t); \boldsymbol{\theta})$ typical time events result from switching processes with jumping system inputs $\boldsymbol{u}(t)$ and from contact problems including stick-slip phenomena with force elements that change their force parameters $\boldsymbol{\theta}$ or even their internal structure during simulation (unilateral force elements, varying friction coefficients $\boldsymbol{\theta}$, ...), see (Eich-Soellner and Führer, 1998; Pfeiffer and Glocker, 1996; Pfeiffer, 2008).

For switching inputs $\boldsymbol{u}(t)$ the time T^* is known in advance but, in general, T^* has to be determined during simulation. State-of-the-art ODE solvers locate time events automatically using *switching functions* that have to be provided in addition to the model equations (1), see (Brenan et al., 1996).

In case of a time event the time integration is stopped at $t = T^*$, system variables $\boldsymbol{q}$, $\boldsymbol{c}$, $\boldsymbol{r}_j$ and system parameters $\boldsymbol{\theta}$ are updated (if necessary) and the simulation is continued after a re-initialization of the solver. Again, frequent time events may slow down the time integration substantially.

Kinematically closed loops In the setup of multibody system models it is important to select an appropriate set of coordinates $\boldsymbol{q}$. For tree structured systems joint coordinates $\boldsymbol{q}$ that describe the degrees of freedom in the joints of the multibody system form a minimum set of generalized coordinates resulting in equations of motion in ODE form (1).

If, however, the multibody system has kinematically closed loops then the joint coordinates $\boldsymbol{q}$ are not longer independent of each other. Loop closing joints restrict the system's configuration to joint coordinates $\boldsymbol{q}$ satisfying n_g *constraints*

$$\mathbf{0} = \boldsymbol{g}(\boldsymbol{q}, t)\,. \tag{29}$$

Coordinate partitioning methods (Shabana, 2001; Wehage and Haug, 1982) locally select a sub-set of linearly independent joint coordinates and compute the remaining (dependent) joint coordinates solving the system of non-linear equations (29). From the numerical viewpoint it proved to be more favourable to keep *all* joint coordinates in the equations of motion that get the form

$$\begin{aligned} \boldsymbol{M}(\boldsymbol{q})\,\ddot{\boldsymbol{q}}(t) &= \boldsymbol{f}(\boldsymbol{q}, \dot{\boldsymbol{q}}, t) - \boldsymbol{G}^\top(\boldsymbol{q}, t)\boldsymbol{\lambda}\,, && \text{(30a)} \\ \mathbf{0} &= \boldsymbol{g}(\boldsymbol{q}, t) && \text{(30b)} \end{aligned}$$

with constraint forces $-\boldsymbol{G}^\top(\boldsymbol{q}, t)\boldsymbol{\lambda}$ that are determined by the constraint matrix $\boldsymbol{G}(\boldsymbol{q}, t) := (\partial \boldsymbol{g}/\partial \boldsymbol{q})(\boldsymbol{q}, t)$ and the Lagrangian multipliers $\boldsymbol{\lambda}(t) \in \mathbb{R}^{n_g}$, see (Roberson and Schwertassek, 1988). The common assumption

$$\operatorname{rank} \boldsymbol{G}(\boldsymbol{q}, t) = n_g$$

(*Grübler* condition) excludes redundant constraints (30b). Eqs. (30) form a second order differential-algebraic equation (DAE), the *descriptor form* of the equations of motion (Brenan et al., 1996).

The descriptor form is not restricted to joint coordinates $\boldsymbol{q}$. Using absolute (Cartesian) coordinates $\boldsymbol{q}$ *any* joint results in a constraint (30b). The resulting equations of motion (30) have a much larger dimension than in the joint coordinate case but $\boldsymbol{M}(\boldsymbol{q})$, $\partial \boldsymbol{f}/\partial \boldsymbol{q}$, $\partial \boldsymbol{f}/\partial \dot{\boldsymbol{q}}$, $\boldsymbol{G}^\top$ are sparse matrices with a structure that reflects the topology of the multibody system (Orlandea, 1973). The counterpart to $\mathcal{O}(N)$–formalisms that eliminate redundant coordinates in a joint coordinate formulation (Brandl et al., 1988; Roberson and Schwertassek, 1988) are *topological solvers* for the absolute coordinate formulation that allow an efficient elimination of redundant coordinates in (17), see (Orlandea, 1973; von Schwerin, 1999).

In DAE terminology the descriptor form (30) is an index-3 system, see (Hairer and Wanner, 1996; Brenan et al., 1996). The constraints (30b) imply additional restrictions on the state variables $\boldsymbol{q}(t)$ and $\boldsymbol{\lambda}(t)$. These *hidden* constraints are obtained by differentiation of (30b):

$$\begin{aligned} \mathbf{0} &= \frac{\mathrm{d}}{\mathrm{d}t}\boldsymbol{g}(\boldsymbol{q}(t),t) = \frac{\partial \boldsymbol{g}}{\partial \boldsymbol{q}}(\boldsymbol{q},t)\cdot\frac{\mathrm{d}\boldsymbol{q}}{\mathrm{d}t}(t) + \frac{\partial \boldsymbol{g}}{\partial t}(\boldsymbol{q},t) \\ &= \boldsymbol{G}(\boldsymbol{q},t)\,\dot{\boldsymbol{q}}(t) + \boldsymbol{g}^{(I)}(\boldsymbol{q},t)\,, \end{aligned} \tag{31a}$$

$$\mathbf{0} = \frac{\mathrm{d}^2}{\mathrm{d}t^2}\boldsymbol{g}(\boldsymbol{q}(t),t) = \boldsymbol{G}(\boldsymbol{q},t)\,\ddot{\boldsymbol{q}}(t) + \boldsymbol{g}^{(II)}(\boldsymbol{q},\dot{\boldsymbol{q}},t) \tag{31b}$$

with functions $\boldsymbol{g}^{(I)}$, $\boldsymbol{g}^{(II)}$ that summarize higher order derivatives of $\boldsymbol{g}$ and the partial derivatives w.r.t. t. The solution of (30) has to satisfy the constraints (30b) on position level as well as the hidden constraints (31a) and (31b) on velocity and acceleration level.

Initial values $\boldsymbol{q}_0$, $\dot{\boldsymbol{q}}_0$, $\boldsymbol{\lambda}_0$ have to be *consistent* with all these constraints:

$$\mathbf{0} = \boldsymbol{g}(\boldsymbol{q}_0,t_0)\,, \tag{32a}$$

$$\mathbf{0} = \boldsymbol{G}(\boldsymbol{q}_0,t_0)\,\dot{\boldsymbol{q}}_0 + \boldsymbol{g}^{(I)}(\boldsymbol{q}_0,t_0)\,, \tag{32b}$$

$$\mathbf{0} = \boldsymbol{G}(\boldsymbol{q}_0,t_0)\,\ddot{\boldsymbol{q}}_0 + \boldsymbol{g}^{(II)}(\boldsymbol{q}_0,\dot{\boldsymbol{q}}_0,t_0) \tag{32c}$$

with

$$\boldsymbol{M}(\boldsymbol{q}_0)\,\ddot{\boldsymbol{q}}_0(t) = \boldsymbol{f}(\boldsymbol{q}_0,\dot{\boldsymbol{q}}_0,t_0) - \boldsymbol{G}^\top(\boldsymbol{q}_0,t_0)\boldsymbol{\lambda}_0. \tag{32d}$$

The computation of consistent initial values $\boldsymbol{q}_0$, $\dot{\boldsymbol{q}}_0$, $\boldsymbol{\lambda}_0$ is part of the initialization in the time integration of (30). If initial values q_{0,i_j} are given for $n_q - n_g$ independent position coordinates q_{i_j}, $(j = 1,\dots,n_q - n_g)$, the remaining initial values q_{0,i_j}, $(j = n_q - n_g + 1,\dots,n_q)$ are uniquely determined by (32a) and may be computed by Newton's method. Furthermore, (32b) defines a system of linear equations for the initial values $\dot{q}_{0,i_j}$, $(j = n_q - n_g + 1,\dots,n_q)$ of n_g dependent velocity components. The initial values $\ddot{\boldsymbol{q}}_0, \boldsymbol{\lambda}_0$ are obtained as solution of the system of linear equations (32c)/(32d).

Advanced DAE models The descriptor form (30) is linear w.r.t. $\boldsymbol{\lambda}$. This structural property is lost if the model has friction forces that depend on the constraint forces $-\boldsymbol{G}^\top\boldsymbol{\lambda}$. The resulting force vector $\boldsymbol{f} = \boldsymbol{f}(\boldsymbol{q},\dot{\boldsymbol{q}},\boldsymbol{\lambda},t)$ has to satisfy the generalized Grübler condition

$$\operatorname{rank}\boldsymbol{G}(\boldsymbol{q},t) = n_g, \tag{33a}$$

$$\operatorname{rank}\Big(\boldsymbol{G}(\boldsymbol{q},t)\,\boldsymbol{M}^{-1}(\boldsymbol{q})\,\Big(\frac{\partial \boldsymbol{f}}{\partial \boldsymbol{\lambda}}(\boldsymbol{q},\dot{\boldsymbol{q}},\boldsymbol{\lambda},t) - \boldsymbol{G}^\top(\boldsymbol{q},t)\Big)\Big) = n_g \tag{33b}$$

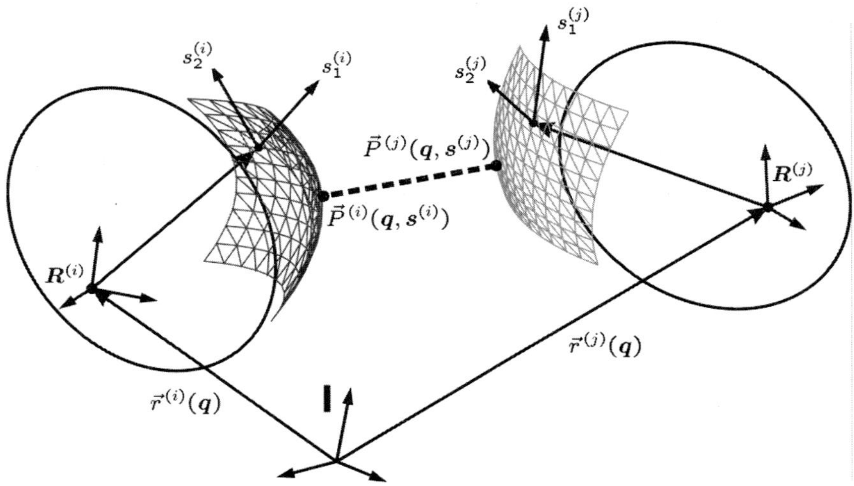

Figure 7. Formulation of contact conditions using surface parameters $\boldsymbol{s}$ as additional algebraic variables.

to guarantee the unique solvability of (30).

Another extension of (30) shows the large potential of DAE formulations in view of user-friendly algorithms for model setup. Consider rigid bodies $(.)^{(i)}$ and $(.)^{(j)}$ that are in contact. To formulate the contact conditions parametrizations $\boldsymbol{s}^{(i)}, \boldsymbol{s}^{(j)} \in \mathbb{R}^2$ of the body surfaces are introduced. Then the position of the points $\vec{P}^{(i)}$ on the surface of Body $(.)^{(i)}$ in $\mathbb{R}^3$ is determined by the position and orientation of this body, i.e., by the multibody system coordinates $\boldsymbol{q}$, and by the surface parameter $\boldsymbol{s}^{(i)}$: $\vec{P}^{(i)} = \vec{P}^{(i)}(\boldsymbol{q}, \boldsymbol{s}^{(i)})$, see Fig. 7.

If both bodies are strictly convex then there is a uniquely defined contact point $\vec{P}^{(i)}(\boldsymbol{q}, \boldsymbol{s}^{(i)}) = \vec{P}^{(j)}(\boldsymbol{q}, \boldsymbol{s}^{(j)})$ that belongs to the surfaces of both bodies. The contact point is characterized by the condition that the normal vector of Body $(.)^{(i)}$ at $\vec{P}^{(i)}$ is parallel to the normal vector of Body $(.)^{(j)}$ at $\vec{P}^{(j)}$, see (Roberson and Schwertassek, 1988). This geometrical condition defines implicitly the contact point coordinates $\boldsymbol{s} = (\boldsymbol{s}^{(i)}, \boldsymbol{s}^{(j)})$ as solution $\boldsymbol{s} = \boldsymbol{s}(\boldsymbol{q}, t)$ of a non-linear system

$$\mathbf{0} = \boldsymbol{h}(\boldsymbol{q}, \boldsymbol{s}, t) \tag{34}$$

with non-singular Jacobian $\partial \boldsymbol{h} / \partial \boldsymbol{s}$, see (Eich-Soellner and Führer, 1998; Simeon et al., 1991).

Contact and friction forces act at the contact point. Furthermore, the contact point coordinates $\boldsymbol{s} = \boldsymbol{s}(\boldsymbol{q}, t)$ are used in the formulation of the contact condition

$$\mathbf{0} = \vec{P}^{(i)}(\boldsymbol{q}, \boldsymbol{s}^{(i)}) - \vec{P}^{(j)}(\boldsymbol{q}, \boldsymbol{s}^{(j)}) \tag{35}$$

that defines constraints $\mathbf{0} = \boldsymbol{g}(\boldsymbol{q}, t)$ in (30). In a classical implementation each evaluation of $\boldsymbol{g}(\boldsymbol{q}, t)$ would require the solution of (34) to get the contact point coordinates $\boldsymbol{s} = \boldsymbol{s}(\boldsymbol{q}, t)$ that have to be inserted in the contact condition (35).

As an alternative, the DAE framework allows to add simply the nonlinear equations (34) to the equations of motion (30) and to append the contact conditions (35) directly to the constraints:

$$\begin{aligned} \boldsymbol{M}(\boldsymbol{q})\,\ddot{\boldsymbol{q}}(t) &= \boldsymbol{f}(\boldsymbol{q}, \boldsymbol{s}, \dot{\boldsymbol{q}}, \boldsymbol{\lambda}, t) - \boldsymbol{G}^\top(\boldsymbol{q}, \boldsymbol{s}, t)\boldsymbol{\lambda}\,, && \text{(36a)}\\ \mathbf{0} &= \boldsymbol{h}(\boldsymbol{q}, \boldsymbol{s}, t)\,, && \text{(36b)}\\ \mathbf{0} &= \boldsymbol{g}(\boldsymbol{q}, \boldsymbol{s}, t)\,. && \text{(36c)} \end{aligned}$$

With this formulation, the solution of (34) w.r.t. $\boldsymbol{s} = \boldsymbol{s}(\boldsymbol{q}, t)$ is left to the DAE solver (Rulka, 1998; Simeon et al., 1991). Note, that the constraint matrix is now

$$\begin{aligned} \boldsymbol{G}(\boldsymbol{q}, \boldsymbol{s}, t) &= \frac{\mathrm{D}}{\mathrm{D}\boldsymbol{q}}\boldsymbol{g}(\boldsymbol{q}, \boldsymbol{s}(\boldsymbol{q}, t), t) = \left[\frac{\partial \boldsymbol{g}}{\partial \boldsymbol{q}} + \frac{\partial \boldsymbol{g}}{\partial \boldsymbol{s}}\frac{\partial \boldsymbol{s}}{\partial \boldsymbol{q}}\right](\boldsymbol{q}, \boldsymbol{s}, t)\\ &= \left[\frac{\partial \boldsymbol{g}}{\partial \boldsymbol{q}} - \frac{\partial \boldsymbol{g}}{\partial \boldsymbol{s}}\left(\frac{\partial \boldsymbol{h}}{\partial \boldsymbol{s}}\right)^{-1}\frac{\partial \boldsymbol{h}}{\partial \boldsymbol{q}}\right](\boldsymbol{q}, \boldsymbol{s}, t) \end{aligned} \tag{37}$$

because of

$$\mathbf{0} = \frac{\mathrm{D}}{\mathrm{D}\boldsymbol{q}}\boldsymbol{h}(\boldsymbol{q}, \boldsymbol{s}(\boldsymbol{q}, t), t) = \left[\frac{\partial \boldsymbol{h}}{\partial \boldsymbol{q}} + \frac{\partial \boldsymbol{h}}{\partial \boldsymbol{s}}\frac{\partial \boldsymbol{s}}{\partial \boldsymbol{q}}\right](\boldsymbol{q}, \boldsymbol{s}, t)\,. \tag{38}$$

From the mathematical viewpoint, Eq. (34) is just a special case of algebraic equations

$$\mathbf{0} = \boldsymbol{b}(\boldsymbol{w}, \boldsymbol{q}, \dot{\boldsymbol{q}}, \boldsymbol{\lambda}, t) \tag{39}$$

with variables $\boldsymbol{w}(t) \in \mathbb{R}^{n_b}$ that are defined implicitly by (39) if $\partial \boldsymbol{b}/\partial \boldsymbol{w}$ is non-singular. The mathematical modelling of force elements may be simplified substantially using such auxiliary variables $\boldsymbol{w}$. A typical example is the modelling of joint friction, see (Rulka, 1998) for a detailed discussion.

With variables $\boldsymbol{w}$ the force vector in (36) gets the general form $\boldsymbol{f} = \boldsymbol{f}(\boldsymbol{q}, \boldsymbol{s}, \dot{\boldsymbol{q}}, \boldsymbol{\lambda}, \boldsymbol{w}, t)$ and the Jacobian $\partial \boldsymbol{f}/\partial \boldsymbol{\lambda}$ in the generalized Grübler condition (33b) has to be substituted by

$$\left[\frac{\partial \boldsymbol{f}}{\partial \boldsymbol{\lambda}} - \frac{\partial \boldsymbol{f}}{\partial \boldsymbol{w}}\left(\frac{\partial \boldsymbol{b}}{\partial \boldsymbol{w}}\right)^{-1}\frac{\partial \boldsymbol{b}}{\partial \boldsymbol{\lambda}}\right](\boldsymbol{q}, \boldsymbol{s}, \dot{\boldsymbol{q}}, \boldsymbol{\lambda}, \boldsymbol{w}, t)\,.$$

Summary Based on the multibody system approach the model equations in applied dynamics are formulated as hybrid system of differential-algebraic equations

$$\begin{aligned} \boldsymbol{M}(\boldsymbol{q})\,\ddot{\boldsymbol{q}}(t) &= \boldsymbol{f}(\boldsymbol{q},\boldsymbol{s},\dot{\boldsymbol{q}},\boldsymbol{c},\boldsymbol{r}_j,\boldsymbol{\lambda},\boldsymbol{w},t) - \boldsymbol{G}^\top(\boldsymbol{q},\boldsymbol{s},t)\boldsymbol{\lambda}\,, &\text{(40a)}\\ \dot{\boldsymbol{c}}(t) &= \boldsymbol{d}(\boldsymbol{c},\boldsymbol{q},\boldsymbol{s},\dot{\boldsymbol{q}},\boldsymbol{r}_j,\boldsymbol{\lambda},\boldsymbol{w},t)\,. &\text{(40b)}\\ \mathbf{0} &= \boldsymbol{b}(\boldsymbol{w},\boldsymbol{q},\boldsymbol{s},\dot{\boldsymbol{q}},\boldsymbol{c},\boldsymbol{r}_j,\boldsymbol{\lambda},t)\,, &\text{(40c)}\\ \mathbf{0} &= \boldsymbol{h}(\boldsymbol{q},\boldsymbol{s},\boldsymbol{r}_j,t)\,, &\text{(40d)}\\ \mathbf{0} &= \boldsymbol{g}(\boldsymbol{q},\boldsymbol{s},\boldsymbol{r}_j,t) &\text{(40e)} \end{aligned}$$

with $t \in [T_j, T_{j+1})$ and discrete state equations

$$\boldsymbol{r}_{j+1} = \boldsymbol{k}(\boldsymbol{r}_j,\boldsymbol{r}_{j-1},\ldots,\boldsymbol{q},\boldsymbol{s},\dot{\boldsymbol{q}},\boldsymbol{c},\boldsymbol{\lambda},\boldsymbol{w},T_{j+1})\,. \tag{41}$$

The continuous part (40) of the model equations forms an index-3 DAE with a complex but characteristic structure that will be exploited in the numerical solution, see Section 5.

It is important that the contact point coordinates $\boldsymbol{s}$ in the constraints (40e) are defined by an equation that is independent of $\dot{\boldsymbol{q}}$ and $\boldsymbol{\lambda}$, see (40d). Otherwise the model equations would not be solvable unless strong additional regularity conditions were satisfied.

5 DAE Time Integration in Applied Dynamics

The time integration of DAE model equations (40) is based on classical ODE methods combined with *index reduction* and *projection* techniques. The time integration methods will be discussed in detail for DAEs (30) that are re-written as first order systems in residual form (19) with

$$\boldsymbol{x}(t) = \begin{pmatrix} \boldsymbol{q}(t) \\ \boldsymbol{v}(t) \\ \boldsymbol{\lambda}(t) \end{pmatrix},$$

and

$$\mathbf{0} = \boldsymbol{F}(\boldsymbol{x},\dot{\boldsymbol{x}},t) = \begin{pmatrix} \dot{\boldsymbol{q}} - \boldsymbol{v} \\ \boldsymbol{M}(\boldsymbol{q})\dot{\boldsymbol{v}} - \boldsymbol{f}(\boldsymbol{q},\boldsymbol{v},t) + \boldsymbol{G}^\top(\boldsymbol{q},t)\boldsymbol{\lambda} \\ \boldsymbol{g}(\boldsymbol{q},t) \end{pmatrix}. \tag{42}$$

At the end of the section we will come back to DAEs of the more general form (40).

The first generation of DAE time integration methods failed in the application to constrained mechanical systems, see (Petzold, 1982). Very similar

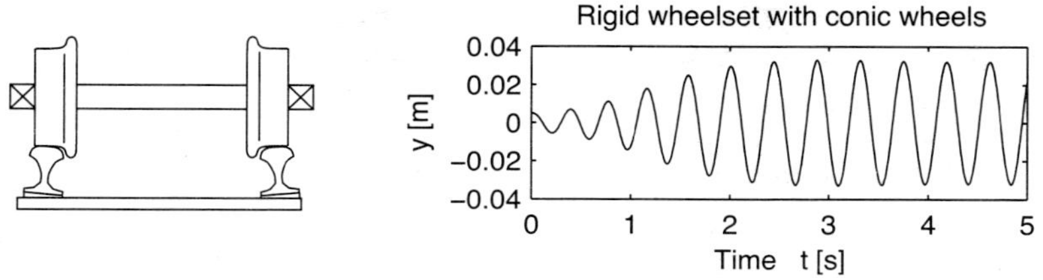

Figure 8. Lateral displacement y of a rigid wheelset performing a hunting motion.

numerical problems were observed in quite different practical applications like robotics, road vehicle simulation and railway vehicle simulation. To discuss these numerical effects in more detail, one of the best analysed benchmark problems from the literature will be used throughout this section (Simeon et al., 1991; Arnold and Netter, 1997). This multibody system model comes from railway dynamics, but in time integration it shows all those numerical effects that are typical of any constrained systems (42) including the ones from road vehicle dynamics.

The benchmark describes a rigid wheelset with conic wheels moving with constant speed along a straight track. Starting with a small initial lateral displacement y the wheelset oscillates in lateral direction, see Fig. 8. This so called hunting motion is a well known phenomenon in railway vehicle dynamics.

The equations of motion are formulated in DAE form (36) with the position coordinates $\boldsymbol{q}(t) \in \mathbb{R}^6$ of the wheelset that correspond to its six degrees of freedom. Two constraints (36c) guarantee permanent contact between the two wheels and the rails (Simeon et al., 1991). Exploiting symmetries in the geometry of wheel and rail the contact point between a single wheel and the rail may be described by a scalar position coordinate $s(t)$ resulting in two equations (36b) that determine implicitly $\boldsymbol{s} = (s_l, s_r)^\top \in \mathbb{R}^2$ for the contact point coordinates on left and right wheel (Arnold and Netter, 1997).

DAE time integration by ODE methods Eqs. (42) suggest to apply BDF (20) and other ODE methods straightforwardly to the time integration of DAEs. Simple test problems like (42) with $\boldsymbol{M} = \boldsymbol{I}_{n_q}$, $\boldsymbol{f} = \boldsymbol{0}$ and linear constraints

$$\boldsymbol{0} = \boldsymbol{g}(\boldsymbol{q}, t) = \boldsymbol{C}\boldsymbol{q} - \boldsymbol{z}(t)$$

show that this approach fails in general: According to (30a) and (31b) we get $\ddot{\boldsymbol{q}}(t) = -\boldsymbol{C}^\top \boldsymbol{\lambda}$ and $\boldsymbol{C}\ddot{\boldsymbol{q}}(t) = \ddot{\boldsymbol{z}}(t)$ resulting in Lagrangian multipliers $\boldsymbol{\lambda}(t) =$

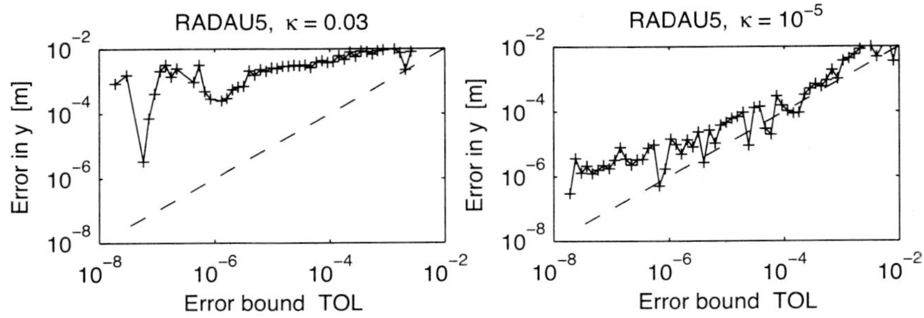

Figure 9. ODE solver RADAU5 applied to the differential-algebraic equations of motion (30) of a rigid wheelset.

$-(\boldsymbol{C}\boldsymbol{C}^\top)^{-1}\ddot{\boldsymbol{z}}(t)$. The BDF solution $\boldsymbol{x}_n = (\boldsymbol{q}_n, \boldsymbol{v}_n, \boldsymbol{\lambda}_n)^\top$ satisfies (20) and hence also

$$\mathbf{0} = \boldsymbol{g}(\boldsymbol{q}_n, t_n) = \boldsymbol{C}\boldsymbol{q}_n - \boldsymbol{z}(t_n), \quad (n > 0)$$

since the constraints (30b) are part of the residual $\boldsymbol{F}$ in (42). A simple computation shows that the application of backward Euler method to (42), i.e., BDF (20) with $k_n = 1$, yields

$$\boldsymbol{\lambda}_{n+1} = \frac{h_n + h_{n-1}}{2h_n}\boldsymbol{\lambda}(t_{n+1}) + \mathcal{O}(h_n) + \mathcal{O}(\frac{h_{n-1}^2}{h_n}).$$

For varying stepsizes the numerical solution $\boldsymbol{\lambda}_{n+1}$ is completely wrong since it does not converge to the analytical solution $\boldsymbol{\lambda}(t_{n+1})$ if $h_n \to 0$ and h_{n-1} is fixed, see also (Petzold, 1982).

For some higher order methods convergence can still be guaranteed (Hairer and Wanner, 1996) but the errors in the numerical solution are typically substantially larger than in the ODE case. This is illustrated by simulation results for the hunting motion of the rigid wheelset in Fig. 8. In this application, the BDF integrator DASSL applied to (36) fails completely. As an alternative the implicit Runge–Kutta solver RADAU5 was used in the numerical tests (Hairer and Wanner, 1996). For RADAU5, experience shows that the error of the numerical solution in ODE applications is usually kept well below the user-defined error tolerances TOL. But in the application to DAE (36) the relative error remains even for extremely small error bounds always in the size of $0.1 \ldots 1.0\,\%$. As a typical example, the left plot of Fig. 9 shows the error in the lateral displacement y for various values of TOL.

To analyse these unsatisfactory results of RADAU5 we modify one of the internal parameters of the solver. BDF (20) and implicit Runge–Kutta methods define $\boldsymbol{x}_{n+1}$ as solution of a system of non-linear equations. In the practical implementation this system is solved iteratively by Newton's method that is stopped if the iteration error is less than $\kappa \cdot \mathrm{TOL}$. In this stopping criterion the user-defined error tolerance for time integration (TOL) is scaled by a constant $\kappa \leq 1$ that is a free control parameter of the solver. Default values are $\kappa = 0.33$ in DASSL (Brenan et al., 1996) and $\kappa = 0.03$ in RADAU5 (Hairer and Wanner, 1996).

The right plot of Fig. 9 shows that the error in time integration is reduced drastically and remains now roughly below the error bounds if κ is set to the very small value $\kappa = 10^{-5}$. The comparison of left and right plot in Fig. 9 illustrates that the direct application of RADAU5 to DAEs (36) makes the solver very sensitive w.r.t. (small) iteration errors in Newton's method.

An analytical perturbation analysis for (42) shows that the errors in $\boldsymbol{q}_n$, $\boldsymbol{v}_n$ and $\boldsymbol{\lambda}_n$ are of size $\mathcal{O}(\theta) + \mathcal{O}(\theta^2/h^2)$, $\mathcal{O}(\theta/h)$ and $\mathcal{O}(\theta/h^2)$, respectively, if the errors in the constraints (30b) are bounded by some small positive constant $\theta \ll 1$, see (Arnold, 1995). Note, that these error bounds *grow* unboundedly for decreasing time stepsize $h := \min_n h_n$.

This error amplification is typical of the direct application of ODE solvers to DAE (42). It may be considered as discrete analogue of corresponding error bounds $\mathcal{O}(\max_t \|\boldsymbol{\theta}(t)\|) + \mathcal{O}(\max_t \|\dot{\boldsymbol{\theta}}(t)\|^2)$, $\mathcal{O}(\max_t \|\dot{\boldsymbol{\theta}}(t)\|)$ and $\mathcal{O}(\max_t \|\ddot{\boldsymbol{\theta}}(t)\|)$ for the analytical solution $\boldsymbol{q}(t)$, $\boldsymbol{v}(t)$ and $\boldsymbol{\lambda}(t)$ of (42) with perturbed constraints

$$\boldsymbol{\theta}(t) = \boldsymbol{g}(\boldsymbol{q}(t), t)$$

and $\max_t \|\boldsymbol{\theta}(t)\| \ll 1$, see (Arnold, 1995).

For perturbations like $\boldsymbol{\theta}(t) = \varepsilon \sin \omega t$ that oscillate with small amplitude $\varepsilon \ll 1$ and high frequency $\omega \gg 1$ the errors in the solution $(\boldsymbol{q}(t), \boldsymbol{v}(t), \boldsymbol{\lambda}(t))$ are much larger than the perturbation itself since $\|\dot{\boldsymbol{\theta}}(t)\| = \mathcal{O}(\omega\varepsilon)$, $\|\ddot{\boldsymbol{\theta}}(t)\| = \mathcal{O}(\omega^2\varepsilon)$. A mechanical interpretation of these results is the well known fact that high frequency kinematic excitations $\boldsymbol{\theta}(t)$ in loop closing joints may cause large joint and constraint forces in a multibody system.

Index reduction and projection In the direct application of ODE methods to DAE (42) the robustness of the solvers may be improved substantially by small values of κ that result in (very) small iteration errors in Newton's method, see the right plot of Fig. 9. On the other hand values of κ that are less than 10^{-3} increase the number of Newton steps per time step substantially and may slow down the ODE solver dramatically.

Instead of applying ODE solvers directly to (42) it proved to be much more advantageous to transform the equations of motion *analytically* before time integration. These analytical transformations are guided by the perturbation analysis. In (30) the Lagrangian multipliers $\boldsymbol{\lambda}(t)$ are defined implicitly by the constraints (30b). Differentiating these constraints $\mathbf{0} = \boldsymbol{g}(\boldsymbol{q}(t), t)$ twice the acceleration constraints (31b) are obtained that define together with $\ddot{\boldsymbol{q}}(t) = \boldsymbol{M}^{-1}(\boldsymbol{q})(\boldsymbol{f} - \boldsymbol{G}^\top \boldsymbol{\lambda})$, see (30a), the system of linear equations

$$\mathbf{0} = -[\boldsymbol{G}\boldsymbol{M}^{-1}\boldsymbol{G}^\top](\boldsymbol{q}, t)\,\boldsymbol{\lambda} + [\boldsymbol{G}\boldsymbol{M}^{-1}\boldsymbol{f}](\boldsymbol{q}, \dot{\boldsymbol{q}}, t) + \boldsymbol{g}^{(II)}(\boldsymbol{q}, \dot{\boldsymbol{q}}, t) \tag{43}$$

for the Lagrangian multipliers $\boldsymbol{\lambda}(t)$. The $r = 2$ differentiation steps introduce the large error terms $\mathcal{O}(\|\ddot{\boldsymbol{\theta}}(t)\|)$ and $\mathcal{O}(\theta/h^2)$ in the analytical and numerical solution, respectively. In DAE terminology the system (30) forms a DAE of index $r + 1 = 3$ that is called the *index-3 formulation* of the equations of motion (Brenan et al., 1996).

For consistent initial values satisfying (32a) the solution of the equations of motion (30) remains unchanged if the constraints (30b) on position level are substituted by the constraints (31a) on velocity level since

$$\boldsymbol{g}(\boldsymbol{q}(t), t) = \underbrace{\boldsymbol{g}(\boldsymbol{q}_0, t_0)}_{=0,\text{ see (32a)}} + \int_{t_0}^{t} \underbrace{\frac{\mathrm{d}}{\mathrm{d}t}\boldsymbol{g}(\boldsymbol{q}(\tau), \tau)}_{=0,\text{ see (31a)}}\,\mathrm{d}\tau = 0\,. \tag{44}$$

From the numerical viewpoint the *index-2 formulation* (30a)/(31a) of the equations of motion is attractive because only $r = 1$ differentiation of the velocity constraints (31a) is necessary to get the system (43) of linear equations defining $\boldsymbol{\lambda}(t)$.

With a perturbation $\boldsymbol{\theta}(t)$ in the velocity constraints (31a) the error terms for this index-2 DAE are of size $\mathcal{O}(\max_t \|\boldsymbol{\theta}(t)\|\ \|\dot{\boldsymbol{\theta}}(t)\|)$ for $\boldsymbol{q}(t)$, $\boldsymbol{v}(t)$ and of size $\mathcal{O}(\max_t \|\dot{\boldsymbol{\theta}}(t)\|)$ for $\boldsymbol{\lambda}(t)$. The error bounds for the numerical solution are $\mathcal{O}(\theta) + \mathcal{O}(\theta^2/h)$ for $\boldsymbol{q}_n$, $\boldsymbol{v}_n$ and $\mathcal{O}(\theta/h)$ for $\boldsymbol{\lambda}_n$. They are smaller by the (small) factor h than the corresponding error bounds for the original index-3 DAE (30a,b), see (Arnold, 1995). Applying RADAU5 with the standard setting $\kappa = 0.03$ to the wheelset example of Fig. 8 in index-2 formulation (30a)/(31a) the error of the numerical solution is kept well below the user-defined error tolerances TOL.

In the same way the *index-1 formulation* (30a)/(31b) of the equations of motion is obtained substituting (30b) by the constraints (31b) on acceleration level. The error bounds for $\boldsymbol{\lambda}(t)$ and $\boldsymbol{\lambda}_n$ are slightly improved to $\mathcal{O}(\max_t \|\boldsymbol{\theta}(t)\|)$ and $\mathcal{O}(\theta)$, respectively. BDF and implicit Runge–Kutta solvers applied to the index-1 formulation are as robust and efficient as in the classical ODE case.

In DAE terminology the substitution of the constraints (30b) by one of its time derivatives is called *index reduction.* Index reduction by differenti-

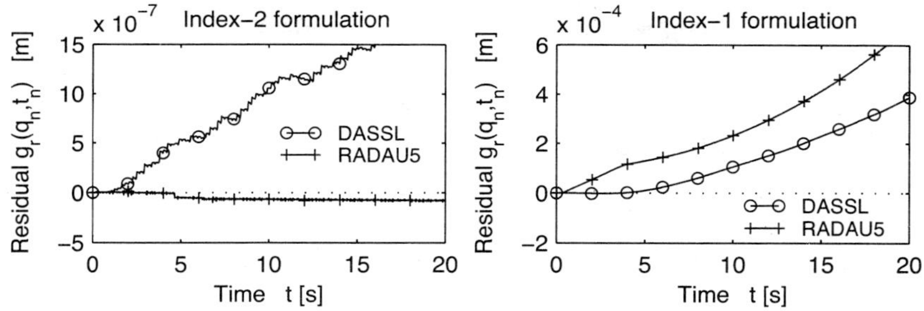

Figure 10. Drift-off effect in the dynamical simulation of the rigid wheelset of Fig. 8 resulting in an increasing distance $g_r(\boldsymbol{q}, t)$ between the right wheel and the rail, i.e., in an increasing error in the constraints $\mathbf{0} = \boldsymbol{g} = (g_l, g_r)^\top$ that are defined by the contact conditions for left and right wheel.

ation improves the solver's robustness substantially but suffers in long-term simulations from the *drift-off effect* that is illustrated by Fig. 10.

Because of (44) the *analytical* solution of the index-2 formulation satisfies the original constraints (30b) exactly for all $t \geq t_0$. In the *numerical* solution the integrand $(\mathrm{d}\boldsymbol{g}/\mathrm{d}t)(\boldsymbol{q}(\tau), \tau)$ in (44) is still bounded by a small constant $\epsilon > 0$ but it does *not* vanish identically. Therefore, the error in (30b) may increase linearly in time t:

$$\|\boldsymbol{g}(\boldsymbol{q}_n, t_n)\| \leq \|\boldsymbol{g}(\boldsymbol{q}_0, t_0)\| + \int_{t_0}^{t_n} \epsilon \,\mathrm{d}t = \epsilon \cdot (t_n - t_0)\,; \tag{45}$$

the numerical solution $\boldsymbol{q}_n$ *drifts* off the manifold $\{\, (\boldsymbol{\eta}, t) \,:\, \boldsymbol{g}(\boldsymbol{\eta}, t) = \mathbf{0} \,\}$ that is defined by the constraints (30b) on position level. The error bound ϵ summarizes discretization and round-off errors and the iteration errors of Newton's method.

For the index-1 formulation a quadratic error growth

$$\|\boldsymbol{g}(\boldsymbol{q}_n, t_n)\| \leq \epsilon \cdot (t_n - t_0)^2$$

has to be expected since the constraints (30b) on position level are substituted by their second derivatives (31b). Practical experience shows that ϵ depends on the solver and on the user-defined error tolerances TOL. In general, however, there is always a linear drift in the time integration of the index-2 formulation and a quadratic drift for the index-1 formulation, see Fig. 10 for a typical example.

An early attempt to avoid both the numerical problems for the index-3 formulation (30) and the drift-off effect in the index-2 and index-1 formula-

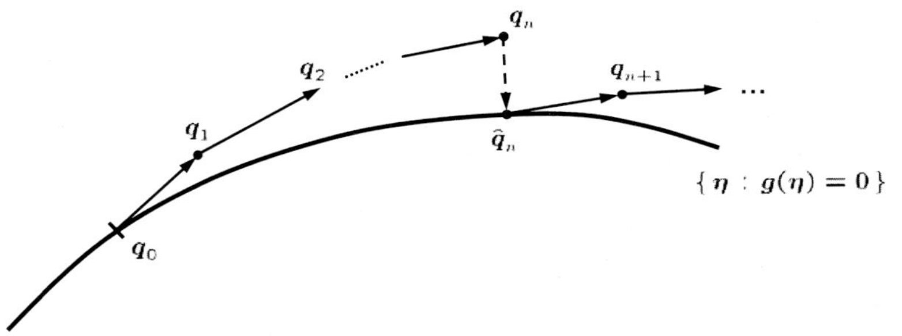

Figure 11. Time integration with projection steps for DAEs (30) with constraints $\boldsymbol{g} = \boldsymbol{g}(\boldsymbol{q})$.

tion goes back to the work of Baumgarte (Baumgarte, 1972) who substituted the constraints (30b) by a linear combination of all three constraints (30b), (31a) and (31b). Because of the problems to select suitable coefficients for this linear combination (*Baumgarte coefficients*) the practical use of Baumgarte's approach is restricted to small scale models, see also (Ascher et al., 1994).

Recently, a Baumgarte like approach that substitutes the original constraints (30b) by a linear combination of (30b) and (31a) was used successfully in real-time applications (*index-2 Baumgarte approach*), see (Arnold et al., 2007a; Valášek et al., 2007).

In off-line simulation, it is today state-of-the-art to avoid the drift-off effect by *projection* techniques (Eich, 1993; Lubich et al., 1995). During the time integration of the index-2 formulation the residual $\|\boldsymbol{g}(\boldsymbol{q}_n, t_n)\|$ in the constraints (30b) is monitored. If the residual exceeds some user-defined small error bound $\varepsilon > 0$ then $\boldsymbol{q}_n$ is projected onto the manifold $\{\boldsymbol{\eta} : \boldsymbol{g}(\boldsymbol{\eta}, t_n) = \mathbf{0}\}$ and the time integration is continued with the projected position coordinates $\hat{\boldsymbol{q}}_n$ instead of $\boldsymbol{q}_n$, see Fig. 11.

Mathematically, the projection defines a minimization problem

$$\|\boldsymbol{\eta} - \boldsymbol{q}_n\| \to \min_{\{\boldsymbol{\eta} : \boldsymbol{g}(\boldsymbol{\eta}, t_n) = \mathbf{0}\}} \tag{46}$$

that can be solved efficiently by simplified Newton iterations to get $\hat{\boldsymbol{q}}_n$, see (Lubich et al., 1995).

For the index-1 formulation the projection of both $\boldsymbol{q}_n$ and $\boldsymbol{v}_n$ helps to avoid the drift in the constraints (30b) and (31a) on position and velocity level (Eich, 1993).

Gear–Gupta–Leimkuhler formulation In complex applications, the use of classical explicit projection methods like (46) is restricted to Runge–Kutta and other one-step methods since the efficient implementation in advanced BDF solvers with order and stepsize control is non-trivial.

Instead of implementing explicit projection steps in the solver, the equations of motion (30) are reformulated in a way that contains implicitly the projection onto the constraint manifold $\{\boldsymbol{\eta} : \boldsymbol{g}(\boldsymbol{\eta}, t_n) = \mathbf{0}\}$. The *Gear–Gupta–Leimkuhler formulation* (or *stabilized index-2 formulation*) of the equations of motion considers the constraints (30b) and (31a) on position and velocity level simultaneously (Gear et al., 1985):

$$\begin{aligned} \dot{\boldsymbol{q}}(t) &= \boldsymbol{v} - \boldsymbol{G}^\top(\boldsymbol{q}, t)\boldsymbol{\mu}, \\ \boldsymbol{M}(\boldsymbol{q})\,\dot{\boldsymbol{v}}(t) &= \boldsymbol{f}(\boldsymbol{q}, \boldsymbol{v}, t) - \boldsymbol{G}^\top(\boldsymbol{q}, t)\boldsymbol{\lambda}, \\ \mathbf{0} &= \boldsymbol{g}(\boldsymbol{q}, t), \\ \mathbf{0} &= \boldsymbol{G}(\boldsymbol{q}, t)\boldsymbol{v} + \boldsymbol{g}^{(I)}(\boldsymbol{q}, t). \end{aligned} \tag{47}$$

The increasing number of equations is compensated by a correction term $-\boldsymbol{G}^\top\boldsymbol{\mu}$ with auxiliary variables $\boldsymbol{\mu}(t) \in \mathbb{R}^{n_g}$. The correction term vanishes identically for the analytical solution ($\boldsymbol{\mu}(t) \equiv \mathbf{0}$) and remains in the size of the user-defined error tolerances TOL for the numerical solution.

Eqs. (47) form an index-2 DAE that may be solved robustly and efficiently by BDF (Gear et al., 1985) and implicit Runge–Kutta methods (Hairer and Wanner, 1996). However, the error estimates in classical ODE solvers tend to overestimate the local errors in the algebraic components $\boldsymbol{\lambda}$, $\boldsymbol{\mu}$ of DAE (47), see (Petzold, 1982). Therefore the components $\boldsymbol{\lambda}$ and $\boldsymbol{\mu}$ should not be considered in the automatic stepsize control of BDF solvers (Petzold and Lötstedt, 1986). For implicit Runge–Kutta solvers the error estimates for $\boldsymbol{\lambda}$ and $\boldsymbol{\mu}$ are scaled by the small factor h_n.

The efficient evaluation of the correction term $\boldsymbol{G}^\top(\boldsymbol{q}, t)\boldsymbol{\mu}$ for given $\boldsymbol{\mu}, \boldsymbol{q}, t$ is non-trivial since multibody formalisms evaluate matrix-vector products $\boldsymbol{G}^\top\boldsymbol{\lambda}$ and $\boldsymbol{G}\boldsymbol{v}$ with $\mathcal{O}(N)$ complexity but do not evaluate the matrix $\boldsymbol{G}(\boldsymbol{q}, t)$ itself.

The key to the efficient solution of (47) was found in the analysis of a special class of overdetermined DAEs (Führer, 1988) that are closely related to DAEs (47), see (Führer and Leimkuhler, 1991). BDF solvers and implicit Runge–Kutta solvers for DAEs (19) approximate the Jacobian

$$\frac{\alpha_{n,0}}{h_n}\frac{\partial \boldsymbol{F}}{\partial \dot{\boldsymbol{x}}} + \frac{\partial \boldsymbol{F}}{\partial \boldsymbol{x}}$$

to compute $\boldsymbol{x}_{n+1}$ iteratively by Newton's method, see (17) and (21). In (47) this Jacobian has a 4×4 block structure reflecting the four equations in (47) and the partitioning $\boldsymbol{x} = (\boldsymbol{q}, \boldsymbol{v}, \boldsymbol{\lambda}, \boldsymbol{\mu})^\top$. Block $(3, 1)$ is an approximation of

$$\frac{\partial \boldsymbol{g}}{\partial \boldsymbol{q}}(\boldsymbol{q}, t) = \boldsymbol{G}(\boldsymbol{q}, t),$$

its transpose may be used to get the matrix-vector product $\boldsymbol{G}^\top \boldsymbol{\mu}$. This algorithm was implemented in the BDF solver ODASSL (Führer, 1988) that exploits furthermore the special structure of (47) to compute $\boldsymbol{\mu}_{n+1}$ more efficiently than in the standard DASSL implementation.

Advanced DAE models The numerical algorithms for DAE time integration that we discussed in detail for classical constrained mechanical systems with equations of motion (30) may be applied as well to the more complex hybrid differential-algebraic model equations (40)/(41) that are typical of applied dynamics.

In the model equations in residual form (42) the vector $\boldsymbol{x}(t)$ is now given by $\boldsymbol{x} = (\boldsymbol{q}, \boldsymbol{s}, \boldsymbol{v}, \boldsymbol{c}, \boldsymbol{\lambda}, \boldsymbol{w})^\top$. Time events and the discrete state equations (41) are handled by the methods that are known from ODE theory, see Section 3.

The differential equations (40b) and the algebraic equations (40c) and (40d) are added to the residual $\boldsymbol{F}(\boldsymbol{x}, \dot{\boldsymbol{x}}, t)$, see (20). As a consequence, the BDF solution $\boldsymbol{x}_{n+1}$ of (40) satisfies the non-linear equations (40d) exactly:

$$\mathbf{0} = \boldsymbol{h}(\boldsymbol{q}_{n+1}, \boldsymbol{s}_{n+1}, \boldsymbol{r}_j, t_{n+1})\,.$$

The BDF solution $\boldsymbol{s}_{n+1}$ coincides with the function value $\boldsymbol{s}(\boldsymbol{q}_{n+1}, t_{n+1})$ of the function $\boldsymbol{s}(\boldsymbol{q}, t)$ that is implicitly defined by

$$\mathbf{0} = \boldsymbol{h}(\boldsymbol{q}, \boldsymbol{s}(\boldsymbol{q}, t), \boldsymbol{r}_j, t)$$

if $\partial \boldsymbol{h} / \partial \boldsymbol{s}$ is non-singular, see (34) and (40d). BDF methods compute for a given geometric configuration $\boldsymbol{q}$ of the multibody system the contact point coordinates $\boldsymbol{s}$ exactly up to the (small) iteration errors in Newton's method. The same result may be shown for the algebraic variables $\boldsymbol{w}$ in (40c).

With the regularity assumptions of Section 4 the model equations (40) form an index-3 DAE. Index reduction and projection techniques have to be applied. It is important to note that the non-linear equations (40c) and (40d) may be solved directly w.r.t. the algebraic variables $\boldsymbol{w}$ and $\boldsymbol{s}$. In contrast to the constraints (40e) the non-linear equations (40c) and (40d) do *not* imply hidden constraints on the state variables like (31).

The index reduction is based on time derivatives of the constraints $\mathbf{0} = \boldsymbol{g}(\boldsymbol{q}, \boldsymbol{s}, t)$. It does not involve any derivatives of the non-linear equations $\mathbf{0} = \boldsymbol{b}(\boldsymbol{w}, \ldots)$ and $\mathbf{0} = \boldsymbol{h}(\boldsymbol{q}, \boldsymbol{s}, t)$. The hidden constraints on the level of velocity coordinates are

$$\mathbf{0} = \frac{\mathrm{d}}{\mathrm{d}t}\boldsymbol{g}(\boldsymbol{q}(t), \boldsymbol{s}(t), t) = \boldsymbol{G}(\boldsymbol{q}, \boldsymbol{s}, t)\,\boldsymbol{v} + \boldsymbol{g}^{(I)}(\boldsymbol{q}, \boldsymbol{s}, t)$$

with matrix $\boldsymbol{G}$ of (37) since $\dot{\boldsymbol{s}}(t)$ is a linear function of $\dot{\boldsymbol{q}}(t) = \boldsymbol{v}(t)$ that may be obtained differentiating $\mathbf{0} = \boldsymbol{h}(\boldsymbol{q}(t), \boldsymbol{s}(t), t)$ implicitly w.r.t. t, see also (38).

Using a rather general concept of DAE index reduction (Kunkel et al., 1997) the Gear–Gupta–Leimkuhler formulation (47) is generalized to complex DAE models (40), see (Arnold, 1996; Arnold et al., 2001). Here, the Gear–Gupta–Leimkuhler formulation (47) is extended by (40b,c,d). To compensate the simultaneous consideration of the constraints on position level and the hidden constraints on velocity level there is, as before, just one correction term. It has to be added to the kinematic equations $\dot{\boldsymbol{q}}(t) = \boldsymbol{v}$ that take the form

$$\dot{\boldsymbol{q}}(t) = \boldsymbol{v} - \boldsymbol{G}^{\top}(\boldsymbol{q}, \boldsymbol{s}, t)\boldsymbol{\mu}$$

with the constraint matrix $\boldsymbol{G}$ of (37) and auxiliary variables $\boldsymbol{\mu}(t) \in \mathbb{R}^{n_g}$. As in the classical approach of Gear, Gupta and Leimkuhler (Gear et al., 1985), the correction term vanishes identically for the analytical solution and remains in the size of the user-defined error tolerances TOL for the numerical solution.

For this extended system the Jacobian $\boldsymbol{J}$ in (21) has a 7×7 block structure with $(\partial \boldsymbol{g}/\partial \boldsymbol{q})$, $(\partial \boldsymbol{g}/\partial \boldsymbol{s})$, $(\partial \boldsymbol{h}/\partial \boldsymbol{s})$, $(\partial \boldsymbol{h}/\partial \boldsymbol{q})$ as blocks $(6,1)$, $(6,2)$, $(5,2)$ and $(5,1)$, respectively. In a modified version of the BDF solver ODASSL these blocks are used for the efficient evaluation of the correction term $\boldsymbol{G}^{\top}\boldsymbol{\mu}$ with matrix $\boldsymbol{G}$ of (37), see (Arnold, 1996).

Summary The robust and efficient dynamical simulation in applied dynamics may be based on DAE time integration methods with DASSL like BDF solvers that are applied to the Gear–Gupta–Leimkuhler formulation of the model equations. The approach is used successfully in industrial multibody system simulation packages (Rulka, 1998).

6 Algorithms and Tools for the Simulation of Multidisciplinary Problems

Typical problems of applied dynamics are far beyond the field of classical multibody analysis. Nevertheless, the methods and software tools of multibody dynamics have been used very successfully as integration platform for large, complex models in applied dynamics including mechanical, hydraulic and electronic components.

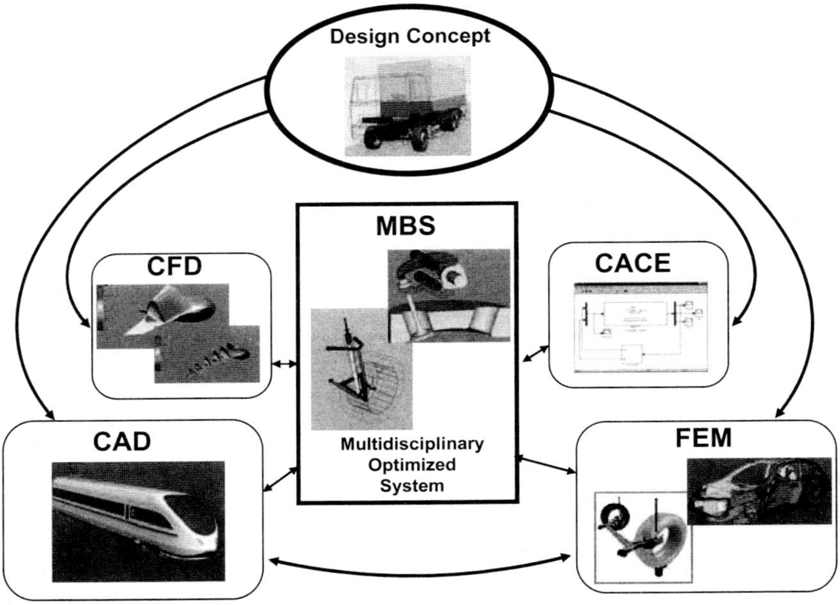

Figure 12. Integrated virtual design using multibody system simulation software and its bi-directional interfaces, see (Veitl, 2001).

Alternatives are general purpose simulation tools like Simulink[2], modelling languages for complex physical systems like ACSL[2] or Modelica[2] and general port-based modelling techniques (Breedveld, 2008). Furthermore, the coupling of two or more monodisciplinary simulation tools from different fields of application has proved to be very useful in industrial applications (*simulator coupling* or *co-simulation*).

Extensions of multibody system simulation packages In the early days of industrial multibody system simulation the focus was on specialized force elements with or without inner state variables $\boldsymbol{c}$ and $\boldsymbol{r}_j$, see Section 4 and (Kortüm et al., 2001).

Today, it might be even more important to embed the dynamical simulation of the mechanical system components efficiently in process chains for the simulation of the overall behaviour of engineering systems. Fig. 12

[2]Simulink is a trademark of The MathWorks, Inc., ACSL is a trademark of The AEgis Technologies Group, Inc., and Modelica is a trademark of the Modelica Association.

shows typical bi-directional interfaces between specialised simulation tools. The interfaces are used for data exchange in the pre- and post-processing of dynamical simulation.

The simulation of flexible multibody systems relies on the interfaces to finite element (FE) tools of structural dynamics. The interface from FE tool to multibody system tool provides mass, damping and stiffness data for the pre-processing of modes $\vec{w}^{(i)}$ in the Rayleigh–Ritz approach (25), see (Wallrapp, 1994). The interface from multibody system tool to FE tool supports the post-processing of multibody system simulation data. Load vectors are transferred to the fatigue analysis in standard FE tools (Dietz et al., 1997).

Modal reduction techniques are also used successfully in the analysis of certain aerodynamic effects. For this *close coupling* of computational fluid dynamics (CFD) and (flexible) multibody dynamics the classical interfaces between FE tools and multibody system tools have been extended by an interface to CFD tools (Krüger and Spieck, 2004).

The interfaces between CAD and multibody system tools support an efficient and fail-safe model setup using the geometrical data of CAD models. The opposite interface transfers all relevant geometrical modifications of the multibody system model automatically back to the CAD model (Trautenberg, 1999).

Controller synthesis is a central problem in applied dynamics. Close links between the methods and tools for the dynamical simulation of mechanical components and the tools of Computer aided control engineering (CACE) are therefore essential, see M. Valášek's contribution to this proceedings volume (Valášek, 2008).

Already in the early 1990's the increasing complexity of controllers in mechatronic system components motivated the integration of multibody system models in CACE tools starting with the export of the system matrices $\boldsymbol{A}$, $\boldsymbol{B}$, $\boldsymbol{C}$ and $\boldsymbol{D}$ describing the linearized equations of motion (6), see (Kortüm and Lugner, 1994). Later, this interface was extended by the export of FORTRAN source code for non-linear equations of motion (2). In both cases the time integration for the overall system including multibody system model and controller is performed in the CACE tool.

Since the solvers of CACE tools are not tailored to the differential-algebraic model equations (40) it is often more attractive to perform the time integration of the overall system including the controller in the multibody system tool. State-of-the-art CACE tools support this approach by a code export interface. A typical example is the Real-Time Workshop[3]

[3] Real-Time Workshop is a trademark of The MathWorks, Inc..

of The MathWorks, Inc., that generates and executes stand-alone C code for developing and testing algorithms modelled in Simulink (Vaculín et al., 2004; Valášek, 2008).

As an alternative to code export the tools of multibody system analysis and CACE may be coupled as well by a function call interface that evaluates during time integration repeatedly the right hand sides of the model equations for various given input data, see (Vaculín et al., 2004) for a more detailed discussion.

General purpose simulation tools and modelling languages Using interfaces between well developed monodisciplinary simulation tools the experienced engineer may study multidisciplinary problems without leaving his usual software environment. Latest new developments of the powerful specialized tools are immediately available and the access to highly developed specialized solvers is straightforward.

Alternative approaches resolutely focus on the multi-domain aspect of complex multidisciplinary applications. The most widespread general purpose simulation tool from system theory is Simulink that has been developed as platform for multi-domain simulation and model-based design of dynamic systems. It has a block-diagram user interface for model setup and a number of toolboxes for special fields of application like SimMechanics and SimDriveline.

The increasing complexity of industrial multidisciplinary applications results in a revival of modelling languages that are designed to bundle the experience of specialists from various engineering domains. Modern modelling languages like Modelica (`http://www.modelica.org/`) are object oriented and do not follow the classical input–output scheme of traditional block-oriented tools (*acausal* modelling).

Instead of classical ODE model equations (2) the model equations get the form of differential-algebraic equations. Modelica is a rapidly developing non-proprietary modelling language that covers a wide range of applications like electrical circuits, vehicle dynamics, hydraulic and pneumatic systems and automotive powertrains, see (Tiller, 2001) and the up-to-date material at the Modelica website `http://www.modelica.org/` for more details of this ongoing development.

Simulator coupling Specialized monodisciplinary tools, general purpose tools, code export / code import interfaces and function call interfaces offer the user a great variety of simulation techniques for multidisciplinary applications (Valášek, 2008). They follow completely different strategies in the design of user interfaces for model setup and in the formulation of mathematical model equations. From the numerical viewpoint, however, all these approaches have in common that finally the full set of model equations is solved in *one* simulation tool by *one* solver.

Advanced code export interfaces are not restricted to the classical export of source code describing model equations for some system components but may include as well the source code for an appropriate time integration method as *embedded solver*. (In practice, the traditional term "code export" is not only used for the export of readable C or FORTRAN source code but also for the export of compiled object code in dynamically linked libraries that are preferred by providers of commercial simulation tools.) With an embedded DAE time integration method, the export of multibody system models to CACE tools is, e.g., not longer restricted to tree structured systems. Another typical application are detailed tyre models that require specially adapted low order fixed stepsize time integration methods to handle the frequent discontinuities in a stick-slip contact model tyre–road. For the dynamical analysis of road vehicles, the tyre model with its embedded solver is imported in industrial multibody system simulation packages (Lugner and Plöchl, 2005).

From the practical viewpoint, the embedding of a solver in exported code may just be another option for the analysis of multidisciplinary problems. The use of an embedded solver in imported code for some subsystem makes, however, a great difference in time integration since the classical model coupling is extended to a solver coupling that may strongly influence accuracy, stability and robustness of the methods that were discussed in Sections 3 and 5. On the other hand, solver coupling offers an attractive way for adapting time integration methods to the specific solution behaviour in some subsystems, see, e.g., Dymola's extended code export interface for real-time applications (Elmqvist et al., 1995) and SIMPACK's Linear subsystem solver for flexible bodies (Dietz, 2005).

Code export / code import interfaces and function call interfaces are tailored to a hierarchical modelling of multidisciplinary problems since finally the simulation of the coupled problem is performed in one master tool that imports the other model components (with or without embedded solver). This hierarchical approach is less appropriate whenever all subsystems of the coupled problem are of equal importance.

For coupled problems of this type the combination of several monodisciplinary simulation tools should not be restricted to the model setup of all system components but should be used as well during time integration. In such a *co-simulation* approach, the subsystems are handled by different solvers and each solver is tailored to the corresponding subsystem. The data exchange is restricted to discrete synchronization points with sampling rates in the range of 1.0 kHz for applications from vehicle dynamics.

The development of interfaces for data exchange and coupling of simulation tools was driven by the needs of practical application. Over the years, a large variety of point-to-point solutions has been implemented that address similar problems but differ in many details. The functionality of S-Functions in MATLAB / Simulink may be considered as a quasi-standard for code export / code import interfaces that is supported by most commercial simulation tools in system dynamics. In vehicle system dynamics, the standardization of co-simulation interfaces in a *Functional mockup interface* is a central goal of the European ITEA 2 project MODELISAR (2008–2011, see `http://www.itea2.org/`).

Case study: Coupled simulation of mechanical and hydraulic components of a servo-hydraulic steering Today, there is no common classification of methods and interfaces for the coupling of simulation tools but detailed investigations have been performed for some typical fields of application (Vaculín et al., 2004; Valášek, 2008) and for several case studies from applied dynamics. As a typical example, Fig. 13 illustrates the coupling between mechanics and hydraulics in the simulation of mechatronic systems (Busch, 2007; Busch et al., 2007).

The mechanical part of a car with servo-hydraulic steering system is modelled in the 3D multibody system simulation package SIMPACK. The very detailed model of the steering (Fig. 14) is clearly beyond the range of standard applications in multibody dynamics. Mechanical, hydraulic and controller components of the servo-hydraulic steering are modelled in the modelling language Modelica. The simulation tool Dymola (`http://www.dynasim.com/`) is used for generating the C code for this subsystem.

The physical interface between the two subsystems "car" and "servo-hydraulic steering" is the rack-and-pinion steering. The excitation of the rack (position, velocity, acceleration) defines the input of the controlled hydraulic system that generates as output an actuator force acting on the mechanical steering components, see Fig. 14. As a result of the steering wheel excitation the actuator force signal reflects the feedback of the roll movement of the car body to the steering system.

Extending the state space form (2) of model equations by system inputs $\boldsymbol{u}(t)$ and system outputs $\boldsymbol{y}(t)$, the coupled subsystems **c**ar and **s**teering are described by

$$\left.\begin{array}{rcl} \dot{\boldsymbol{x}}^{[\mathbf{c}]}(t) &=& \boldsymbol{\varphi}^{[\mathbf{c}]}(\boldsymbol{x}^{[\mathbf{c}]}(t), \boldsymbol{u}^{[\mathbf{c}]}(t)) \\ \boldsymbol{y}^{[\mathbf{c}]}(t) &=& \boldsymbol{\gamma}^{[\mathbf{c}]}(\boldsymbol{x}^{[\mathbf{c}]}(t)) \end{array}\right\}, \quad \left.\begin{array}{rcl} \dot{\boldsymbol{x}}^{[\mathbf{s}]}(t) &=& \boldsymbol{\varphi}^{[\mathbf{s}]}(\boldsymbol{x}^{[\mathbf{s}]}(t), \boldsymbol{u}^{[\mathbf{s}]}(t)) \\ \boldsymbol{y}^{[\mathbf{s}]}(t) &=& \boldsymbol{\gamma}^{[\mathbf{s}]}(\boldsymbol{x}^{[\mathbf{s}]}(t)) \end{array}\right\} \tag{48a}$$

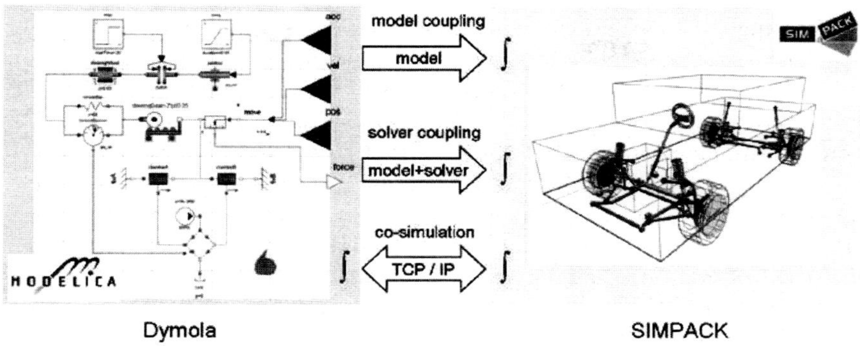

Figure 13. Coupling of simulation tools in the dynamical simulation of a car with servo-hydraulic steering system (Busch, 2007; Busch et al., 2007).

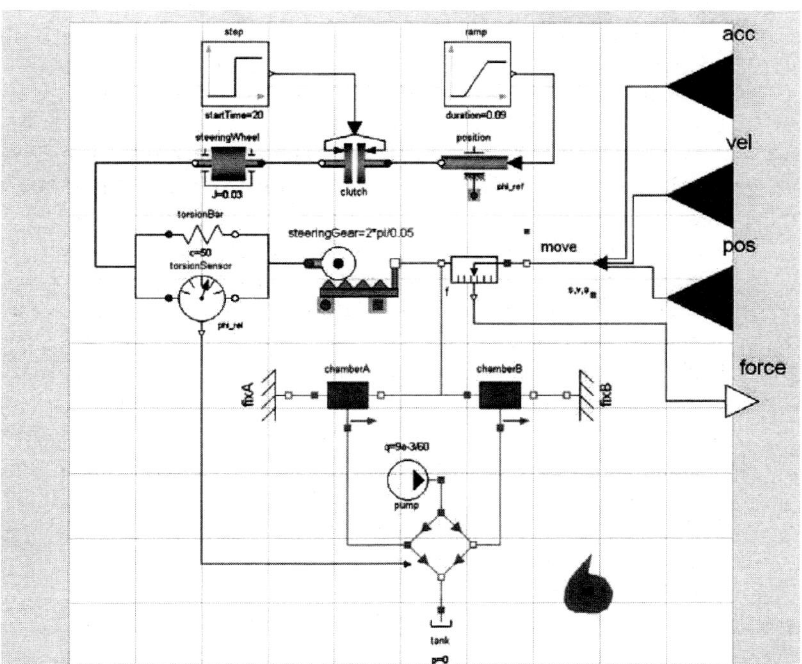

Figure 14. Block diagram for the Modelica model of a servo-hydraulic steering (Busch et al., 2007).

and the coupling conditions

$$\left.\begin{array}{rcll} \boldsymbol{u}^{[\mathrm{s}]}(t) & = & \boldsymbol{y}^{[\mathrm{c}]}(t)\,, & \text{(excitation of the rack)},\\ \boldsymbol{u}^{[\mathrm{c}]}(t) & = & \boldsymbol{y}^{[\mathrm{s}]}(t)\,, & \text{(actuator force)}. \end{array}\right\} \tag{48b}$$

Fig. 13 summarizes basic strategies for coupling mechanical and hydraulic components during simulation (Busch et al., 2007) that go back to the previous work of one of the co-authors who studied in his PhD thesis the simulation of tilting trains with hydraulic actuators (Dronka, 2004), see also (Liebig et al., 2001). For model coupling, see Fig. 13, the C code describing the hydraulic part $(\boldsymbol{.})^{[\mathrm{s}]}$ defines a user force element in SIMPACK. In this approach, SIMPACK is the master tool and the time integration of the overall system is performed by SIMPACK's default integrator SODASRT.

As an alternative, Dymola offers the joint export of C code describing the physical Modelica model with C code of an embedded time integration method which is to be selected out of several fixed stepsize methods (Elmqvist et al., 1995). In the master tool SIMPACK, the steering model with embedded solver is imported as user force element, see Fig. 13. In time integration, SIMPACK's SODASRT solver for the mechanical part $(\boldsymbol{.})^{[\mathrm{c}]}$ is combined with the embedded fixed stepsize solver of the hydraulic force element $(\boldsymbol{.})^{[\mathrm{s}]}$ (*solver coupling*). The solver coupling results in several non-trivial algorithmic problems (Busch, 2007), see also (Veitl, 2001; Dietz, 2005), but offers the opportunity to use larger time stepsizes $h_n^{[\mathrm{c}]}$ in the mechanical part if the (fixed) stepsize $h^{[\mathrm{s}]}$ in the hydraulic subsystem is restricted for stability or accuracy reasons (Busch, 2007), see also (Hippmann et al., 2005).

Since Dymola exports C code for *parametrized* Modelica models, system parameters of the steering model $(\boldsymbol{.})^{[\mathrm{s}]}$ may later be updated conveniently. Changes of the model structure would, however, require a new code export by Dymola for updating the user force element in SIMPACK in a model coupling or solver coupling approach. These nested model updates involve the risk of inconsistent models. Furthermore, the exported code does not consider future version updates of the exporting tool resulting in an additional risk of inconsistency. These limitations of model and solver coupling are avoided by a co-simulation that combines methods and solvers of both simulation tools during time integration using a TCP/IP interface for data exchange at the synchronization points, see Fig. 13.

The co-simulation between Dymola and SIMPACK proceeds in *macro steps* $T_j \to T_{j+1}$ of length $H := T_{j+1} - T_j$ with typical macro stepsizes in the range of $H = 1.0\,\mathrm{ms}$, see Fig. 15. At the synchronization point T_j, the subsystem outputs $\boldsymbol{y}^{[\mathrm{c}]}(T_j)$, $\boldsymbol{y}^{[\mathrm{s}]}(T_j)$ are exchanged between Dymola and

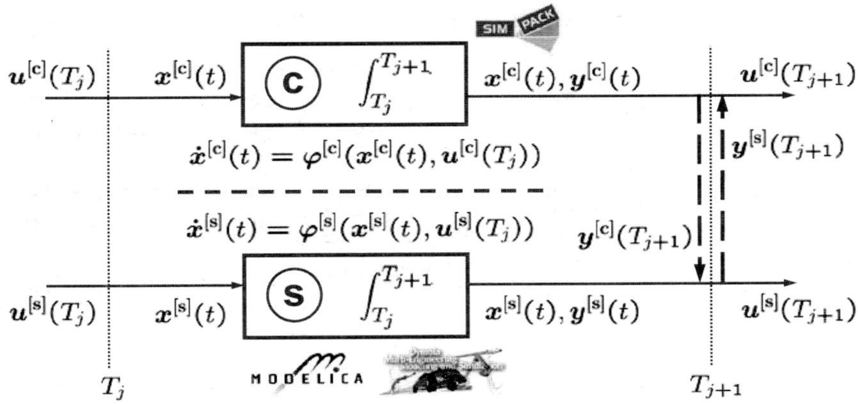

Figure 15. Schematic view of a macro step in a co-simulation coupling SIMPACK for subsystem $(.)^{[c]}$ (car) with Dymola for subsystem $(.)^{[s]}$ (steering).

SIMPACK. Approximating the physical coupling conditions (48b) in the macro step $T_j \to T_{j+1}$ by

$$u^{[s]}(t) := u^{[s]}(T_j) = y^{[c]}(T_j), \quad u^{[c]}(t) := u^{[c]}(T_j) = y^{[s]}(T_j), \tag{49}$$

($t \in [T_j, T_{j+1}]$), the subsystems car and steering are completely decoupled.

They are integrated independently from each other by SIMPACK's SODASRT solver (subsystem $(.)^{[c]}$) and by the DASSL solver in Dymola (subsystem $(.)^{[s]}$). There is no communication between both solvers before the next synchronization point T_{j+1} that defines the end of the current macro step. The co-simulation is continued with the next macro step $T_{j+1} \to T_{j+2}$ starting with a data exchange at $t = T_{j+1}$ and so on.

From the viewpoint of the multibody system tool, input variables from other simulation tools like $u^{[c]}(T_j) = y^{[s]}(T_j)$ in Fig. 15 may be considered as a special type of discrete variables r_j in (40): In the multibody system tool the values $u^{[c]}(T_j)$ of the coupling variables r_j are kept constant during the whole macro step $T_j \to T_{j+1}$. For the other simulation tool in the co-simulation environment (Dymola) the update at $t = T_{j+1}$, that is formally written as (41), involves the time integration of subsystem $(.)^{[s]}$ from synchronization point $t = T_j$ to synchronization point $t = T_{j+1}$ to get $y^{[s]}(T_{j+1}) = r_{j+1} = k(\ldots)$.

Modular time integration With an independent time integration of all subsystems in each macro step and frozen subsystem inputs $\boldsymbol{u}^{[\cdot]}(t) := \boldsymbol{u}^{[\cdot]}(T_j)$, $(\, t \in [T_j, T_{j+1}]\,)$, co-simulation may be extended straightforwardly to coupled problems that are composed of more than $r = 2$ subsystems. There is, however, a large potential for improving the performance of co-simulation algorithms by an optimized sequential treatment of the subsystems and by more sophisticated approximations for the system inputs $\boldsymbol{u}^{[\cdot]}(t)$, see (Kübler and Schiehlen, 2000). Improved co-simulation algorithms for typical applied problems have been studied by many authors, see, e.g., (Busch, 2007) for a detailed investigation of the Dymola–SIMPACK co-simulation for mechatronic systems like the servo-hydraulic steering in Fig. 13.

For a more general analysis of numerical aspects we use in the following a condensed notation of coupled problems like (48): In the right hand sides of (48a), the subsystem inputs $\boldsymbol{u}^{[c]}, \boldsymbol{u}^{[s]}$ are substituted by $\boldsymbol{y}^{[s]}, \boldsymbol{y}^{[c]}$, see (48b). Since $\boldsymbol{y}^{[s]}, \boldsymbol{y}^{[c]}$ may be expressed by $\boldsymbol{x}^{[s]}, \boldsymbol{x}^{[c]}$, see (48a), we get finally a coupled problem of the general form

$$
\begin{aligned}
\dot{\boldsymbol{x}}^{[1]}(t) &= \boldsymbol{\varphi}^{[1]}(\boldsymbol{x}^{[1]}(t), \boldsymbol{x}^{[2]}(t), \ldots, \boldsymbol{x}^{[r]}(t), t)\,, \\
&\;\;\vdots \\
\dot{\boldsymbol{x}}^{[r]}(t) &= \boldsymbol{\varphi}^{[r]}(\boldsymbol{x}^{[1]}(t), \boldsymbol{x}^{[2]}(t), \ldots, \boldsymbol{x}^{[r]}(t), t)
\end{aligned}
\tag{50}
$$

with $r \geq 2$ subsystems.

From the numerical viewpoint, the important difference between co-simulation and classical simulation techniques is the restriction of data exchange between subsystems to the synchronization points T_j that results in r independent time integration problems in each macro step:

$$
\dot{\boldsymbol{x}}^{[i]}(t) = \boldsymbol{\varphi}^{[i]}({}^{i}\tilde{\boldsymbol{x}}^{[1]}(t), \ldots, {}^{i}\tilde{\boldsymbol{x}}^{[i-1]}(t), \boldsymbol{x}^{[i]}(t), {}^{i}\tilde{\boldsymbol{x}}^{[i+1]}(t), \ldots, {}^{i}\tilde{\boldsymbol{x}}^{[r]}(t), t)\,,
\tag{51}
$$

$(\, t \in [T_j, T_{j+1}],\ i = 1, \ldots, r\,)$ with approximations

$$
{}^{i}\tilde{\boldsymbol{x}}^{[k]}(t) \approx \boldsymbol{x}^{[k]}(t)\,, \quad (\, k \neq i\,)\,,
$$

for the state vectors of other subsystems. The combination of r possibly different time integration methods for the r subsystems (51) is called a *modular* time integration method for the coupled problem (50). For more than 20 years, modular time integration methods have found much interest in the numerical analysis community because of their large potential for parallelization (Jackson, 1991).

For the decoupled time integration of the r subsystems (51), the coupling terms may simply be frozen in each macro step, see Fig. 15:

$$ {}^{i}\tilde{\boldsymbol{x}}^{[k]}(t) := \boldsymbol{x}^{[k]}(T_j)\,, \quad (t \in [T_j, T_{j+1}],\ k \neq i)\,. \tag{52}$$

This constant extrapolation of $\boldsymbol{x}^{[k]}$ from $t = T_j$ to the macro step $T_j \to T_{j+1}$ is very popular in practical applications. Eqs. (51) get the form

$$\dot{\boldsymbol{x}}^{[i]}(t) = \boldsymbol{\varphi}^{[i]}(\boldsymbol{x}^{[1]}(T_j), \dots, \boldsymbol{x}^{[i-1]}(T_j), \boldsymbol{x}^{[i]}(t), \boldsymbol{x}^{[i+1]}(T_j), \dots, \boldsymbol{x}^{[r]}(T_j), t)$$

and show in a comparison with (8) that constant extrapolation handles all coupling terms between subsystems similar to the explicit Euler method in classical time integration. Therefore, constant extrapolation shares all benefits and drawbacks of the classical explicit Euler method: it is robust and easy to implement and may be used successfully in applications with low accuracy requirements. On the other hand, constant extrapolation may result in a strong restriction of the macro stepsize H for accuracy reasons (low approximation order) and for stability reasons (stiff coupling terms).

The accuracy of modular time integration may be improved using higher order approximations ${}^{i}\tilde{\boldsymbol{x}}^{[k]}(t)$ in each macro step $T_j \to T_{j+1}$. If all r subsystems (51) are solved at the same time, the definition of ${}^{i}\tilde{\boldsymbol{x}}^{[k]}(t)$ has to be based on function values $\boldsymbol{x}^{[k]}(t)$ with $t \leq T_j$. Linear extrapolation uses $\boldsymbol{x}^{[k]}(T_{j-1})$ and $\boldsymbol{x}^{[k]}(T_j)$ to get for all $t \in [T_j, T_{j+1}]$, $k \neq i$:

$$ {}^{i}\tilde{\boldsymbol{x}}^{[k]}(t) := \boldsymbol{x}^{[k]}(T_j) + \frac{t - T_j}{T_j - T_{j-1}}(\boldsymbol{x}^{[k]}(T_j) - \boldsymbol{x}^{[k]}(T_{j-1}))\,. \tag{53}$$

In the same way, quadratic extrapolation of $\boldsymbol{x}^{[k]}(t)$ is based on $\boldsymbol{x}^{[k]}(T_{j-2})$, $\boldsymbol{x}^{[k]}(T_{j-1})$ and $\boldsymbol{x}^{[k]}(T_j)$ and so on.

In a sequential implementation, the r subsystems (51) are again solved independently from each other but now the macro steps $T_j \to T_{j+1}$ are already completed for all subsystems $(\boldsymbol{.})^{[k]}$ with $k < i$ before starting the time integration in subsystem $(\boldsymbol{.})^{[i]}$. Similar to a slowest first strategy in multirate time integration (Gear and Wells, 1984) the approximations ${}^{i}\tilde{\boldsymbol{x}}^{[k]}(t)$ with $k < i$ may therefore be defined by *interpolation* using $\boldsymbol{x}^{[k]}(T_{j+1})$ and the linear approximation gets the form

$$ {}^{i}\tilde{\boldsymbol{x}}^{[k]}(t) := \boldsymbol{x}^{[k]}(T_j) + \frac{t - T_j}{T_{j+1} - T_j}(\boldsymbol{x}^{[k]}(T_{j+1}) - \boldsymbol{x}^{[k]}(T_j))\,,\ (k < i)\,, \tag{54}$$

see also (53).

Such a sequential staggered algorithm is used successfully in the dynamical simulation of combustion engines with chain drives, see Fig. 16 and

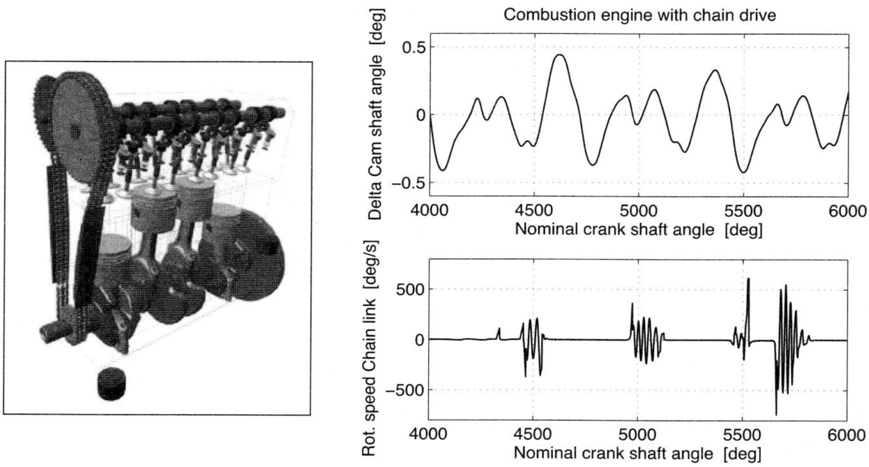

Figure 16. Dynamical simulation of a combustion engine with chain drive (Hippmann et al., 2005), cited from (Arnold, 2007).

(Hippmann et al., 2005). Very stiff contact forces between the chain links introduce high frequency oscillations in the subsystem "chain drive", see the lower right plot in Fig. 16 that shows the velocity of one of the chain links. The upper plot illustrates that the coupled system engine / chain drive is a multiscale problem with clearly separated time scales. The deviation Δ_c of the cam shaft angle from its nominal position is a typical state variable of the slow subsystem "engine". Both subsystems are coupled by contact forces F_c acting between gear wheels and chain links.

In (Hippmann et al., 2005), the coupled system is solved by a modular time integration method. In the first stage of each macro step, subsystem $(\boldsymbol{.})^{[1]}$ (engine) is solved by SIMPACK's SODASRT solver with time stepsizes $h_n^{[1]}$ of moderate size. The input from the chain drive (subsystem $(\boldsymbol{.})^{[2]}$) is approximated by a quadratic extrapolation ${}^{\mathbf{1}}\tilde{\boldsymbol{x}}^{[2]}(t)$ using the function values at $t = T_{j-2}, T_{j-1}, T_j$. If macro step $T_j \to T_{j+1}$ is completed for subsystem $(\boldsymbol{.})^{[1]}$, the simulation results $\boldsymbol{x}^{[1]}(T_j)$ and $\boldsymbol{x}^{[1]}(T_{j+1})$ are used to get an approximation ${}^{\mathbf{2}}\tilde{\boldsymbol{x}}^{[1]}(t) \approx \boldsymbol{x}^{[1]}(t)$ by linear interpolation (54).

In the second stage of the macro step, subsystem $(\boldsymbol{.})^{[2]}$ (chain drive) has to be integrated with very small time stepsizes $h_n^{[2]}$ because of the high frequency oscillations in the chain drive, see the lower right plot of Fig. 16. Here, an explicit Runge–Kutta method is used that avoids the large numerical effort for Jacobian evaluations in SODASRT. Compared with classical simulation techniques (SODASRT applied to the complete coupled system engine / chain drive), the numerical effort was reduced by a factor of 20 using the modular approach. For this application, macro stepsizes in the range of $H = 1.0\,\mu$s were found to be appropriate (Hippmann et al., 2005).

In industrial applications, it is state-of-the-art to use fixed macro stepsizes H in co-simulation to avoid problems with nested stepsize control algorithms in a combination of a global macro stepsize control with r local (micro) stepsize and order control algorithms for the r subsystems (51). A macro stepsize control based on an error estimate by Richardson extrapolation is proposed in (Kübler, 2000).

The accuracy of modular time integration may be studied along the lines of the convergence analysis for classical ODE time integration methods (Ascher and Petzold, 1998; Hairer et al., 1993). The asymptotic error analysis for $H \to 0$ results in a bound for the global error ϵ that depends on the global errors $\epsilon^{[i]}$, $(i = 1, \ldots, r)$, of the time integration methods in the r subsystems (51), on the macro stepsize H and on the approximation order of ${}^{i}\tilde{\boldsymbol{x}}^{[k]}(t)$:

$$\epsilon := \max_{\substack{0 \le n \le N \\ i=1,\ldots,r}} \|\boldsymbol{x}_n^{[i]} - \boldsymbol{x}^{[i]}(t_n)\| \le C_0(\epsilon^{[1]} + \ldots + \epsilon^{[r]}) + CH^{q+1} . \qquad (55)$$

Errors $\epsilon^{[i]}$ from the time integration in the subsystems may slightly be amplified by a factor $C_0 = \mathcal{O}(1)$. The main new contribution to the error of modular time integration methods results from the use of extrapolated or interpolated approximations ${}^{i}\tilde{\boldsymbol{x}}^{[k]}(t)$ in (51). The error constant $C > 0$ in (55) reflects the physical influence of the coupling terms on the overall system behaviour. If the subsystems are only weakly coupled, constant C may be very small, but in general we have $C = \mathcal{O}(1)$. The approximation order of ${}^{i}\tilde{\boldsymbol{x}}^{[k]}(t)$ is characterized by the exponent $q + 1$ in (55) with $q = 0$ for constant extrapolation (52), $q = 1$ for linear extrapolation (53) and linear interpolation (54), $q = 2$ for quadratic approximations and so on.

In (Arnold, 2007), the influence of the approximation order on the global error ϵ in (55) was studied for a simple test model that describes a mathematical pendulum being coupled to a damped oscillator with one horizontal degree of freedom, see Fig. 17. The system is composed of $r = 2$ subsystems that are both solved by the backward Euler method (14) with fixed (micro) stepsizes $h^{[1]} = h^{[2]} = h$ resulting in error terms $\epsilon^{[1]} = \mathcal{O}(h)$, $\epsilon^{[2]} = \mathcal{O}(h)$, see (55).

In Fig. 18, that shows the global error ϵ vs. macro stepsize H, the error terms $\epsilon^{[1]}$ and $\epsilon^{[2]}$ dominate for small macro stepsizes H. The minimum value of ϵ is reduced by a factor of 2 if the micro stepsize h is reduced

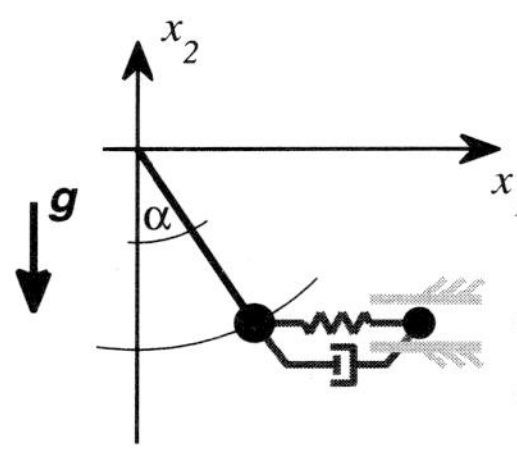

$$
\begin{aligned}
m_{\text{pend}}\, l\ddot{\alpha} &= -m_{\text{pend}}\, g \sin\alpha + \cos\alpha\, F_{\text{c}}(\alpha, \dot{\alpha}, x_1, \dot{x}_1) \\
m_{\text{osc}}\, \ddot{x}_1 &= -F_{\text{c}}(\alpha, \dot{\alpha}, x_1, \dot{x}_1) \\
F_{\text{c}} &= k\,(x_1 - l\sin\alpha) + d\,(\dot{x}_1 - l\dot{\alpha}\cos\alpha)
\end{aligned}
$$

$$
\begin{aligned}
&l = 1.0\,\text{m}\,,\ g = 9.81\,\text{m/s}^2,\ m_{\text{pend}} = m_{\text{osc}} = 1.0\,\text{kg}\,,\\
&\alpha(0) = 5^\circ,\ \ x_1(0) = 0.1\,\text{m}\ +\ 1.0\,\text{m} \times \sin\alpha(0)\,,\\
&\dot{\alpha}(0) = 0\,\text{rad/s}\,,\ \ \dot{x}_1(0) = 0\,\text{m/s}\,.
\end{aligned}
$$

Figure 17. Test problem: Mathematical pendulum coupled to an oscillator.

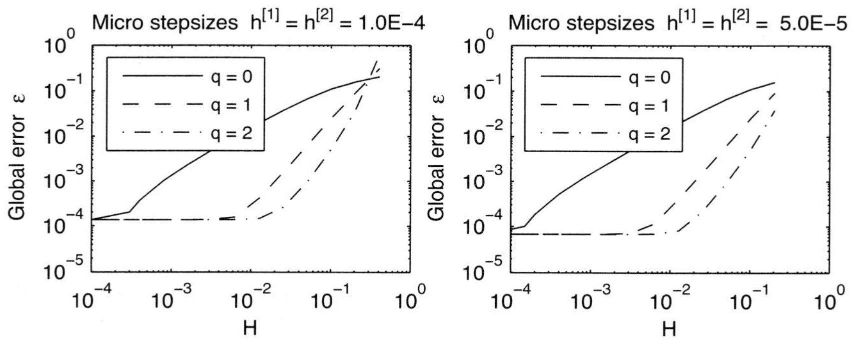

Figure 18. Error of modular time integration methods: The nonstiff case ($k = 10.0\,\text{N/m}$, $d = 1.0\,\text{Ns/m}$), $t \in [0\,\text{s}, 4\,\text{s}]$, see (Arnold, 2007).

from $h = 0.1\,\text{ms}$ (left plot) to $h = 0.05\,\text{ms}$ (right plot). For larger macro stepsizes, the error term CH^{q+1} that is introduced by the approximations ${}^{i}\tilde{\boldsymbol{x}}^{[k]}(t) \approx \boldsymbol{x}^{[k]}(t)$ in (51) dominates the overall global error ϵ. Quadratic ($q = 2$) and linear ($q = 1$) extrapolation are clearly superior to the classical constant extrapolation (52).

The simple test problem of Fig. 17 may be used as well to illustrate a typical stability problem of all classical co-simulation techniques. The use of extrapolated function values ${}^{i}\tilde{\boldsymbol{x}}^{[k]}(t)$ in (51) makes the modular time integration method in part to an explicit method. For coupled problems (50) with stiff coupling terms, the modular method behaves similar to an explicit Runge–Kutta or multistep method applied to a stiff ODE (2), see Section 3. For macro stepsizes H beyond a small stability bound, the modular time integration method gets unstable.

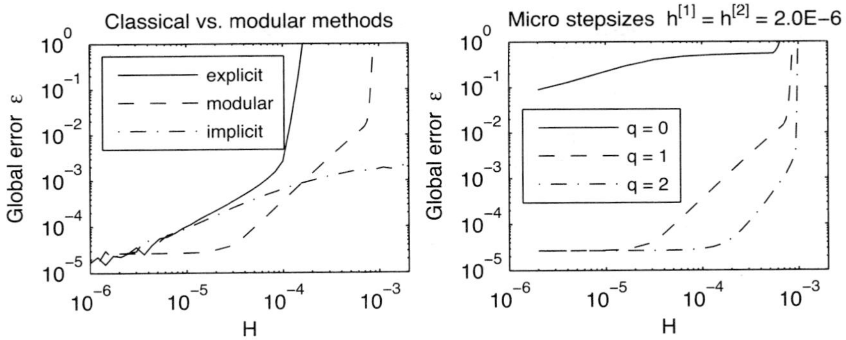

Figure 19. Error of modular time integration methods: The stiff case ($k = 10^6\,\mathrm{N/m}$, $d = 10^2\,\mathrm{Ns/m}$), $t \in [0.1\,\mathrm{s}, 0.2\,\mathrm{s}]$. Left plot: Modular time integration with $q = 1$ vs. explicit Euler method (8) and backward Euler method (14). Right plot: Modular time integration with extrapolation orders $q = 0$, $q = 1$ and $q = 2$, see (Arnold, 2007).

For the test problem of Fig. 17 with a stiff spring in the oscillator ($k = 10^6\,\mathrm{N/m}$, $d = 10^2\,\mathrm{Ns/m}$), the left plot of Fig. 19 shows a comparison between a classical ODE time integration of the overall system by the explicit and by the backward Euler method and a modular time integration combining a backward Euler method in both subsystems with a linear extrapolation of coupling terms according to (53). This modular time integration method was used successfully in the nonstiff case, see Fig. 18, but fails in the stiff case unless a stability condition $H \leq H^*$ is satisfied. The combination of stiff solvers in the subsystems and extrapolation of coupling terms makes the modular method (dashed line in the left plot of Fig. 19) more robust than the explicit Euler method (8) applied to the overall system (solid line). The comparison with the dash dotted line, that corresponds to backward Euler method (14) applied to the overall system, shows, however, that the excellent stability of stiff ODE solvers is lost in the modular approach because of the (explicit) extrapolation steps.

The right plot of Fig. 19 shows that the stability problem may be observed for any extrapolation order but the stability bound H^* depends on q. For this test problem, the results for constant extrapolation are completely useless. A detailed stability analysis for modular time integration methods is technically very complicated since it has to take into account several types of stiff coupling terms and different extrapolation and interpolation methods for the definition of ${}^{i}\tilde{\boldsymbol{x}}^{[k]}(t)$ in (51), see (Skelboe and Andersen,

1989) for a discussion of multirate methods for stiff ODEs. In practice, the stability bound H^* is determined by some numerical tests and a macro stepsize $H := \alpha_s H^*$ with a safety factor $\alpha_s < 1$ is used since there is even today no other general strategy to guarantee the stability of modular time integration methods in the stiff case.

The stability of modular time integration methods for coupled DAEs cannot be achieved in such a straightforward way since a stepsize independent contractivity condition has to be satisfied, see (Deuflhard et al., 1987) for a closely related stability analysis of half-explicit DAE time integration methods. To avoid these stability problems, nonlinear projection steps (Kübler and Schiehlen, 2000), linear stabilization steps based on Jacobians of coupling terms (Arnold and Günther, 2001) and an "ideal gluing algorithm" for the coupling of time integration methods (Tseng and Hulbert, 1999) were proposed in the literature. All these methods were successfully used in nontrivial case studies from applied dynamics but the development of an easy-to-use, robust and efficient general purpose method for the numerically stable co-simulation of coupled DAEs is still a topic of ongoing research.

Bibliography

M. Arnold. A perturbation analysis for the dynamical simulation of mechanical multibody systems. *Applied Numerical Mathematics*, 18:37–56, 1995.

M. Arnold. Numerical problems in the dynamical simulation of wheel-rail systems. *Z. Angew. Math. Mech.*, Proceedings of ICIAM 95, Issue 3: 151–154, 1996.

M. Arnold. Multi-rate time integration for large scale multibody system models. In P. Eberhard, editor, *IUTAM Symposium on Multiscale Problems in Multibody System Contacts*, pages 1–10. Springer, 2007.

M. Arnold and M. Günther. Preconditioned dynamic iteration for coupled differential-algebraic systems. *BIT Numerical Mathematics*, 41:1–25, 2001.

M. Arnold and H. Netter. Wear profiles and the dynamical simulation of wheel-rail systems. In M. Brøns, M.P. Bendsøe, and M.P. Sørensen, editors, *Progress in Industrial Mathematics at ECMI 96*, pages 77–84. Teubner, Stuttgart, 1997.

M. Arnold, V. Mehrmann, and A. Steinbrecher. Index reduction in industrial multibody system simulation. Technical Report IB 532–01–01, DLR German Aerospace Center, Institute of Aeroelasticity, Vehicle System Dynamics Group, 2001.

M. Arnold, A. Fuchs, and C. Führer. Efficient corrector iteration for DAE time integration in multibody dynamics. *Comp. Meth. Appl. Mech. Eng.*, 195:6958–6973, 2006.

M. Arnold, B. Burgermeister, and A. Eichberger. Linearly implicit time integration methods in real-time applications: DAEs and stiff ODEs. *Multibody System Dynamics*, 17:99–117, 2007a.

M. Arnold, Ch. Weidemann, and L. Mauer. Stepsize control versus fixed stepsize time integration: Theory and practical experience. In C.L. Bottasso, P. Masarati, and L. Trainelli, editors, *Proc. of Multibody Dynamics 2007 (ECCOMAS Thematic Conference)*, Milan, Italy, 2007b.

U. Ascher and L.R. Petzold. *Computer Methods for Ordinary Differential Equations and Differential–Algebraic Equations.* SIAM, Philadelphia, 1998.

U.M. Ascher, H. Chin, and S. Reich. Stabilization of DAEs and invariant manifolds. *Numer. Math.*, 67:131–149, 1994.

J. Baumgarte. Stabilization of constraints and integrals of motion in dynamical systems. *Computer Methods in Applied Mechanics and Engineering*, 1:1–16, 1972.

H. Brandl, R. Johanni, and M. Otter. A very efficient algorithm for the simulation of robots and similar multibody systems without inversion of the mass matrix. In P. Kopacek, I. Troch, and K. Desoyer, editors, *Theory of Robots*, pages 95–100, Oxford, 1988. Pergamon Press.

P.C. Breedveld. Port based modelling of multidomain physical systems in terms of bond graphs. – *Published in this volume*, 2008.

K.E. Brenan, S.L. Campbell, and L.R. Petzold. *Numerical solution of initial–value problems in differential–algebraic equations.* SIAM, Philadelphia, 2nd edition, 1996.

O. Brüls, A. Cardona, and M. Géradin. Modelling, simulation and control of flexible multibody systems. – *Published in this volume*, 2008.

M. Busch. Entwicklung einer SIMPACK–Modelica/Dymola–Schnittstelle. Diploma Thesis, Martin Luther University Halle-Wittenberg, Institute of Computer Science, 2007.

M. Busch, M. Arnold, A. Heckmann, and S. Dronka. Interfacing SIMPACK to Modelica / Dymola for multi-domain vehicle system simulations. *SIMPACK News (INTEC GmbH, Wessling, Germany)*, 11(2):01–03, 2007.

T.F. Coleman, B.S. Garbow, and J.J. Moré. Software for estimating sparse Jacobian matrices. *ACM Transactions on Mathematical Software*, 10: 329–345, 1984.

P. Deuflhard, E. Hairer, and J. Zugck. One–step and extrapolation methods for differential–algebraic systems. *Numer. Math.*, 51:501–516, 1987.

S. Dietz. *Vibration and Fatigue Analysis of Vehicle Systems using Component Modes.* Fortschritt-Berichte VDI Reihe 12, Nr. 401. VDI–Verlag, Düsseldorf, 1999.

S. Dietz. The new powerful linear subsystem solver for flexible bodies in multi-body systems. In J.M. Goicolea, J. Cuadrado, and J.C. García Orden, editors, *Proc. of Multibody Dynamics 2005 (ECCOMAS Thematic Conference)*, Madrid, Spain, 2005.

S. Dietz, H. Netter, and D. Sachau. Fatigue life prediction by coupling finite-element and multibody systems calculations. In *Proceedings of DETC'97, ASME Design Engineering Technical Conferences DETC/VIB-4229*, Sacramento, California, 1997.

J.R. Dormand and P.J. Prince. A family of embedded Runge–Kutta formulae. *J. Comp. Appl. Math.*, 6:19–26, 1980.

S. Dronka. *Die Simulation gekoppelter Mehrkörper- und Hydraulik–Modelle mit Erweiterung für Echtzeitsimulation.* Shaker Verlag, Aachen, 2004.

E. Eich. Convergence results for a coordinate projection method applied to mechanical systems with algebraic constraints. *SIAM J. Numer. Anal.*, 30:1467–1482, 1993.

E. Eich-Soellner and C. Führer. *Numerical Methods in Multibody Dynamics.* Teubner–Verlag, Stuttgart, 1998.

A. Eichberger. *Simulation von Mehrkörpersystemen auf parallelen Rechnerarchitekturen.* Fortschritt-Berichte VDI Reihe 8, Nr. 332. VDI–Verlag, Düsseldorf, 1993.

H. Elmqvist, M. Otter, and F.E. Cellier. Inline integration: A new mixed symbolic / numeric approach for solving differential–algebraic equation systems. In *Proc. ESM'95, European Simulation Multiconference, Prague, Czech Republic, June 5–8, 1995*, pages xxiii–xxxiv, 1995.

C. Führer. Differential-algebraische Gleichungssysteme in mechanischen Mehrkörpersystemen. Theorie, numerische Ansätze und Anwendungen. Technical report, TU München, Mathematisches Institut und Institut für Informatik, 1988.

C. Führer and B. Leimkuhler. Numerical solution of differential–algebraic equations for constrained mechanical motion. *Numer. Math.*, 59:55–69, 1991.

C.W. Gear and R.R. Wells. Multirate linear multistep methods. *BIT*, 24: 484–502, 1984.

C.W. Gear, B. Leimkuhler, and G.K. Gupta. Automatic integration of Euler–Lagrange equations with constraints. *J. Comp. Appl. Math.*, 12&13:77–90, 1985.

E. Hairer and G. Wanner. *Solving Ordinary Differential Equations. II. Stiff and Differential-Algebraic Problems.* Springer–Verlag, Berlin Heidelberg New York, 2nd edition, 1996.

E. Hairer, S.P. Nørsett, and G. Wanner. *Solving Ordinary Differential Equations. I. Nonstiff Problems.* Springer–Verlag, Berlin Heidelberg New York, 2nd edition, 1993.

A. Heckmann, M. Arnold, and O. Vaculín. A modal multifield approach for an extended flexible body description in multibody dynamics. *Multibody System Dynamics*, 13:299–322, 2005.

G. Hippmann, M. Arnold, and M. Schittenhelm. Efficient simulation of bush and roller chain drives. In J.M. Goicolea, J. Cuadrado, and J.C. García Orden, editors, *Proc. of Multibody Dynamics 2005 (ECCOMAS Thematic Conference)*, Madrid, Spain, 2005.

S. Iwnicki, editor. *The Manchester Benchmarks for Rail Vehicle Simulation.* Supplement to Vehicle System Dynamics, Vol. 31, Swets & Zeitlinger, Lisse, 1999.

K. Jackson. A survey of parallel numerical methods for initial value problems for ordinary differential equations. *IEEE Transactions on Magnetics*, 27:3792–3797, 1991.

J.J. Kalker. *Three-Dimensional Elastic Bodies in Rolling Contact.* Kluwer Academic Publishers, Dordrecht Boston London, 1990.

C.T. Kelley. *Iterative Methods for Optimization.* SIAM, Philadelphia, 1999.

C.T. Kelley. *Solving Nonlinear Equations with Newton's Method.* SIAM, Philadelphia, 2003.

W. Kortüm and P. Lugner. *Systemdynamik und Regelung von Fahrzeugen.* Springer–Verlag, Berlin Heidelberg New York, 1994.

W. Kortüm and W. Schiehlen. General purpose vehicle system dynamics software based on multibody formalisms. *Vehicle System Dynamics*, 14: 229–263, 1985.

W. Kortüm, W.O. Schiehlen, and M. Arnold. Software tools: From multibody system analysis to vehicle system dynamics. In H. Aref and J.W. Phillips, editors, *Mechanics for a New Millennium*, pages 225–238, Dordrecht, 2001. Kluwer Academic Publishers.

W.R. Krüger and M. Spieck. Aeroelastic effects in multibody dynamics. *Vehicle System Dynamics*, 41:383–399, 2004.

R. Kübler. *Modulare Modellierung und Simulation mechatronischer Systeme.* Fortschritt-Berichte VDI Reihe 20, Nr. 327. VDI–Verlag GmbH, Düsseldorf, 2000.

R. Kübler and W. Schiehlen. Two methods of simulator coupling. *Mathematical and Computer Modelling of Dynamical Systems*, 6:93–113, 2000.

P. Kunkel, V. Mehrmann, W. Rath, and J. Weickert. GELDA: A software package for the solution of general linear differential algebraic equations. *SIAM J. Sci. Comp.*, 18:115–138, 1997.

S. Liebig, S. Helduser, M. Stüwing, and S. Dronka. Die Modellierung und Simulation gekoppelter mechanischer und hydraulischer Systeme (`http://vwitme011.vkw.tu-dresden.de/TrafficForum/vwt_2001/beitraege/VWT18proceedings_pages505-524.pdf`). In *Proc. of the 18th Dresden Conference on Traffic and Transportation Sciences, September 17-18, 2001*, pages 505–524, 2001.

Ch. Lubich, Ch. Engstler, U. Nowak, and U. Pöhle. Numerical integration of constrained mechanical systems using MEXX. *Mech. Struct. Mach.*, 23:473–495, 1995.

P. Lugner and M. Plöchl, editors. *Tyre Models for Vehicle Dynamics Analysis.* Supplement 1 to Vol. 43 of *Vehicle System Dynamics*, 2005.

C.B. Moler. *Numerical Computing with MATLAB.* SIAM, Philadelphia, 2004.

J.J. Moré, D.C. Sorensen, B.S. Garbow, and K.E. Hillstrom. The MINPACK project. In W.J. Cowell, editor, *Sources and Development of Mathematical Software*, pages 88–111. Prentice–Hall, Englewood Cliffs, N.J., 1984.

N. Orlandea. *Development and Application of Node–Analogous Sparsity–Oriented Methods for Simulation of Mechanical Dynamic Systems.* PhD thesis, University of Michigan, 1973.

L.R. Petzold. Differential/algebraic equations are not ODEs. *SIAM J. Sci. Stat. Comput.*, 3:367–384, 1982.

L.R. Petzold and P. Lötstedt. Numerical solution of nonlinear differential equations with algebraic constraints II: practical implications. *SIAM J. Sci. Stat. Comput.*, 7:720–733, 1986.

A. Pfeiffer. *Numerische Sensitivitätsanalyse unstetiger multidisziplinärer Modelle mit Anwendungen in der gradientenbasierten Optimierung.* Fortschritt-Berichte VDI Reihe 20, Nr. 417. VDI–Verlag, Düsseldorf, 2008.

F. Pfeiffer and Ch. Glocker. *Multibody Dynamics with Unilateral Contacts.* Wiley & Sons, New York, 1996.

G. Rill. *Simulation von Kraftfahrzeugen.* Fundamentals and Advances in the Engineering Sciences. Vieweg, Braunschweig Wiesbaden, 1994.

R.E. Roberson and R. Schwertassek. *Dynamics of Multibody Systems.* Springer–Verlag, Berlin Heidelberg New York, 1988.

W. Rulka. *Effiziente Simulation der Dynamik mechatronischer Systeme für industrielle Anwendungen.* PhD thesis, Vienna University of Technology, Department of Mechanical Engineering, 1998.

W. Schiehlen. Multibody system dynamics: roots and perspectives. *Multibody System Dynamics*, 1:149–188, 1997.

W. Schiehlen and P. Eberhard. Multibody systems and applied dynamics. – *Published in this volume*, 2008.

R. Schwertassek and O. Wallrapp. *Dynamik flexibler Mehrkörpersysteme.* Vieweg, 1999.

A.A. Shabana. *Dynamics of Multibody Systems.* Cambridge University Press, Cambridge, 2nd edition, 1998.

A.A. Shabana. *Computational Dynamics.* John Wiley & Sons, Inc., New York, 2nd edition, 2001.

B. Simeon. *Numerische Simulation gekoppelter Systeme von partiellen und differential-algebraischen Gleichungen in der Mehrkörperdynamik.* Fortschritt-Berichte VDI Reihe 20, Nr. 325. VDI–Verlag, Düsseldorf, 2000.

B. Simeon, C. Führer, and P. Rentrop. Differential-algebraic equations in vehicle system dynamics. *Surveys on Mathematics for Industry*, 1:1–37, 1991.

S. Skelboe and P.U. Andersen. Stability properties of backward Euler multirate formulas. *SIAM J. Sci. Stat. Comp.*, 10:1000–1009, 1989.

M.M. Tiller. *Introduction to Physical Modeling with Modelica.* Kluwer Academic Publishers, Boston Dordrecht London, 2001.

W. Trautenberg. *Bidirektionale Kopplung zwischen CAD und Mehrkörpersimulationssystemen.* PhD thesis, Munich University of Technology, Department of Mechanical Engineering, 1999.

F.C. Tseng and G.M. Hulbert. Network–distributed multibody dynamics simulation — gluing algorithm. In J.A.C. Ambrósio and W.O. Schiehlen, editors, *Advances in Computational Multibody Dynamics*, pages 521–540, IDMEC/IST Lisbon, Portugal, 1999.

O. Vaculín, M. Valášek, and W.R. Krüger. Overview of coupling of multibody and control engineering tools. *Vehicle System Dynamics*, 41:415–429, 2004.

M. Valášek. Modelling, simulation and control of mechatronical systems. – *Published in this volume*, 2008.

M. Valášek, Z. Šika, and O. Vaculín. Multibody formalism for real-time application using natural coordinates and modified state space. *Multibody System Dynamics*, 17:209–227, 2007.

A. Veitl. *Integrierter Entwurf innovativer Stromabnehmer.* Fortschritt-Berichte VDI Reihe 12, Nr. 449. VDI–Verlag, Düsseldorf, 2001.

A. Veitl, T. Gordon, A. van de Sand, M. Howell, M. Valášek, O. Vaculín, and P. Steinbauer. Methodologies for coupling simulation models and codes in mechatronic system analysis and design. In *Proc. of the 16th IAVSD–Symposium on Dynamics of Vehicles on Roads and Tracks*, pages 231–243. Supplement to Vehicle System Dynamics, Vol. 33, Swets & Zeitlinger, 1999.

R. von Schwerin. *MultiBody System SIMulation – Numerical Methods, Algorithms, and Software*, volume 7 of *Lecture Notes in Computational Science and Engineering*. Springer, Berlin Heidelberg, 1999.

O. Wallrapp. Linearized flexible multibody dynamics including geometric stiffening effects. *Mechanics of Structures and Machines*, 19:385–409, 1991.

O. Wallrapp. Standardization of flexible body modeling in multibody system codes, Part I: Definition of standard input data. *Mechanics of Structures and Machines*, 22:283–304, 1994.

R.A. Wehage and E.J. Haug. Generalized coordinate partitioning for dimension reduction in analysis of constrained dynamic systems. *J. Mech. Design*, 104:247–255, 1982.

Ch. Weidemann, L. Mauer, and M. Arnold. Improving the calculation speed for time domain integration of complex railway vehicles. In *Proceedings of ACMD2006, The Third Asian Conference on Multibody Dynamics, August 1-4, 2006*, Komaba, Tokyo, Japan, 2006.

Real-Time Multibody Dynamics and Applications

Javier Cuadrado, Daniel Dopico, Miguel A. Naya, and M. Gonzalez

Dep. de Ingenieria Industrial II, University of La Coruña, Ferrol, Spain

Abstract The simulation of multibody dynamics in real-time is becoming more and more relevant, since it is required by the growing and technologically advanced hardware- and human-in-the-loop applications. Both the efficient calculation of multibody dynamics and the adequate consideration of mechanical phenomena, like flexibility or contact, are essential for such applications. Accordingly, this contribution presents the following contents: a) two variants of a real-time formalism, capable of performing very fast calculations of the dynamics of complex multibody systems; b) the experimental validation of the formalism in the case of a prototype car; c) the use of the formalism for the design of vehicle controllers through a human-in-the-loop application. As conclusion, the work provides a reliable tool for the real-time simulation of multibody dynamics, which can be used to develop practical industrial applications.

1 Introduction

The simulation of multibody dynamics in real-time is becoming more and more relevant, since it is required by the growing and technologically advanced hardware- and human-in-the-loop applications. An illustrative example of hardware-in-the-loop application is the testing and tuning of automotive Electronics Control Units (ECU), which nowadays is carried out against a computer model of the vehicle rather than against its physical counterpart. Examples of human-in-the-loop applications are the different types of simulators (cars, planes, cranes, excavators, etc.) used for personnel training. Furthermore, the emerging field of Virtual Reality (VR), which possesses a huge potential of industrial applicability, can also lead to mixed human- and hardware-in-the-loop applications, as long as it introduces the user to an immersive and interactive physic-based simulation through hardware interfaces like sensoring devices and haptics.

In order to face the abovementioned challenging applications, computational methods for the dynamics of multibody systems are demanded which encompass the two following characteristics:
a) Efficient calculation, affected by all the problem stages: modeling, coordinates selection, formulation of the equations of motion, numerical integration, and implementation.
b) Adequate consideration of mechanical phenomena, like flexibility, contact, impact, friction, damping, etc., since more and more detailed models are demanded.

According to what has been exposed, the following contents are presented in this work:
a) Two variants of a method for real-time simulation of multibody dynamics, one in absolute coordinates and another in relative coordinates, whose efficiency is compared through several examples. Some numerical integrators are proposed for the method. Finally, criteria of use of both the two variants and the different integrators are provided.
b) The experimental validation of the formalism through comparison of calculated and measured stresses on a prototype car, so as to verify that the method provides accurate results, and can deal with both rigid and flexible links.
c) The use of the method for the design of vehicle controllers through a human-in-the-loop application, in order to show the aptitude of the formalism for addressing an industrially meaningful application of real-time multibody dynamics.

2 Real-Time Method

During the last years, the authors have worked on efficient formulations for the dynamics of multibody systems. The result has been an index-3 augmented Lagrangian formulation with projections of velocities and accelerations, which features the trapezoidal rule (or any other Newmark-type integrator) as numerical integrator. The method has shown to be robust and accurate, successfully facing singular configurations, changing topologies and stiff systems, as well as efficient, achieving real-time performance on a conventional PC when simulating detailed models of complex multibody systems, like cars or excavators.

Commonly, real-time methods use non-iterative schemes provided by explicit or semi-implicit integrators, and seek to perform the function evaluation (calculation of the accelerations for a given time and given positions and velocities) as fast as possible. However, the proposed formalism features an iterative scheme, obtained from the combination of the equations of motion and the equations of an implicit integrator; the resulting nonlinear system of equations is solved by means of the Newton-Raphson procedure, leading to a highly stable and robust algorithm. Two variants of the described formulation have been developed.

In the first variant, known as method in absolute coordinates, the multibody system is described by means of natural or fully-Cartesian coordinates, the kinematics of the system are established through constraint equations, and the force and inertia terms are built by superposition of the corresponding terms at body level.

The second variant, known as method in relative coordinates, requires that, first of all, the spanning tree version of the multibody system is obtained by means of the cut joint method, so that closed loops are opened. Then, the resulting open loop mechanism is described by a set of relative coordinates, but a set of Cartesian body coordinates is also defined. The relation between both sets of coordinates is established at velocity level. The kinematics of the problem is addressed by what may be called "classical relations", coming from the fundamental mechanics. The force and inertia terms are built recursively.

In what follows, the two variants of the formalism will be described in detail. Then, some examples will be analyzed in order to show the performance of the method and compare the two variants. Later, alternative structural integrators to the basic trapezoidal rule will be considered, and the performance of the two variants when combined with such integrators will be examined again. Finally, criteria of use of both the two variants and the different integrators are provided.

2.1 The Method in Absolute Coordinates

As indicated above, this method employs natural or fully-Cartesian coordinates to describe the multibody system. In general, the number of natural coordinates required to define a mechanism is lower than the number of reference point coordinates needed to model the same mechanism, which means that the former modeling technique produces a smaller problem size and, therefore, is advantageous from the efficiency point of view. Natural coordinates were introduced in 1982 by Garcia de Jalon; a detailed explanation about their use can be found in Garcia de Jalon and Bayo (1994). However, in order to provide the reader with an idea about these coordinates, an example is shown here.

Natural coordinates are the Cartesian coordinates of some points of the mechanism, and the Cartesian components of some unit vectors (axes) of the mechanism. In Fig. 1, fixed points and unit vectors are represented in

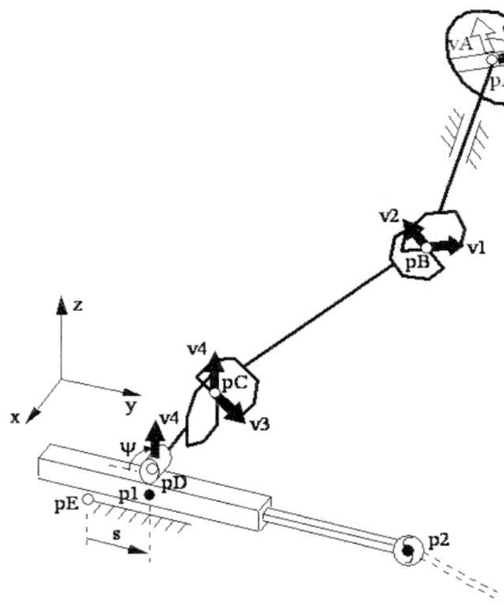

Figure 1. A steering mechanism modeled in natural coordinates.

white, while mobile points and unit vectors are represented in black. Only the mobile entities are problem variables. Sometimes, as in the example, it is helpful to include some relative coordinates (angles or distances) in the set of variables. In this case, the modeling is said to be done in mixed coordinates.

In the example, the mobile entities are two points (**p1** and **p2**) and four unit vectors (**v1**, **v2**, **v3** and **v4**), although only the y-coordinate of points **p1** and **p2** is considered as variable (x- and z-coordinate of these two points are constant during the motion of the mechanism). Moreover, angles φ and ψ and distance s, have been included as problem variables. Therefore, the set of problem variables is

$$\begin{aligned}\mathbf{q}^{\mathrm{T}} = \{ & \; y_1 \quad y_2 \quad v_{1x} \quad v_{1y} \quad v_{1z} \quad v_{2x} \quad v_{2y} \quad v_{2z} \\ & v_{3x} \quad v_{3y} \quad v_{3z} \quad v_{4x} \quad v_{4y} \quad v_{4z} \quad \varphi \quad \psi \quad s \; \}\end{aligned} \tag{1}$$

whose size is 17. Since the mechanism has got one single degree of freedom, 16 constraint equations must relate the variables. They are the following:

$$(x_B - x_A)v_{1x} + (y_B - y_A)v_{1y} + (z_B - z_A)v_{1z} = 0 \tag{2a}$$

$$v_{1x}^2 + v_{1y}^2 + v_{1z}^2 - 1 = 0 \tag{2b}$$

$$v_{1x}v_{2x} + v_{1y}v_{2y} + v_{1z}v_{2z} = 0 \tag{2c}$$

$$v_{2x}^2 + v_{2y}^2 + v_{2z}^2 - 1 = 0 \tag{2d}$$

$$v_{3x}^2 + v_{3y}^2 + v_{3z}^2 - 1 = 0 \tag{2e}$$

$$(x_C - x_B)v_{2x} + (y_C - y_B)v_{2y} + (z_C - z_B)v_{2z} = 0 \tag{2f}$$

$$(x_C - x_B)v_{3x} + (y_C - y_B)v_{3y} + (z_C - z_B)v_{3z} = 0 \tag{2g}$$

$$v_{2x}v_{3x} + v_{2y}v_{3y} + v_{2z}v_{3z} = 0 \tag{2h}$$

$$v_{3x}v_{4x} + v_{3y}v_{4y} + v_{3z}v_{4z} = 0 \tag{2i}$$

$$v_{4x}^2 + v_{4y}^2 + v_{4z}^2 - 1 = 0 \tag{2j}$$

$$(x_D - x_C)v_{4x} + (y_D - y_C)v_{4y} + (z_D - z_C)v_{4z} = 0 \tag{2k}$$

$$v_{4y} + \cos\psi = 0 \tag{2l}$$

$$(x_1 - x_E)^2 + (y_1 - y_E)^2 + (z_1 - z_E)^2 - s^2 = 0 \tag{2m}$$

$$s + R\psi - c = 0 \tag{2n}$$

$$(x_2 - x_1)^2 + (y_2 - y_1)^2 + (z_2 - z_1)^2 - L_{12}^2 = 0 \tag{2o}$$

$$v_{1x}v_{Ax} + v_{1y}v_{Ay} + v_{1z}v_{Az} - \cos\varphi = 0 \tag{2p}$$

Equations (2a) and (2b) impose the rigid character of the first body of the kinematic chain (the one with the steering wheel). Equation (2c) corresponds to the first universal joint. Equations (2d) to (2h) establish the rigid body conditions for the next body in the chain. Equation (2i) corresponds to the second universal joint. Equations (2j) and (2k) are the rigid body conditions of the body containing the pinion. Equation (2l) defines the angle rotated by the pinion, Eq. (2m) defines the distance at the rack, and Eq. (2n) provides the relation between both relative coordinates, thus serving to define the rack and pinion pair. Equation (2o) is a constant distance condition between two points of the rack. Finally, Eq. (2p) serves to define the angle rotated by the steering wheel.

Summarizing, the constraint equations among natural coordinates arise from three sources: rigid body conditions, joint conditions (not always necessary), and relative coordinates (in case they are introduced).

When using absolute coordinates, the kinematics of the mechanism is easily stated through the constraint equations and their time derivatives. The position problem consists of getting the position vector $\mathbf{q}$ for a known value of the variables representing the degrees of freedom of the mechanism. If the set of constraints is denoted by $\mathbf{\Phi}$, the position problem is written as

$$\mathbf{\Phi}(\mathbf{q}) = \mathbf{0} \tag{3}$$

Since this system of equations is usually nonlinear, it is linearized with respect to a certain guess value of the positions, $\mathbf{q}_i$, thus leading to

$$\mathbf{\Phi}(\mathbf{q}) \simeq \mathbf{\Phi}(\mathbf{q}_i) + \mathbf{\Phi_q}(\mathbf{q}_i)(\mathbf{q} - \mathbf{q}_i) = \mathbf{0} \tag{4}$$

so that the following linear system should be solved iteratively until Eq. (3) is satisfied within a prescribed tolerance (Newton-Raphson iteration),

$$\mathbf{\Phi_q}(\mathbf{q}_i)(\mathbf{q}_{i+1} - \mathbf{q}_i) = -\mathbf{\Phi}(\mathbf{q}_i) \tag{5}$$

The velocity problem consists of obtaining the velocity vector $\dot{\mathbf{q}}$ for a known value of the first time derivative of the variables representing the degrees of freedom of the mechanism. Differentiating Eq. (3) with respect to the time gives,

$$\mathbf{\Phi_q}(\mathbf{q})\dot{\mathbf{q}} = \mathbf{0} \tag{6}$$

which is a linear system in the velocities.

The acceleration problem consists of obtaining the acceleration vector $\ddot{\mathbf{q}}$ for a known value of the second time derivative of the variables representing the degrees of freedom of the mechanism. Differentiating Eq. (6) with respect to the time yields,

$$\mathbf{\Phi_q}(\mathbf{q})\ddot{\mathbf{q}} = -\dot{\mathbf{\Phi}}_\mathbf{q}(\mathbf{q}, \dot{\mathbf{q}})\dot{\mathbf{q}} \tag{7}$$

which is a linear system in the accelerations, whose leading matrix is the same already evaluated for the velocity problem.

Going now to the dynamic terms, mass matrix and force vector, they are obtained by superposition of the corresponding terms at body level. A generic three-dimensional body is considered in Fig. 2, modeled in natural coordinates by means of one point and three non-coplanar vectors (this group of entities will be called the basic set). By the moment, it is not required that the vectors have unit norm.

The position of any material point of the body can be expressed as function of the positions of the point and the three vectors in the form,

$$\bar{\mathbf{r}} = \bar{\mathbf{r}}_1 + c_1\bar{\mathbf{u}} + c_2\bar{\mathbf{v}} + c_3\bar{\mathbf{w}} = \bar{\mathbf{r}}_1 + \begin{bmatrix} \bar{\mathbf{u}} & \bar{\mathbf{v}} & \bar{\mathbf{w}} \end{bmatrix}\mathbf{c} = \bar{\mathbf{r}}_1 + \bar{\mathbf{X}}\mathbf{c} \tag{8}$$

where the coefficients c_1, c_2 and c_3 remain constant during the motion, due to the rigid character of the body.

Differentiating Eq. (8) with respect to time provides the velocity of any material point of the body and, then, the kinetic energy of the body can be obtained. By doing so, the following expression results for the mass matrix

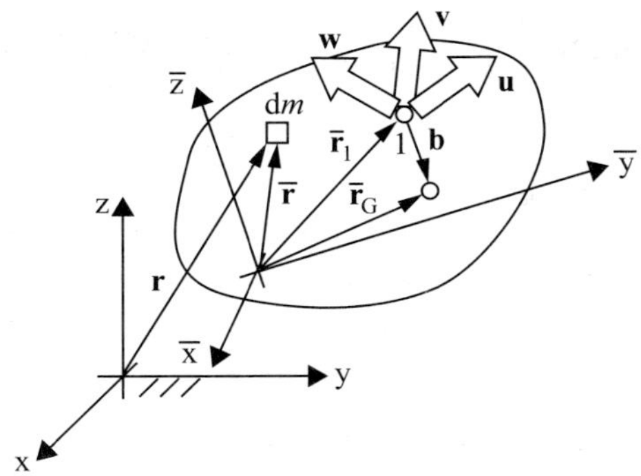

Figure 2. General three-dimensional body modeled in natural coordinates.

of the body,

$$\mathbf{M}=\left[\begin{array}{cc} m & m\left(\overline{\mathbf{r}}_{\mathrm{G}}-\overline{\mathbf{r}}_1\right)^{\mathrm{T}}\overline{\mathbf{X}}^{-\mathrm{T}} \\ m\overline{\mathbf{X}}^{-1}\left(\overline{\mathbf{r}}_{\mathrm{G}}-\overline{\mathbf{r}}_1\right) & \overline{\mathbf{X}}^{-1}\left(\mathbf{\Pi}_{\mathrm{G}}+m\left(\overline{\mathbf{r}}_{\mathrm{G}}-\overline{\mathbf{r}}_1\right)\left(\overline{\mathbf{r}}_{\mathrm{G}}-\overline{\mathbf{r}}_1\right)^{\mathrm{T}}\right)\overline{\mathbf{X}}^{-\mathrm{T}} \end{array}\right] \tag{9}$$

where m is the mass of the body, $\overline{\mathbf{r}}_1$ is the vector of local coordinates of point 1, $\overline{\mathbf{r}}_{\mathrm{G}}$ is the vector of local coordinates of the mass center of the body, and

$$\mathbf{\Pi}_{\mathrm{G}}=\left[\begin{array}{ccc} \frac{1}{2}\left(-I_x+I_y+I_z\right) & I_{xy} & I_{xz} \\ I_{xy} & \frac{1}{2}\left(I_x-I_y+I_z\right) & I_{yz} \\ I_{xz} & I_{yz} & \frac{1}{2}\left(I_x+I_y-I_z\right) \end{array}\right] \tag{10}$$

being the elements of this matrix the inertia moments and products of the body with respect to a reference frame parallel to the local one, with its origin at the mass center of the body.

Each element of the matrix given in Eq. (9) must be multiplied by the identity matrix of size 3x3. In this way, the resulting matrix has got the size 12x12, corresponding to the 12 variables which have served to describe the body, i.e. the coordinates of point 1 and the components of vectors **u**, **v**, and **w**.

It is not common that a body is defined with one point and three vectors (the basic set), as in the described case. If the body is modeled with a different set of entities, the procedure to get the mass matrix is quite easy. In the following, the case of a body described with two points and two unit vectors is explained (the two unit vectors and the vector defined by the two points, are assumed not to be coplanar).

If the modeling points are called 1 and 2, and the modeling unit vectors are called $\mathbf{u}$ and $\mathbf{v}$, the relation between the basic and the real set of entities is,

$$\begin{Bmatrix} \mathbf{r}_1 \\ \mathbf{u} \\ \mathbf{v} \\ \mathbf{w} \end{Bmatrix} = \begin{bmatrix} \mathbf{I} & \mathbf{0} & \mathbf{0} & \mathbf{0} \\ -\mathbf{I} & \mathbf{I} & \mathbf{0} & \mathbf{0} \\ \mathbf{0} & \mathbf{0} & \mathbf{I} & \mathbf{0} \\ \mathbf{0} & \mathbf{0} & \mathbf{0} & \mathbf{I} \end{bmatrix} \begin{Bmatrix} \mathbf{r}_1 \\ \mathbf{r}_2 \\ \mathbf{u} \\ \mathbf{v} \end{Bmatrix} \tag{11}$$

or, in compact form,

$$\mathbf{q}_0 = \mathbf{T}\mathbf{q} \tag{12}$$

Since matrix $\mathbf{T}$ is constant, identical relation is established between the time derivatives of $\mathbf{q}_0$ and $\mathbf{q}$. Then, if the mass matrix obtained for the basic set in Eq. (9) is called $\mathbf{M}_0$, substitution of the basic velocities by the real ones according to Eq. (12) in the kinetic energy of the body leads to the following expression of the mass matrix for the real set,

$$\mathbf{M} = \mathbf{T}^{\mathrm{T}}\mathbf{M}_0\mathbf{T} \tag{13}$$

For other sets of entities defining the body, the procedure is analogous. If, as in the described cases, the number of variables selected to model the body is enough to represent a three-dimensional space (one point and three non-coplanar directions), the resulting mass matrix is constant and, therefore, no vector of velocity dependent forces exists. However, when less variables are selected, such a vector must be calculated.

Regarding the force vector, the case of a force acting on a material point of a body, illustrated in Fig. 3, is addressed.

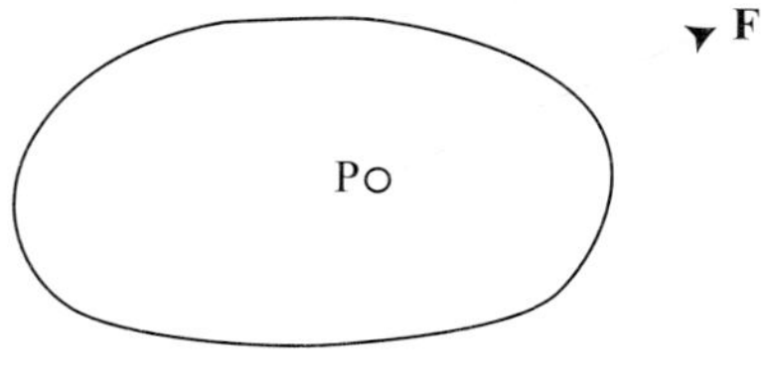

Figure 3. Force acting on a point of a body.

Particularizing Eq. (8) for the point P, and writing the expression in a more compact form yields,

$$\mathbf{r}_{\mathrm{P}} = \begin{bmatrix} \mathbf{I} & c_1\mathbf{I} & c_2\mathbf{I} & c_3\mathbf{I} \end{bmatrix} \begin{Bmatrix} \mathbf{r}_1 \\ \mathbf{u} \\ \mathbf{v} \\ \mathbf{w} \end{Bmatrix} = \mathbf{C}_{\mathrm{P}}\mathbf{q} \tag{14}$$

Then, substitution of Eq. (14) in the power given by the force leads to the following value for the force vector,

$$\mathbf{Q} = \mathbf{C}_{\mathrm{P}}^{\mathrm{T}}\mathbf{F} \tag{15}$$

Once the kinematic and dynamic terms have been derived, the formalism in absolute coordinates is exposed. A more detailed explanation is provided in Cuadrado et al. (2000). The equations of motion of the whole multibody system are given by an index-3 augmented Lagrangian formulation in the form

$$\mathbf{M}\ddot{\mathbf{q}} + \mathbf{\Phi}_{\mathbf{q}}^{\mathrm{T}}\alpha\mathbf{\Phi} + \mathbf{\Phi}_{\mathbf{q}}^{\mathrm{T}}\boldsymbol{\lambda}^{*} = \mathbf{Q} \tag{16}$$

where $\mathbf{M}$ is the mass matrix, $\ddot{\mathbf{q}}$ are the accelerations, $\mathbf{\Phi}_{\mathbf{q}}$ the Jacobian matrix of the constraint equations, α the penalty factor, $\mathbf{\Phi}$ the constraints vector, $\boldsymbol{\lambda}^{*}$ the Lagrange multipliers and $\mathbf{Q}$ the vector of applied and velocity dependent inertia forces. The Lagrange multipliers are obtained from the following iteration process (given by sub-index i, while sub-index n stands for time-step),

$$\boldsymbol{\lambda}_{i+1}^{*} = \boldsymbol{\lambda}_{i}^{*} + \alpha\mathbf{\Phi}_{i+1}\ ,\ i = 0{,}1{,}2,... \tag{17}$$

where the value of $\boldsymbol{\lambda}_{0}^{*}$ is taken equal to the $\boldsymbol{\lambda}^{*}$ obtained in the previous time-step.

As integration scheme, the implicit single-step trapezoidal rule has been adopted. The corresponding difference equations in velocities and accelerations are:

$$\dot{\mathbf{q}}_{n+1} = \frac{2}{\Delta t}\mathbf{q}_{n+1} + \hat{\dot{\mathbf{q}}}_{n} \text{ with } \hat{\dot{\mathbf{q}}}_{n} = -\left(\frac{2}{\Delta t}\mathbf{q}_{n} + \dot{\mathbf{q}}_{n}\right) \tag{18}$$

$$\ddot{\mathbf{q}}_{n+1} = \frac{4}{\Delta t^{2}}\mathbf{q}_{n+1} + \hat{\ddot{\mathbf{q}}}_{n} \text{ with } \hat{\ddot{\mathbf{q}}}_{n} = -\left(\frac{4}{\Delta t^{2}}\mathbf{q}_{n} + \frac{4}{\Delta t}\dot{\mathbf{q}}_{n} + \ddot{\mathbf{q}}_{n}\right) \tag{19}$$

being Δt the time-step. Dynamic equilibrium can be established at time-step n+1 by introducing the difference equations (18) and (19) into the equations of motion (16), leading to a nonlinear system of equations in which the positions at time-step n+1 are the unknowns,

$$\mathbf{f}\left(\mathbf{q}_{n+1}\right) = 0 \tag{20}$$

Such a system can be solved through the Newton-Raphson iterative procedure, being the residual vector,

$$\mathbf{f}\left(\mathbf{q}\right) = \frac{\Delta t^{2}}{4}\left(\mathbf{M}\ddot{\mathbf{q}} + \mathbf{\Phi}_{\mathbf{q}}^{\mathrm{T}}\alpha\mathbf{\Phi} + \mathbf{\Phi}_{\mathbf{q}}^{\mathrm{T}}\boldsymbol{\lambda}^{*} - \mathbf{Q}\right) \tag{21}$$

and the approximated tangent matrix,

$$\frac{\partial \mathbf{f}(\mathbf{q})}{\partial \mathbf{q}} = \mathbf{M} + \frac{\Delta t}{2}\mathbf{C} + \frac{\Delta t^2}{4}\left(\mathbf{\Phi}_\mathbf{q}^{\mathrm{T}}\alpha\mathbf{\Phi}_\mathbf{q} + \mathbf{K}\right) \tag{22}$$

where $\mathbf{C}$ and $\mathbf{K}$ represent the contribution of damping and elastic forces of the system provided they exist. The equivalent meaning of the term $\mathbf{\Phi}_\mathbf{q}^{\mathrm{T}}\alpha\mathbf{\Phi}_\mathbf{q}$ and the stiffness matrix $\mathbf{K}$ should be noted.

A closer look at the tangent matrix reveals that ill-conditioning may appear when the time-step becomes small. It may be seen in Eq. (22) that $\mathbf{K}$ and the constraint terms are multiplied by Δt^2, $\mathbf{C}$ by Δt and $\mathbf{M}$ is not affected by the step size. As a consequence when Δt reaches small values, large round-off errors will occur. In fact, it has been demonstrated in Brenan et al. (1989), that for an index-3 differential-algebraic equation, the tangent matrix has a condition number of order $1/\Delta t^3$. Consequently, the method is bound to have round-off errors for step sizes smaller than 10^{-5}, which lets a sufficient range for solving practical problems.

The procedure explained above yields a set of positions $\mathbf{q}_{n+1}$ that not only satisfies the equations of motion (16), but also the constraint conditions $\mathbf{\Phi} = 0$. However, it is not expected that the corresponding sets of velocities and accelerations satisfy $\dot{\mathbf{\Phi}} = 0$ and $\ddot{\mathbf{\Phi}} = 0$, because these conditions have not been imposed in the solution process. To overcome this difficulty, mass-damping-stiffness-orthogonal projections in velocities and accelerations are performed. It can be seen that the projections leading matrix is the same tangent matrix appearing in equation (22). Therefore, triangularization is avoided and projections in velocities and accelerations are carried out with just forward recursions and back substitutions.

If $\dot{\mathbf{q}}^*$ and $\ddot{\mathbf{q}}^*$ are the velocities and accelerations obtained after convergence has been achieved in the Newton-Raphson iteration, their cleaned counterparts $\dot{\mathbf{q}}$ and $\ddot{\mathbf{q}}$ are calculated from

$$\frac{\partial \mathbf{f}(\mathbf{q})}{\partial \mathbf{q}}\dot{\mathbf{q}} = \left[\mathbf{M} + \frac{\Delta t}{2}\mathbf{C} + \frac{\Delta t^2}{4}\mathbf{K}\right]\dot{\mathbf{q}}^* - \frac{\Delta t^2}{4}\mathbf{\Phi}_\mathbf{q}^{\mathrm{T}}\alpha\mathbf{\Phi}_t \tag{23}$$

for the velocities, and

$$\frac{\partial \mathbf{f}(\mathbf{q})}{\partial \mathbf{q}}\ddot{\mathbf{q}} = \left[\mathbf{M} + \frac{\Delta t}{2}\mathbf{C} + \frac{\Delta t^2}{4}\mathbf{K}\right]\ddot{\mathbf{q}}^* - \frac{\Delta t^2}{4}\mathbf{\Phi}_\mathbf{q}^{\mathrm{T}}\alpha\left(\dot{\mathbf{\Phi}}_\mathbf{q}\dot{\mathbf{q}} + \dot{\mathbf{\Phi}}_t\right) \tag{24}$$

for the accelerations.

Since matrices tend to be sparse when applying this formulation, the use of sparse matrix technology is recommended to solve all the linear sets of equations and to evaluate the product $\mathbf{\Phi}_\mathbf{q}^{\mathrm{T}}\mathbf{\Phi}_\mathbf{q}$.

2.2 The Method in Relative Coordinates

As the name of the method indicates, in this case relative coordinates are used to describe the mechanism. However, the mechanism which is described is not the original one, but its spanning tree version, obtained once all the closed kinematic loops have been opened by the cut joint method, i.e. some joints are eliminated in order to open the corresponding closed loops. For example, Fig. 4 shows a four-bar mechanism, in which the cut joint method is applied to open the single existing closed loop, yielding a three degrees of freedom spanning tree version of the mechanism, described by the three relative coordinates z_1, z_2 and z_3.

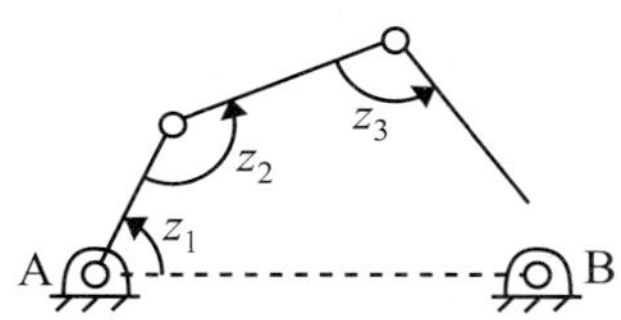

Figure 4. Spanning tree version of a four-bar mechanism.

Of course, the two constraint equations (in the planar case) imposing that the point at the end of the third link is coincident with the fixed point B must be considered in order to recover the original mechanism.

Besides the mentioned set of relative coordinates, a second set of coordinates is defined in this method, the so-called body coordinates, which can be expressed at velocity level for each body in the following form,

$$\mathbf{Z} = \left\{ \begin{array}{c} \dot{\mathbf{s}} \\ \boldsymbol{\omega} \end{array} \right\} \tag{25}$$

being $\dot{\mathbf{s}}$ the velocity of the point of the body which in that particular time is coincident with the fixed frame origin, and $\boldsymbol{\omega}$ the angular velocity of the body.

In this method, the set of relative coordinates and their derivatives are the starting point of a new time-step. This means that, first of all, the kinematics of the spanning tree version of the mechanism should be calculated in a recursive form, from the base to the leaves.

At position level, the coordinates of all the relevant points of the mechanism can be obtained by applying the classical kinematic relations,

$$\mathbf{r} = \mathbf{r}_0 + \mathbf{A}\bar{\mathbf{r}} \tag{26}$$

where $\mathbf{r}$ is the vector containing the coordinates of any point of a body, $\mathbf{r}_0$ is the vector containing the coordinates of the origin of the local frame of the body, and $\mathbf{A}$ is the rotation matrix relating the local frame of the body with the fixed frame. Equation (26) is complemented by

$$\mathbf{A}_i = \mathbf{A}_{i-1}\mathbf{A}_{i,i-1} \tag{27}$$

which provides the rotation matrix of body i as a function of the rotation matrix of body *i-1* (the previous body in the kinematic chain), and the rotation matrix due to the joint connecting the two bodies.

Analogously, at velocity level, the velocities of all the relevant points of the mechanism and the angular velocities of all the bodies can be calculated by means of the classical kinematic relations,

$$\mathbf{v} = \mathbf{v}_0 + \boldsymbol{\omega} \times (\mathbf{r} - \mathbf{r}_0) \tag{28}$$

$$\boldsymbol{\omega}_i = \boldsymbol{\omega}_{i-1} + \boldsymbol{\omega}_{i,i-1} \tag{29}$$

where $\mathbf{v}$ is the vector containing the Cartesian velocities of any point of a body, $\mathbf{v}_0$ is the vector containing the Cartesian velocities of any other point of the body, and $\boldsymbol{\omega}$ is the angular velocity of the body. The angular velocity of a body, $\boldsymbol{\omega}_i$, can be obtained as the sum of the angular velocity of the previous body, $\boldsymbol{\omega}_{i-1}$, plus the relative angular velocity, $\boldsymbol{\omega}_{i,i-1}$, due to the joint connecting the two bodies.

At acceleration level, the following expressions are the equivalent ones to the already shown for positions and velocities,

$$\mathbf{a} = \mathbf{a}_0 + \boldsymbol{\alpha} \times (\mathbf{r} - \mathbf{r}_0) + \boldsymbol{\omega} \times (\boldsymbol{\omega} \times (\mathbf{r} - \mathbf{r}_0)) \tag{30}$$

$$\boldsymbol{\alpha}_i = \boldsymbol{\alpha}_{i-1} + \boldsymbol{\alpha}_{i,i-1} + \boldsymbol{\omega}_{i-1} \times \boldsymbol{\omega}_{i,i-1} \tag{31}$$

being, in this case, $\mathbf{a}$ for acceleration of points, and $\boldsymbol{\alpha}$ for angular acceleration of bodies.

Accordingly, the relation between the body coordinates (and their derivatives) of two neighbor bodies can be expressed as

$$\mathbf{Z}_i = \mathbf{Z}_{i-1} + \mathbf{b}_i\dot{\mathbf{z}}_i \tag{32}$$

$$\dot{\mathbf{Z}}_i = \dot{\mathbf{Z}}_{i-1} + \mathbf{b}_i\ddot{\mathbf{z}}_i + \mathbf{d}_i \tag{33}$$

where the form of terms $\mathbf{b}_i$ and $\mathbf{d}_i$ depends on the type of joint that connects the bodies *i-1* and i. In the case of a revolute joint, they take the form

$$\mathbf{b}_i = \begin{Bmatrix} \tilde{\mathbf{r}}_i\mathbf{u}_i \\ \mathbf{u}_i \end{Bmatrix} \tag{34}$$

$$\mathbf{d}_i = \begin{Bmatrix} \tilde{\mathbf{r}}_i\left(\tilde{\boldsymbol{\omega}}_{i-1}\mathbf{u}_i\right)\dot{z}_i + \left(\tilde{\boldsymbol{\omega}}_{i-1}\tilde{\boldsymbol{\omega}}_{i-1} - \tilde{\boldsymbol{\omega}}_i\tilde{\boldsymbol{\omega}}_i\right)\mathbf{r}_i \\ \tilde{\boldsymbol{\omega}}_{i-1}\mathbf{u}_i\dot{z}_i \end{Bmatrix} \tag{35}$$

being $\mathbf{r}_i$ the point where the joint is located, and $\mathbf{u}_i$ a unit vector in the direction of the joint. In the case of a prismatic joint, the terms take the form

$$\mathbf{b}_i = \left\{ \begin{array}{c} \mathbf{u}_i \\ \mathbf{0} \end{array} \right\} \tag{36}$$

$$\mathbf{d}_i = \left\{ \begin{array}{c} 2\tilde{\boldsymbol{\omega}}_{i-1}\mathbf{u}_i\dot{z}_i \\ \mathbf{0} \end{array} \right\} \tag{37}$$

Therefore, at the view of Eq. (32), a matrix $\mathbf{R}$ can be defined so that the following relationship between the relative and the body coordinates stands,

$$\mathbf{Z} = \mathbf{R}\dot{\mathbf{z}} \tag{38}$$

where now $\mathbf{Z}$ includes the body coordinates of all the bodies of the mechanism. Due to the body coordinates adopted, the form of matrix $\mathbf{R}$ is rather special, and can be always decomposed as

$$\mathbf{R} = \mathbf{T}\mathbf{R}_d \tag{39}$$

with $\mathbf{T}$ a connectivity matrix, and $\mathbf{R}_d$ a diagonal matrix.

In order to explain the structure of matrix $\mathbf{R}$, an example of spanning tree mechanism is provided in Fig. 5.

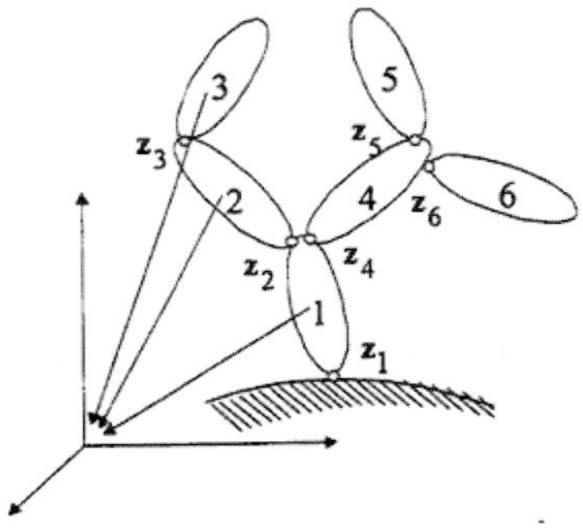

Figure 5. Spanning tree mechanism.

For the example of Fig. 5, Eq. (38) takes the following form,

$$\mathbf{Z} = \begin{Bmatrix} \mathbf{Z}_1 \\ \mathbf{Z}_2 \\ \mathbf{Z}_3 \\ \mathbf{Z}_4 \\ \mathbf{Z}_5 \\ \mathbf{Z}_6 \end{Bmatrix} = \begin{bmatrix} \mathbf{b}_1 & \mathbf{0} & \mathbf{0} & \mathbf{0} & \mathbf{0} & \mathbf{0} \\ \mathbf{b}_1 & \mathbf{b}_2 & \mathbf{0} & \mathbf{0} & \mathbf{0} & \mathbf{0} \\ \mathbf{b}_1 & \mathbf{b}_2 & \mathbf{b}_3 & \mathbf{0} & \mathbf{0} & \mathbf{0} \\ \mathbf{b}_1 & \mathbf{0} & \mathbf{0} & \mathbf{b}_4 & \mathbf{0} & \mathbf{0} \\ \mathbf{b}_1 & \mathbf{0} & \mathbf{0} & \mathbf{b}_4 & \mathbf{b}_5 & \mathbf{0} \\ \mathbf{b}_1 & \mathbf{0} & \mathbf{0} & \mathbf{b}_4 & \mathbf{0} & \mathbf{b}_6 \end{bmatrix} \begin{Bmatrix} \dot{z}_1 \\ \dot{z}_2 \\ \dot{z}_3 \\ \dot{z}_4 \\ \dot{z}_5 \\ \dot{z}_6 \end{Bmatrix} \tag{40}$$

And the matrices $\mathbf{T}$ and $\mathbf{R}_d$ appearing in Eq. (39) result

$$\mathbf{T} = \begin{bmatrix} \mathbf{I}_6 & \mathbf{0} & \mathbf{0} & \mathbf{0} & \mathbf{0} & \mathbf{0} \\ \mathbf{I}_6 & \mathbf{I}_6 & \mathbf{0} & \mathbf{0} & \mathbf{0} & \mathbf{0} \\ \mathbf{I}_6 & \mathbf{I}_6 & \mathbf{I}_6 & \mathbf{0} & \mathbf{0} & \mathbf{0} \\ \mathbf{I}_6 & \mathbf{0} & \mathbf{0} & \mathbf{I}_6 & \mathbf{0} & \mathbf{0} \\ \mathbf{I}_6 & \mathbf{0} & \mathbf{0} & \mathbf{I}_6 & \mathbf{I}_6 & \mathbf{0} \\ \mathbf{I}_6 & \mathbf{0} & \mathbf{0} & \mathbf{I}_6 & \mathbf{0} & \mathbf{I}_6 \end{bmatrix} \tag{41}$$

$$\mathbf{R}_d = \begin{bmatrix} \mathbf{R}_1 & \mathbf{0} & \mathbf{0} & \mathbf{0} & \mathbf{0} & \mathbf{0} \\ \mathbf{0} & \mathbf{R}_2 & \mathbf{0} & \mathbf{0} & \mathbf{0} & \mathbf{0} \\ \mathbf{0} & \mathbf{0} & \mathbf{R}_3 & \mathbf{0} & \mathbf{0} & \mathbf{0} \\ \mathbf{0} & \mathbf{0} & \mathbf{0} & \mathbf{R}_4 & \mathbf{0} & \mathbf{0} \\ \mathbf{0} & \mathbf{0} & \mathbf{0} & \mathbf{0} & \mathbf{R}_5 & \mathbf{0} \\ \mathbf{0} & \mathbf{0} & \mathbf{0} & \mathbf{0} & \mathbf{0} & \mathbf{R}_6 \end{bmatrix} \tag{42}$$

where $\mathbf{I}_6$ is the 6x6 identity matrix and the $\mathbf{R}_i$ are 6x1 vectors.

Differentiating Eq. (38) with respect to time yields,

$$\dot{\mathbf{Z}} = \mathbf{R}\ddot{\mathbf{z}} + \dot{\mathbf{R}}\dot{\mathbf{z}} \tag{43}$$

where the term $\dot{\mathbf{R}}\dot{\mathbf{z}}$ follows a similar structure of that explained for matrix $\mathbf{R}$, as can be deducted by taking into account Eq. (33),

$$\dot{\mathbf{R}}\dot{\mathbf{z}} = \mathbf{T}\dot{\mathbf{R}}_d\dot{\mathbf{z}} = \begin{Bmatrix} \mathbf{d}_1 \\ \mathbf{d}_1 + \mathbf{d}_2 \\ \mathbf{d}_1 + \mathbf{d}_2 + \mathbf{d}_3 \\ \mathbf{d}_1 + \mathbf{d}_4 \\ \mathbf{d}_1 + \mathbf{d}_4 + \mathbf{d}_5 \\ \mathbf{d}_1 + \mathbf{d}_4 + \mathbf{d}_6 \end{Bmatrix} = \begin{Bmatrix} \left(\dot{\mathbf{R}}\dot{\mathbf{z}}\right)_1 \\ \left(\dot{\mathbf{R}}\dot{\mathbf{z}}\right)_2 \\ \left(\dot{\mathbf{R}}\dot{\mathbf{z}}\right)_3 \\ \left(\dot{\mathbf{R}}\dot{\mathbf{z}}\right)_4 \\ \left(\dot{\mathbf{R}}\dot{\mathbf{z}}\right)_5 \\ \left(\dot{\mathbf{R}}\dot{\mathbf{z}}\right)_6 \end{Bmatrix} \tag{44}$$

Once the kinematics has been described, the way to obtain the dynamic terms, mass matrix and force vector, is addressed now. The dynamic terms in body coordinates are, for a single body,

$$\bar{\mathbf{M}}_i = \begin{bmatrix} m\mathbf{I_3} & -m\tilde{\mathbf{g}} \\ m\tilde{\mathbf{g}} & \mathbf{J} - m\tilde{\mathbf{g}}\tilde{\mathbf{g}} \end{bmatrix} \tag{45}$$

$$\bar{\mathbf{Q}}_i = \left\{ \begin{array}{c} \mathbf{f} - \omega \times (\omega \times m\mathbf{g}) \\ \mathbf{n} - \omega \times \mathbf{J}\omega + \mathbf{g} \times (\mathbf{f} - \omega \times (\omega \times m\mathbf{g})) \end{array} \right\} \tag{46}$$

wherein m is the body mass, $\mathbf{I}_3$ is the 3x3 identity matrix, $\mathbf{g}$ is the global position of the mass center of the body, $\tilde{\mathbf{g}}$ is the dual anti-symmetric matrix of $\mathbf{g}$, $\mathbf{J}$ is the inertia tensor of the body with respect to a reference frame parallel to the global one at the mass center of the body, $\mathbf{f}$ is the vector of forces applied to the body, and $\mathbf{n}$ is the vector of applied moments with respect to the mass center of the body. These expressions are obtained from the Newton-Euler equations, taking into account that the center of mass is substituted by the point of the body that, at that particular time, is coincident with the fixed frame origin.

The mass matrix and force vector of the whole mechanism in body coordinates are then,

$$\bar{\mathbf{M}} = \begin{bmatrix} \bar{\mathbf{M}}_1 & \mathbf{0} & \mathbf{0} & \mathbf{0} & \mathbf{0} & \mathbf{0} \\ \mathbf{0} & \bar{\mathbf{M}}_2 & \mathbf{0} & \mathbf{0} & \mathbf{0} & \mathbf{0} \\ \mathbf{0} & \mathbf{0} & \bar{\mathbf{M}}_3 & \mathbf{0} & \mathbf{0} & \mathbf{0} \\ \mathbf{0} & \mathbf{0} & \mathbf{0} & \bar{\mathbf{M}}_4 & \mathbf{0} & \mathbf{0} \\ \mathbf{0} & \mathbf{0} & \mathbf{0} & \mathbf{0} & \bar{\mathbf{M}}_5 & \mathbf{0} \\ \mathbf{0} & \mathbf{0} & \mathbf{0} & \mathbf{0} & \mathbf{0} & \bar{\mathbf{M}}_6 \end{bmatrix} ; \bar{\mathbf{Q}} = \left\{ \begin{array}{c} \bar{\mathbf{Q}}_1 \\ \bar{\mathbf{Q}}_2 \\ \bar{\mathbf{Q}}_3 \\ \bar{\mathbf{Q}}_4 \\ \bar{\mathbf{Q}}_5 \\ \bar{\mathbf{Q}}_6 \end{array} \right\} \tag{47}$$

If the virtual power principle is applied, the dynamic equations of the system can be written as (the asterisk indicates the virtual character of the corresponding velocities),

$$\mathbf{Z}^{*\mathrm{T}} \left(\bar{\mathbf{M}}\dot{\mathbf{Z}} - \bar{\mathbf{Q}} \right) = 0 \tag{48}$$

and substituting now in Eq. (48) the result of Eq. (38) and (43) yields,

$$\dot{\mathbf{z}}^{*\mathrm{T}} \left\{ \mathbf{R}^{\mathrm{T}}\bar{\mathbf{M}}\mathbf{R}\ddot{\mathbf{z}} - \mathbf{R}^{\mathrm{T}} \left(\bar{\mathbf{Q}} - \bar{\mathbf{M}}\dot{\mathbf{R}}\dot{\mathbf{z}} \right) \right\} = 0 \tag{49}$$

or, in a more compact form,

$$\dot{\mathbf{z}}^{*\mathrm{T}} \left\{ \mathbf{M}\ddot{\mathbf{z}} - \mathbf{Q} \right\} = 0 \tag{50}$$

Therefore, according to Eq. (50), the mass matrix in relative coordinates is,

$$\mathbf{M} = \mathbf{R}^{\mathrm{T}}\bar{\mathbf{M}}\mathbf{R} = \mathbf{R}_d^{\mathrm{T}}\left(\mathbf{T}^{\mathrm{T}}\bar{\mathbf{M}}\mathbf{T}\right)\mathbf{R}_d \tag{51}$$

with the following specific values for the example of Fig. 5,

$$\mathbf{T}^{\mathrm{T}}\bar{\mathbf{M}}\mathbf{T} = \begin{bmatrix} \mathbf{M}_1 & \mathbf{M}_2 & \mathbf{M}_3 & \mathbf{M}_4 & \mathbf{M}_5 & \mathbf{M}_6 \\ & \mathbf{M}_2 & \mathbf{M}_3 & \mathbf{0} & \mathbf{0} & \mathbf{0} \\ & & \mathbf{M}_3 & \mathbf{0} & \mathbf{0} & \mathbf{0} \\ & & & \mathbf{M}_4 & \mathbf{M}_5 & \mathbf{M}_6 \\ & sym & & & \mathbf{M}_5 & \mathbf{0} \\ & & & & & \mathbf{M}_6 \end{bmatrix} ; \quad \begin{array}{c} \mathbf{M_6} = \bar{\mathbf{M}}_6 \\ \mathbf{M_5} = \bar{\mathbf{M}}_5 \\ \mathbf{M_4} = \bar{\mathbf{M}}_4 + \mathbf{M}_5 + \mathbf{M}_6 \\ \mathbf{M_3} = \bar{\mathbf{M}}_3 \\ \mathbf{M_2} = \bar{\mathbf{M}}_2 + \mathbf{M}_3 \\ \mathbf{M_1} = \bar{\mathbf{M}}_1 + \mathbf{M}_2 + \mathbf{M}_4 \end{array} \tag{52}$$

while the force vector in relative coordinates is,

$$\mathbf{Q} = \mathbf{R}^{\mathrm{T}}\left(\bar{\mathbf{Q}} - \bar{\mathbf{M}}\dot{\mathbf{R}}\dot{\mathbf{z}}\right) = \mathbf{R}_d^{\mathrm{T}}\left(\mathbf{T}^{\mathrm{T}}\left(\bar{\mathbf{Q}} - \bar{\mathbf{M}}\dot{\mathbf{R}}\dot{\mathbf{z}}\right)\right) \tag{53}$$

with the following specific values for the example of Fig. 5,

$$\mathbf{T}^{\mathrm{T}}\left(\bar{\mathbf{Q}} - \bar{\mathbf{M}}\dot{\mathbf{R}}\dot{\mathbf{z}}\right) = \begin{Bmatrix} \mathbf{Q}_1 \\ \mathbf{Q}_2 \\ \mathbf{Q}_3 \\ \mathbf{Q}_4 \\ \mathbf{Q}_5 \\ \mathbf{Q}_6 \end{Bmatrix} ; \quad \begin{array}{c} \mathbf{Q}_6 = \bar{\mathbf{Q}}_6 - \bar{\mathbf{M}}_6\left(\dot{\mathbf{R}}\dot{\mathbf{z}}\right)_6 \\ \mathbf{Q}_5 = \bar{\mathbf{Q}}_5 - \bar{\mathbf{M}}_5\left(\dot{\mathbf{R}}\dot{\mathbf{z}}\right)_5 \\ \mathbf{Q}_4 = \bar{\mathbf{Q}}_4 - \bar{\mathbf{M}}_4\left(\dot{\mathbf{R}}\dot{\mathbf{z}}\right)_4 + \mathbf{Q}_5 + \mathbf{Q}_6 \\ \mathbf{Q}_3 = \bar{\mathbf{Q}}_3 - \bar{\mathbf{M}}_3\left(\dot{\mathbf{R}}\dot{\mathbf{z}}\right)_3 \\ \mathbf{Q}_2 = \bar{\mathbf{Q}}_2 - \bar{\mathbf{M}}_2\left(\dot{\mathbf{R}}\dot{\mathbf{z}}\right)_2 + \mathbf{Q}_3 \\ \mathbf{Q}_1 = \bar{\mathbf{Q}}_1 - \bar{\mathbf{M}}_1\left(\dot{\mathbf{R}}\dot{\mathbf{z}}\right)_1 + \mathbf{Q}_2 + \mathbf{Q}_4 \end{array} \tag{54}$$

Hence, the mass matrix and force vector in relative coordinates are obtained recursively, from the leaves to the base of the mechanism, by accumulating the mass matrices and force vectors of the bodies expressed in body coordinates.

Once the kinematic and dynamic terms have been derived, the formalism in relative coordinates, which is basically identical to the one already explained for the method in absolute coordinates, is exposed here again for the sake of clarity. For more details, the reader is referred to Cuadrado et al. (2004a). The dynamic equations are stated according to the index-3 augmented Lagrangian formulation in the form,

$$\mathbf{M}\ddot{\mathbf{z}} + \mathbf{\Phi}_{\mathbf{z}}^{\mathrm{T}}\alpha\mathbf{\Phi} + \mathbf{\Phi}_{\mathbf{z}}^{\mathrm{T}}\boldsymbol{\lambda}^* = \mathbf{Q} \tag{55}$$

where $\mathbf{z}$ are the relative coordinates, $\mathbf{M}$ is the mass matrix of the mechanism expressed in terms of the relative coordinates, $\mathbf{\Phi}$ is the constraints vector due to the closure conditions of the loops, $\mathbf{\Phi_z}$ is the Jacobian matrix of the constraints, α is the penalty factor, $\mathbf{Q}$ is the vector of applied and velocity-dependent forces, and $\boldsymbol{\lambda}^*$ is the vector of Lagrange multipliers obtained from the following iteration process (given by sub-index i, while sub-index n stands for the time-step):

$$\boldsymbol{\lambda}^*_{i+1} = \boldsymbol{\lambda}^*_i + \alpha\boldsymbol{\Phi}_{i+1} \tag{56}$$

where the value of $\boldsymbol{\lambda}^*_0$ is taken equal to the $\boldsymbol{\lambda}^*$ worked out in the previous time-step.

The Jacobian matrix of the constraints, $\mathbf{\Phi_z}$, implies the differentiation of the constraints vector with respect to the relative coordinates $\mathbf{z}$. This can be easily done by applying the chain differentiation rule as,

$$\mathbf{\Phi_z} = \mathbf{\Phi_q}\mathbf{q_z} \tag{57}$$

In the proposed method, natural coordinates are used at the cut-points to impose the closure conditions of the loops. This means that the constraints are expressed in terms of such coordinates, named $\mathbf{q}$ in Eq. (57). Therefore, the term $\mathbf{\Phi_q}$ is the Jacobian matrix of the constraints when natural coordinates are used, while the term $\mathbf{q_z}$ simply represents the velocities of the natural coordinates $\mathbf{q}$, when unit velocities are successively given to the relative coordinates $\mathbf{z}$.

As integration scheme, the implicit single-step trapezoidal rule has been adopted. The corresponding difference equations in velocities and accelerations are:

$$\dot{\mathbf{z}}_{n+1} = \frac{2}{\Delta t}\mathbf{z}_{n+1} + \hat{\dot{\mathbf{z}}}_n \;;\; \hat{\dot{\mathbf{z}}}_n = -\left(\frac{2}{\Delta t}\mathbf{z}_n + \dot{\mathbf{z}}_n\right) \tag{58}$$

$$\ddot{\mathbf{z}}_{n+1} = \frac{4}{\Delta t^2}\mathbf{z}_{n+1} + \hat{\ddot{\mathbf{z}}}_n \;;\; \hat{\ddot{\mathbf{z}}}_n = -\left(\frac{4}{\Delta t^2}\mathbf{z}_n + \frac{4}{\Delta t}\dot{\mathbf{z}}_n + \ddot{\mathbf{z}}_n\right) \tag{59}$$

being Δt the time-step. Dynamic equilibrium can be established at time-step $n+1$ by introducing the difference equations (58) and (59) into the equations of motion (55), leading to a non-linear system of algebraic equations, where the $\mathbf{z}_{n+1}$ are the unknowns,

$$\mathbf{f}\left(\mathbf{z}_{n+1}\right) = 0 \tag{60}$$

Such a system can be solved by the Newton-Raphson iteration, being the residual vector

$$\mathbf{f}\left(\mathbf{z}\right) = \frac{\Delta t^2}{4}\left(\mathbf{M}\ddot{\mathbf{z}} + \mathbf{\Phi}_\mathbf{z}^{\mathrm{T}}\alpha\mathbf{\Phi} + \mathbf{\Phi}_\mathbf{z}^{\mathrm{T}}\boldsymbol{\lambda}^* - \mathbf{Q}\right) \tag{61}$$

and the approximated tangent matrix

$$\frac{\partial \mathbf{f}(\mathbf{z})}{\partial \mathbf{z}} = \mathbf{M} + \frac{\Delta t}{2}\mathbf{C} + \frac{\Delta t^2}{4}\left(\mathbf{\Phi}_{\mathbf{z}}^{\mathrm{T}} \alpha \mathbf{\Phi}_{\mathbf{z}} + \mathbf{K}\right) \tag{62}$$

where $\mathbf{C}$ and $\mathbf{K}$ represent the contribution of damping and elastic forces of the system provided they exist. Comments regarding the equivalent meaning of the terms $\mathbf{\Phi}_{\mathbf{z}}^{\mathrm{T}} \alpha \mathbf{\Phi}_{\mathbf{z}}$ and $\mathbf{K}$, as well as ill-conditioning problems of the approximated tangent matrix for time-step sizes under 10^{-5}, are identical to those made above for the method in absolute coordinates.

The explained procedure yields a set of positions $\mathbf{z}_{n+1}$ that not only satisfies the equations of motion (55), but also the constraint conditions $\mathbf{\Phi} = 0$. However, it is not expected that the corresponding sets of velocities and accelerations satisfy $\dot{\mathbf{\Phi}} = \mathbf{0}$ and $\ddot{\mathbf{\Phi}} = \mathbf{0}$, because these conditions have not been imposed in the solution process. To overcome this difficulty, mass-damping-stiffness-orthogonal projections in velocities and accelerations are performed. It can be seen that the projections leading matrix is the same tangent matrix appearing in Eq. (62). Therefore, triangularization is avoided and projections in velocities and accelerations are carried out with just forward reductions and back substitutions.

If $\dot{\mathbf{z}}^*$ and $\ddot{\mathbf{z}}^*$ are the velocities and accelerations obtained after convergence has been achieved in the Newton-Raphson iteration, their cleaned counterparts $\dot{\mathbf{z}}$ and $\ddot{\mathbf{z}}$ are calculated from,

$$\frac{\partial \mathbf{f}(\mathbf{z})}{\partial \mathbf{z}}\dot{\mathbf{z}} = \left[\mathbf{M} + \frac{\Delta t}{2}\mathbf{C} + \frac{\Delta t^2}{4}\mathbf{K}\right]\dot{\mathbf{z}}^* - \frac{\Delta t^2}{4}\mathbf{\Phi}_{\mathbf{z}}^{\mathrm{T}} \alpha \mathbf{\Phi}_t \tag{63}$$

for the velocities, and,

$$\frac{\partial \mathbf{f}(\mathbf{z})}{\partial \mathbf{z}}\ddot{\mathbf{z}} = \left[\mathbf{M} + \frac{\Delta t}{2}\mathbf{C} + \frac{\Delta t^2}{4}\mathbf{K}\right]\ddot{\mathbf{z}}^* - \frac{\Delta t^2}{4}\mathbf{\Phi}_{\mathbf{z}}^{\mathrm{T}} \alpha \left(\dot{\mathbf{\Phi}}_{\mathbf{z}}\dot{\mathbf{z}} + \dot{\mathbf{\Phi}}_t\right) \tag{64}$$

When modeling in relative coordinates, matrices become less sparse than when modeling in absolute coordinates. However, in most cases, the use of sparse matrix techniques is still advantageous.

2.3 Modeling Comparison

In order to illustrate the modeling differences when using the method in absolute or relative coordinates, the example of Fig. 6 is provided. It shows a car suspension mechanism, which is modeled with natural coordinates in Fig. 6a, and with relative coordinates in Fig. 6b.

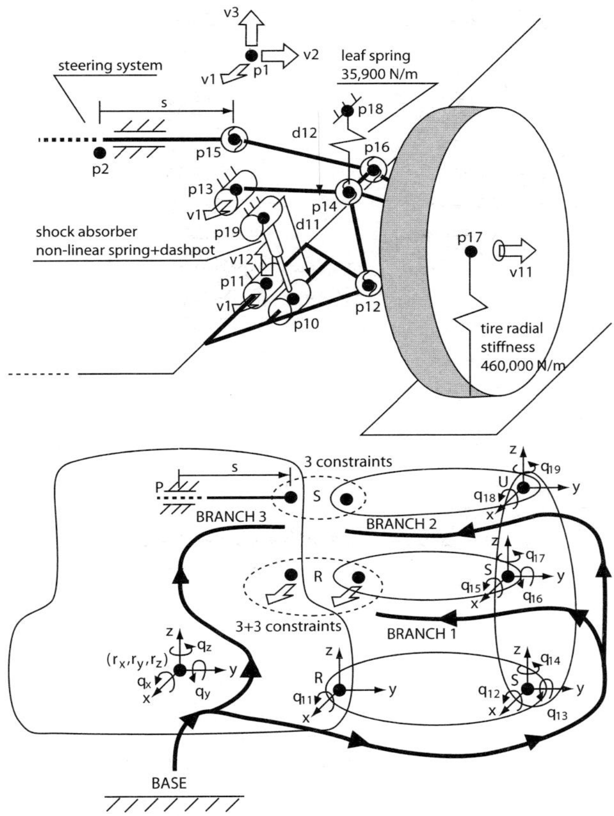

Figure 6. a) Modeling in absolute coordinates; b) Modeling in relative coordinates.

2.4 The Projections

The projections in velocities and accelerations have a stabilizing effect on the solution. Details about the projections applied in the two variants of the method presented above can be found in Bayo and Ledesma (1996). However, in order to provide the reader with an idea about their meaning, the projections in velocities are developed here. The variables will be designated by $\mathbf{q}$, although they can represent either the absolute or relative coordinates.

If $\dot{\mathbf{q}}^*$ are the velocities obtained after convergence has been achieved in the Newton-Raphson iteration, the following optimization problem is set:

$$\begin{aligned} &\min\ V = \frac{1}{2}\left(\dot{\mathbf{q}} - \dot{\mathbf{q}}^*\right)^{\mathrm{T}} \mathbf{P} \left(\dot{\mathbf{q}} - \dot{\mathbf{q}}^*\right) \\ &\text{subject to } \frac{\Delta t^2}{4}\dot{\mathbf{\Phi}}\left(\mathbf{q},\dot{\mathbf{q}},t\right) = 0 \end{aligned} \tag{65}$$

where the weight matrix $\mathbf{P}$ can be set to different values. In the proposed formulation, matrix $\mathbf{P}$ takes the following value,

$$\mathbf{P} = \left[\mathbf{M} + \frac{\Delta t}{2}\mathbf{C} + \frac{\Delta t^2}{4}\mathbf{K}\right] \tag{66}$$

The constraints are imposed to the optimization problem through the penalty technique, thus leading to,

$$\min\ V^* = \frac{1}{2}\left(\dot{\mathbf{q}} - \dot{\mathbf{q}}^*\right)^{\mathrm{T}} \mathbf{P} \left(\dot{\mathbf{q}} - \dot{\mathbf{q}}^*\right) + \frac{\Delta t^2}{8}\dot{\mathbf{\Phi}}^{\mathrm{T}}\alpha\dot{\mathbf{\Phi}} \tag{67}$$

In order to get the minimum of such a function, its derivatives with respect to the velocities $\dot{\mathbf{q}}$ are equated to zero,

$$\frac{\partial V^*}{\partial \dot{\mathbf{q}}} = \mathbf{P}\left(\dot{\mathbf{q}} - \dot{\mathbf{q}}^*\right) + \frac{\Delta t^2}{4}\dot{\mathbf{\Phi}}_{\dot{\mathbf{q}}}^{\mathrm{T}}\alpha\dot{\mathbf{\Phi}} = \mathbf{P}\left(\dot{\mathbf{q}} - \dot{\mathbf{q}}^*\right) + \frac{\Delta t^2}{4}\mathbf{\Phi}_{\mathbf{q}}^{\mathrm{T}}\alpha\left(\mathbf{\Phi}_{\mathbf{q}}\dot{\mathbf{q}} + \mathbf{\Phi}_t\right) = \mathbf{0} \tag{68}$$

so that the final expression for the projections is obtained,

$$\begin{aligned} &\left[\mathbf{M} + \frac{\Delta t}{2}\mathbf{C} + \frac{\Delta t^2}{4}\left(\mathbf{\Phi}_{\mathbf{q}}^{\mathrm{T}}\alpha\mathbf{\Phi}_{\mathbf{q}} + \mathbf{K}\right)\right]\dot{\mathbf{q}} = \\ &\left[\mathbf{M} + \frac{\Delta t}{2}\mathbf{C} + \frac{\Delta t^2}{4}\mathbf{K}\right]\dot{\mathbf{q}}^* - \frac{\Delta t^2}{4}\mathbf{\Phi}_{\mathbf{q}}^{\mathrm{T}}\alpha\mathbf{\Phi}_t \end{aligned} \tag{69}$$

The reason to take the value of the weight matrix $\mathbf{P}$ indicated in Eq. (66) was to obtain, in the final system of Eq. (69), the same leading matrix than that of the Newton-Raphson iteration required to achieve convergence within a time-step with the proposed formulation.

2.5 The Algorithm

In this section, the flow diagram that shows how to implement the proposed method is presented (see Fig. 7). It can be helpful for those readers interested in programming the method in any of its two variants. As in the previous section, the variables will be designated by $\mathbf{q}$, although they can represent either the absolute or relative coordinates.

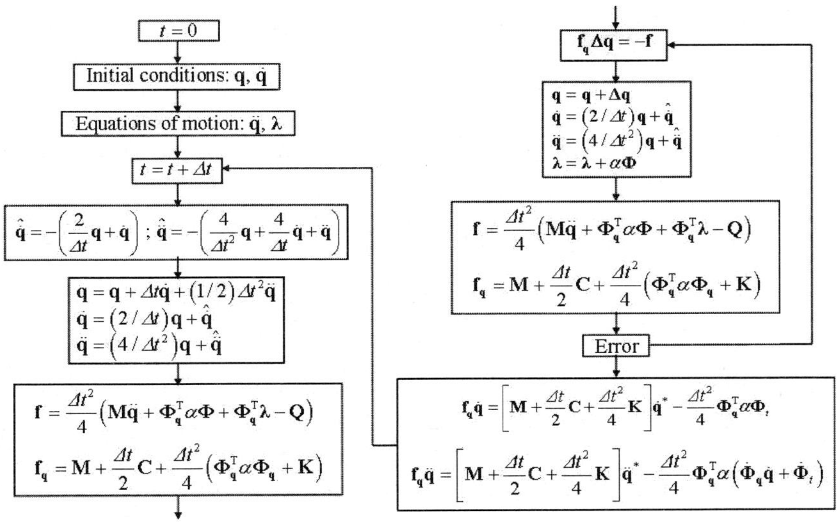

Figure 7. Flow diagram of the method.

2.6 Examples

To analyze the performance of the proposed method, and to compare its two variants, four examples have been solved. The implementation of the three first cases has been carried out in the computing environment Matlab, which means that the reported CPU times would be drastically reduced if the same programs were written in Fortran or C languages. However, in order to show the behavior of the method when facing a complex and realistic problem, and when all the resources available are used to achieve real-time performance, the fourth case has been implemented in Fortran language.

Double four-bar mechanism. The first example is a one degree of freedom assembly of two four-bar linkages, illustrated in Fig. 8, which has been proposed as an example to test the performance in cases where the mechanism undergoes singular configurations. When the mechanism reaches a horizontal position, the number of degrees of freedom instantaneously increases from 1 to 3. All the links have a uniformly distributed unit mass and a unit length. The gravity force acts in the negative vertical direction. At the beginning, the links pinned to the ground are in vertical position and upwards, receiving a unit clockwise angular velocity.

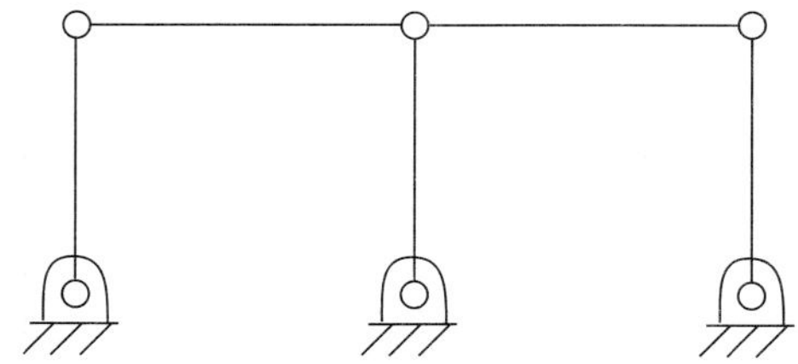

Figure 8. Double four-bar mechanism.

The simulation lasts for 10 s, during which the mechanism goes through the singular position ten times. Table 1 shows the obtained results by using the two variants of the method for different time-steps (Δt). The error is measured as the maximum deviation, in J, suffered by the system energy. The time is the CPU time, in s, required to carry out the simulation. It can

Table 1. Results for the double four-bar mechanism.

	Absolute		Relative	
Δt	Error	Time	Error	Time
0.01	-0.21	0.86	-0.01	3.20
0.03	-0.83	0.37	-0.12	1.32
0.05	-9.03	0.26	-0.34	0.83
0.1	wrong results		-1.34	0.52

be observed that the absolute variant is more efficient than the relative one for the same time-step, while the latter is found to be more accurate and robust, as it shows an smaller error for the same time-step and, moreover, works for larger time-steps than the former. Therefore, if the efficiency of both methods was compared for the same level of error, similar performances would be obtained. It should be noted that, in this example, the number of global coordinates is 6, while the number of relative coordinates is 5. It is not then a case in which the relative variant can get a great advantage due to the reduced number of coordinates with respect to the absolute one (this will occur in large examples).

Car suspension. The second example, shown in Fig. 9, is the suspension system of the Iltis vehicle (see Frik et al. (1993)), which was proposed as a benchmark problem by the European automobile industry to check multibody dynamic codes.

The suspension starts moving from at rest conditions at a speed of 5

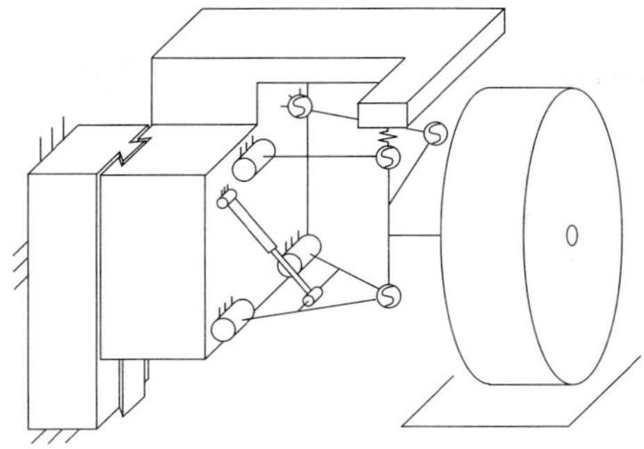

Figure 9. Iltis suspension.

m/s, and its position does not correspond to the static equilibrium. Thus, it freely oscillates until the static equilibrium position is reached. After three seconds, the suspension drops down a curb of 10 cm height, and afterwards oscillates until the equilibrium is reached again. The complete analysis lasts for 7 s. It is carried out using the two variants of the method, the performance results being depicted in Table 2. This time the error is obtained as the difference, in J, between the lost gravitational energy and the energy dissipated by the shock-absorber.

Table 2. Results for the Iltis suspension.

	Absolute		Relative	
Δt	Error	Time	Error	Time
0.001	-43.13	17.8	-42.05	43.2
0.01	-43.45	2.04	-42.80	5.44
0.03	-57.96	0.84	no convergent	

In this case, unlike the previous one, the system shows a stiff behaviour, since the block representing one fourth of the whole vehicle has great inertia and, therefore, moves at slow frequency, while the wheel possesses little mass and, subsequently, moves at high frequency. The absolute variant is less accurate but more robust than the relative one, since it works for larger time-steps. Regarding efficiency, the absolute variant keeps the same advantage with respect to the relative one than in the previous example, since 25 global coordinates have been used face to 8 relative coordinates. It is supposed that many more variables are needed, so that the absolute variant requires a huge effort to solve the equation system, thus balancing its advantage.

Serial robot. The PUMA robot, designed by Unimation and Co. and shown in Fig. 10, is an example of a 6 degrees of freedom serial manipulator. It has been often used by different authors (see, for example, Angeles and Ma (1988)) to illustrate methods and procedures in several areas of robotics. In this work, the robot has been taken as an example of multibody system undergoing changing configurations.

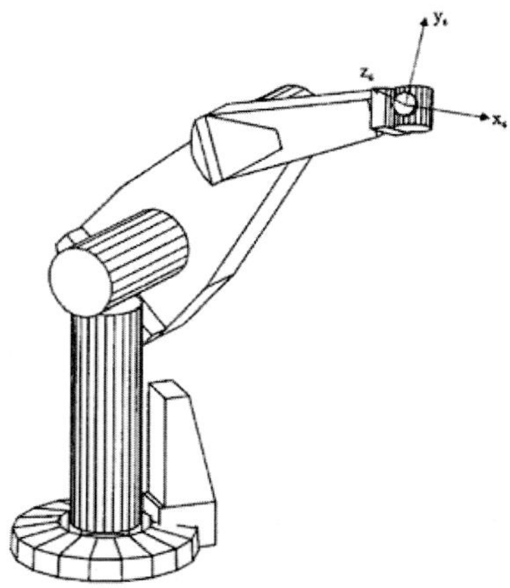

Figure 10. PUMA robot.

Starting from rest, torques are provided at the six hinges of the robot so that, in a time of 2 s, it arrives at a new position in the space, again in rest conditions. Once the new position has been reached, a point of the hand is attached to the ground, so that the robot loses 3 degrees-of-freedom. In this new configuration, torques are applied to the three rotational pairs of the hand, and a second maneuver, which lasts 4 s and ends with rest conditions, is performed. Finally, the robot is released from its attachment and returned in 2 s to the initial position by torques acting at the six revolute joints, once more finishing the maneuver at rest conditions. Therefore, the total simulation time is 8 s. Table 3 illustrates the results obtained when applying the two variants of the method, where the error has been calculated as the distance, in mm, between the hand positions at initial and final times (which, ideally, should be coincident).

This time, the absolute variant shows the worst behavior either in ef-

Table 3. Results for the PUMA robot.

	Absolute		Relative	
Δt	Error	Time	Error	Time
0.01	0.53	6.25	0.34	4.02
0.05	3.40	1.92	1.80	1.74
0.1	no convergent		no convergent	

ficiency, accuracy or robustness, probably because the number of global coordinates defined to model the robot is 45 (sparse matrix techniques have not been used in this case), far from the 6 corresponding relative coordinates needed.

Iltis vehicle. The fourth example, shown in Fig. 11, is the full model of the Iltis vehicle (Frik et al. (1993)), already referenced when describing the second example. The simulation consists of 8 s of motion with the vehicle going up an inclined ramp and then down a series of stairs, at a constant horizontal speed of 5 m/s (the road profile is illustrated in Fig. 12). The simulation leads to a rather violent motion with acceleration peaks of up to $5g$.

Figure 11. Iltis vehicle.

As said above, this time the programs have been implemented in Fortran language and, furthermore, sparse matrix technologies have been applied where needed, so that the potential of the proposed method for achieving real-time performance when dealing with complex and realistic models can be perceived. The simulation has been carried out by means of the two variants of the method studied so far in this paper. Moreover, in order to have a better reference to appraise the attained efficiency, commercial software ADAMS has also been used to perform the simulation. ADAMS default options (see ADAMS (2000)) have been chosen to run the simulation: variable time-step size Gear integrator along with an index-3 formulation, error of 10^{-3} and Jacobian evaluation in one out of four iterations. All the programs have been run on a PC with one AMD Athlon XP processor 1600+@1.4 GHz.

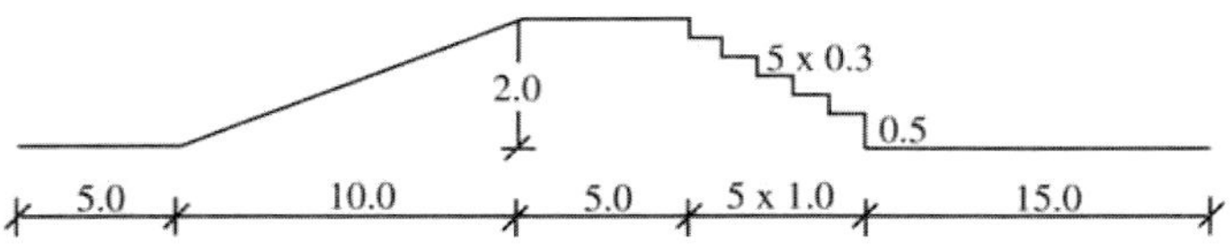

Figure 12. Road profile.

Table 4 illustrates the CPU times, in s, obtained when applying the two variants of the method and the ADAMS solver. Given that the two variants being compared use a fix time-step size integrator scheme while ADAMS features a variable time-step size procedure, the following criterion has been adopted in order to present the results: first, a fix time-step size of 0.01 s has been selected for the two variants of the method and, later on, the largest possible fix time-step size for which good results are obtained with each variant has been also considered, so as to show not only efficiency but robustness as well. Finally, the last row of Table 4 shows how many times each alternative is faster than real-time, which gives a simple and clear information about performance.

Table 4. Results for the Iltis vehicle.

Absolute		Relative		ADAMS
$\Delta t = 0.01$	$\Delta t = 0.0175$	$\Delta t = 0.01$	$\Delta t = 0.035$	Δt variable
5.41	3.80	0.72	0.38	7.62
2.11		21.05		1.05

In this example, it turns out that, definitely, the absolute variant is largely less efficient than the relative one. This can be explained by the fact that the absolute method needs 168 variables to model the vehicle, while the number of relative coordinates required by the relative method is only 26 (from which 10 are the system degrees of freedom). Moreover, the absolute variant is also less robust, since the largest time-step size it can reach while giving correct results is clearly smaller than that of its competitor. Finally, comparison with commercial software ADAMS shows that the proposed formulation can be seen as a good candidate for real-time general-purpose applications, as it encompasses good levels of efficiency, accuracy and robustness: the less performing absolute variant is easier to program, while the more performing relative variant is more involved.

2.7 Alternative Integrators

So far, the two variants of the method being compared (absolute and relative) have been built using the trapezoidal rule as numerical integrator. In this section, the influence of the numerical integrator in the behavior of each method is to be addressed. Three families of integrators, all of them belonging to the so-called structural integrators group, are to be tested: the Newmark family with numerical dissipation, the HHT algorithm, and the Generalized-α method. The reason to use the structural integrators is that they show a completely analogous scheme to the one of the trapezoidal rule (which in fact is a particular case of all of them), so that the general strategy of combining the dynamic and integrator equations can be preserved.

The general form of the dynamic equations is,

$$\mathbf{M}(\mathbf{q}) \cdot \ddot{\mathbf{q}} + \mathbf{P}(\mathbf{q}, \dot{\mathbf{q}}) = \mathbf{0} \tag{70}$$

where $\mathbf{M}(\mathbf{q})$ is the mass matrix and $\mathbf{P}(\mathbf{q}, \dot{\mathbf{q}})$ contains all the terms which are constant or functions of positions and/or velocities. When the above-mentioned integrators are applied, Eq. (70) turns into,

$$\begin{aligned} &[(1-\delta_m) \cdot \mathbf{M}_{n+1} \cdot \ddot{\mathbf{q}}_{n+1} + \delta_m \cdot \mathbf{M}_n \cdot \ddot{\mathbf{q}}_n] + \\ &[(1-\delta_f) \cdot \mathbf{P}(\mathbf{q}_{n+1}, \dot{\mathbf{q}}_{n+1}) + \delta_f \cdot \mathbf{P}(\mathbf{q}_n, \dot{\mathbf{q}}_n)] = \mathbf{0} \end{aligned} \tag{71}$$

which means that the dynamic equilibrium is stated from partial contributions from step n and $n+1$.

On the other hand, the difference equations of Newmark integrator are:

$$\begin{aligned} \dot{\mathbf{q}}_{n+1} &= \frac{\gamma}{\beta \cdot \Delta t} \cdot \mathbf{q}_{n+1} + \hat{\dot{\mathbf{q}}}_n \\ \hat{\dot{\mathbf{q}}}_n &= -\left(\frac{\gamma}{\beta \cdot \Delta t} \cdot \mathbf{q}_n + \left(\frac{\gamma}{\beta} - 1\right) \cdot \dot{\mathbf{q}}_n + \Delta t \left(\frac{\gamma}{2\beta} - 1\right) \ddot{\mathbf{q}}_n\right) \end{aligned} \tag{72}$$

$$\begin{aligned} \ddot{\mathbf{q}}_{n+1} &= \frac{1}{\beta \cdot \Delta t^2} \cdot \mathbf{q}_{n+1} + \hat{\ddot{\mathbf{q}}}_n \\ \hat{\ddot{\mathbf{q}}}_n &= -\left(\frac{\gamma}{\beta \cdot \Delta t^2} \cdot \mathbf{q}_n + \frac{\gamma}{\beta \cdot \Delta t} \cdot \dot{\mathbf{q}}_n + \left(\frac{1}{2\beta} - 1\right) \ddot{\mathbf{q}}_n\right) \end{aligned} \tag{73}$$

Eqs. (71-73) are the basis from which the two variants of the dynamic formulation implementing the three families of integrators here considered are to be obtained. As an example, the expressions for the residual vector, the approximated tangent matrix, and the projections in velocities and accelerations are to be written. They can be compared with their corresponding counterparts when using the trapezoidal rule, given by Eq. (21-24) for the method in absolute coordinates, and by Eq. (61-64) for the method in relative coordinates. As in previous occasions, the variables will be designated by $\mathbf{q}$, although they can represent either the absolute or relative coordinates.

$$\mathbf{f}(\mathbf{q}) = \beta\Delta t^2 \Big[(1-\delta_m)\mathbf{M}_{n+1}\ddot{\mathbf{q}}_{n+1} + (1-\delta_f)\left(\boldsymbol{\Phi}_{\mathbf{q}}^{\mathrm{T}}\alpha\boldsymbol{\Phi} + \boldsymbol{\Phi}_{\mathbf{q}}^{\mathrm{T}}\boldsymbol{\lambda}^* - \mathbf{Q}\right)_{n+1} + \\ +\delta_m\mathbf{M}_n\ddot{\mathbf{q}}_n + \delta_f\left(\boldsymbol{\Phi}_{\mathbf{q}}^{\mathrm{T}}\alpha\boldsymbol{\Phi} + \boldsymbol{\Phi}_{\mathbf{q}}^{\mathrm{T}}\boldsymbol{\lambda}^* - \mathbf{Q}\right)_n\Big] \tag{74}$$

$$\frac{\partial\mathbf{f}(\mathbf{q})}{\partial\mathbf{q}} = (1-\delta_m)\mathbf{M}_{n+1} + (1-\delta_f)\gamma\Delta t\mathbf{C} + (1-\delta_f)\beta\Delta t^2\left(\boldsymbol{\Phi}_{\mathbf{q}}^{\mathrm{T}}\alpha\boldsymbol{\Phi}_{\mathbf{q}} + \mathbf{K}\right) \tag{75}$$

$$\frac{\partial\mathbf{f}(\mathbf{q})}{\partial\mathbf{q}}\dot{\mathbf{q}} = \left[(1-\delta_m)\mathbf{M}_{n+1} + (1-\delta_f)\gamma\Delta t\mathbf{C} + (1-\delta_f)\beta\Delta t^2\mathbf{K}\right]\dot{\mathbf{q}}^* - \\ -(1-\delta_f)\beta\Delta t^2\boldsymbol{\Phi}_{\mathbf{q}}^{\mathrm{T}}\alpha\boldsymbol{\Phi}_t \tag{76}$$

$$\frac{\partial\mathbf{f}(\mathbf{q})}{\partial\mathbf{q}}\ddot{\mathbf{q}} = \left[(1-\delta_m)\mathbf{M}_{n+1} + (1-\delta_f)\gamma\Delta t\mathbf{C} + (1-\delta_f)\beta\Delta t^2\mathbf{K}\right]\ddot{\mathbf{q}}^* - \\ -(1-\delta_f)\beta\Delta t^2\boldsymbol{\Phi}_{\mathbf{q}}^{\mathrm{T}}\alpha\left(\dot{\boldsymbol{\Phi}}_{\mathbf{q}}\dot{\mathbf{q}} + \dot{\boldsymbol{\Phi}}_t\right) \tag{77}$$

Four parameters appear in Eq. (71-73): δ_m, δ_f, γ, β. Different values of these four parameters lead to the different families of integrators. From the studies performed on structural integrators (usually for linear systems) concerning stability and accuracy issues (see, for example, Geradin and Cardona (2001)), it is concluded that the more appropriate values for the parameters are those shown in Table 5. As it can be observed in Table 5, the four parameters of each family are obtained as functions of one single parameter, called in the Table the *basic parameter*. The admissible range of variation of such basic parameters is also illustrated in the Table.

With this choice of the parameters, the three families are unconditionally stable for linear systems. In respect to the accuracy, the HHT and the Generalized-α algorithms are second order, while the Newmark dissipative is only first order accurate.

In order to find out the influence of the numerical integrator selected on the final performance of the two variants previously compared in this paper,

Table 5. Parameters of the integrators.

	Newmark dissip.	HHT	Generalized-α
Basic parameter	$\xi \in [-1, 0]$	$\delta_f \in \left[0, \frac{1}{3}\right]$	$\rho_\infty \in [0, 1]$
δ_m	0	0	$\frac{2\rho_\infty - 1}{\rho_\infty + 1}$
δ_f	0	δ_f	$\frac{\rho_\infty}{\rho_\infty + 1}$
γ	$\frac{(1-2\xi)}{2}$	$\frac{(1+2\delta_f)}{2}$	$\frac{1}{2} - \delta_m + \delta_f$
β	$\frac{(1-\xi)^2}{4}$	$\frac{(1+\delta_f)^2}{4}$	$\frac{1}{4}\left(1 + \delta_f - \delta_m\right)^2$

the three families of integrators (Newmark dissipative, HHT, Generalized-α have been implemented for each formulation (absolute and relative). Then, the already described simulation of the Iltis vehicle has been executed with each combination of dynamic formulation and numerical integrator (i.e. six combinations are possible).

Multiple executions have been carried out for each combination, so as to perform a two-dimensional sweeping: basic parameter and time-step size. Table 6 shows the ranges of variation of the basic parameter and the time-step size for each integrator family, along with the increments considered for these magnitudes to perform the different executions.

Table 6. Range of variation and increment of basic parameter and time-step for each integrator family.

	Newmark dissip.	HHT	Generalized-α
Basic parameter	$\xi \in [-1, 0]$	$\delta_f \in \left[0, \frac{1}{3}\right]$	$\rho_\infty \in [0, 1]$
Increment	0.02	0.01	0.02
Δt (s)	[0.01,0.05]	[0.01,0.05]	[0.01,0.05]
Increment (s)	0.0025	0.0025	0.0025

For all the executions, the CPU-time spent to perform the simulation has been recorded, as well as the solution error. To evaluate such error, the solution attained with the corresponding dynamic formulation and the trapezoidal rule for a time-step of 0.001 s has been taken as reference. Then, the error of a certain combination of dynamic formulation and integrator has been obtained as,

$$error = \frac{1}{(nse + 1)} \sum_{i=0}^{nse} |q_i - q_i^*| \tag{78}$$

where nse stands for the number of time-steps considered to evaluate the error, q_i represents the history of the vertical coordinate of a chassis point in the neighborhood of the mass center, and q_i^* is the solution taken as reference. To make the error highly sensitive to the actual discrepancies between any solution and the reference solution, the proposed error has been evaluated considering only the time-steps going from second 5 to second 8 (end) of the simulation, since the most violent motion takes place in that time interval.

The next two subsections are devoted to present the results obtained from the huge amount of executions carried out, in such a way that discussion can be established, and conclusions extracted. Note that each execution comes from a certain combination of dynamic formulation and numerical integrator, and is defined by the assigned values of basic parameter and time-step size. Therefore, a particular execution can be referred to as (*dynamic formulation*, *numerical integrator*, *basic parameter*, Δt). The following abbreviations have been used: absolute (ABS), relative (REL), trapezoidal rule (TRP), Newmark dissipative (NWD), HHT (HHT), Generalized-α (GNα).

Test of efficiency-robustness. In this test, for each combination of dynamic formulation and integrator, the fastest execution, i.e. the execution with the couple of values (*basic parameter*, Δt) which spent the smallest CPU-time to carry out the simulation, has been recorded. The only condition established is convergence along the whole simulation, but the error has not been limited. A maximum time-step size of 0.05 s has been allowed for all the methods, as indicated in Table 6. The best performance achieved for each formulation when combined with the trapezoidal rule is also repeated here for comparison. The results are listed in Table 7 where, for each dynamic formulation, the integrator which has led to the most efficient method has been boldfaced.

The trend of Newmark dissipative when combined with either the absolute or the relative formulation (i.e. the integration is performed in dependent coordinates through a penalty approach) is that bigger damping produces better convergence but worse accuracy. A careful selection of the basic parameter enables to take larger time-steps than those used for the trapezoidal rule, while keeping a good level of accuracy, thus achieving high efficiency rates.

Regarding the HHT algorithm, when combined with either the absolute or the relative formulation, it is observed that, if small time-steps are used, the trapezoidal rule shows to be more efficient and accurate. However, for large time-steps the trend is just the opposite, and good levels of efficiency are attained while keeping moderate error values if an adequate value of the basic parameter is chosen.

Table 7. The most efficient integration schemes for each dynamic formulation.

	Basic parameter	Δt (s)	CPU-time (s)	$\frac{\text{Real - time}}{\text{CPU - time}}$	Error (x10^{-2})
ABS+TRP	-	0.0175	3.86	2.07	5.671
ABS+NWD	**-0.62**	**0.045**	**1.74**	**4.60**	**9.702**
ABS+HHT	0.15	0.03	2.62	3.05	7.167
ABS+GNα	0	0.025	2.77	2.89	8.186
REL+TRP	-	0.035	0.37	21.62	9.016
REL+NWD	**-0.52**	**0.05**	**0.25**	**32.00**	**9.795**
REL+HHT	0.35	0.0425	0.33	24.24	6.267
REL+GNα	0	0.05	0.28	28.57	6.189

In what respects to the Generalized-α method, when combined with either the absolute or the relative formulation, it comes that the trapezoidal rule is clearly superior for small time-steps. However, the trend can be inverted for large time-steps if a good value of the spectral radius ρ_∞ is selected for the Generalized-α integrator. Such selection should be carried out very carefully, as the algorithm is very sensitive to small changes in the value of the spectral radius. A too high value of ρ_∞ leads to unstable behavior, while a too low value of ρ_∞ eliminates an excessive part of the solution.

Summarizing, both penalty formulations (i.e. absolute and relative) clearly benefit from the use of structural integrators with some numerical damping, so obtaining clear enhancements in efficiency with respect to the use of the trapezoidal rule as integrator. The best performance is achieved with the Newmark dissipative family. Furthermore, the advantage of the relative formulation over the absolute one, already shown when using the trapezoidal rule (see Table 4), is preserved.

For the reader to appraise the meaning of the error values presented in Table 7, the accuracy obtained in the fastest simulation presented ($9.795 \cdot 10^{-2}$ m) has been illustrated in Fig. 13, through comparison between such solution and that used as reference to evaluate the error (trapezoidal rule with time-step of 0.001 s). The plot corresponds to the vertical coordinate of a chassis point in the neighborhood of the mass center, which was used to work out the error (see Eq. 78). The least damped line corresponds to the reference.

It can be seen in Fig. 13 that significant differences are obtained during the final and more violent phase of the motion. The reason is that the

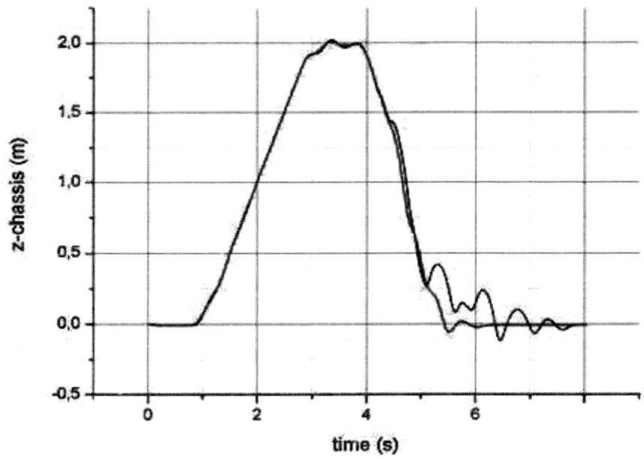

Figure 13. The fastest simulation with time-step under 0.05 s vs. the reference simulation.

method achieves its excellent level of stability, which in turn allows for incredibly large time-steps and very high efficiency rates, at the price of introducing some numerical damping, thus suppressing part of the solution. Moreover, since so large time-steps (0.05 s) are used, many things happen between two consecutive time-steps, which also causes a loss of information about the system response. Therefore, depending on the application, the analyst will have to accept a compromise between the levels of accuracy and efficiency he desires to achieve, but what has been clearly stated is that the proposed method shows excellent stability, thus allowing the analyst to take very large time-step sizes.

Test of efficiency-accuracy. In this test, for each combination of dynamic formulation and integrator, the fastest execution producing an error value under a prescribed limit, i.e. the execution with the couple of values (*basic parameter*, Δt) which spent the smallest CPU-time to carry out the simulation while leading to an error value under $4 \cdot 10^{-2}$ m, has been recorded. The corresponding results for each formulation when combined with the trapezoidal rule have also been obtained for comparison. The results are listed in Table 8 where, for each dynamic formulation, the integrator which has led to the most efficient method has been boldfaced.

Table 8. The most efficient integration schemes under a given error value for each dynamic formulation.

	Basic parameter	Δt (s)	CPU-time (s)	$\frac{\text{Real - time}}{\text{CPU - time}}$	Error (x10^{-2})
ABS+TRP	-	0.015	4.39	1.82	3.310
ABS+NWD	-0.04	0.0175	3.76	2.13	3.595
ABS+HHT	**0.04**	**0.02**	**3.56**	**2.25**	**3.764**
ABS+GNα	0.16	0.0175	4.01	2.00	3.774
REL+TRP	-	0.015	0.62	12.90	2.656
REL+NWD	**-0.08**	**0.0375**	**0.36**	**22.22**	**3.457**
REL+HHT	0.22	0.0375	0.37	21.62	2.981
REL+GNα	0.02	0.0425	0.36	22.22	3.928

In general terms, Table 8 confirms what has been said in the previous subsection: both penalty formulations do benefit from integrators with some numerical damping. However, this time the gain achieved by the relative formulation from the use of dissipative integrators is clearly superior to the one obtained by its absolute competitor. Again the Newmark dissipative family yields the best results and, furthermore, its combination with the relative formulation produces a method which encompasses the highest levels of efficiency, accuracy and robustness.

In order to have a feeling of how good the obtained solutions are, now that a maximum error value of $4 \cdot 10^{-2}$ m has been accepted, the accuracy obtained in the fastest simulation presented ($3.457 \cdot 10^{-2}$ m) has been illustrated in Fig. 14, through comparison between such solution and that used as reference to evaluate the error (trapezoidal rule with time-step of 0.001 s). The plot corresponds to the vertical coordinate of a chassis point in the neighborhood of the mass center, which was used to work out the error (see Eq. 78). Again, the least damped line corresponds to the reference.

It can be perceived that the obtained solution, yielding an error of $3.457 \cdot 10^{-2}$ m, is much more accurate than the one illustrated in Fig. 13, corresponding to an error of $9.795 \cdot 10^{-2}$ m. Accordingly, the efficiency of the most accurate solution comes down to a factor of 22.22, from the 32.00 for the least accurate one. Therefore, it can be stated that the hybrid formulation combined with the Newmark dissipative integrator leads to a method that provides high efficiency along with very acceptable accuracy.

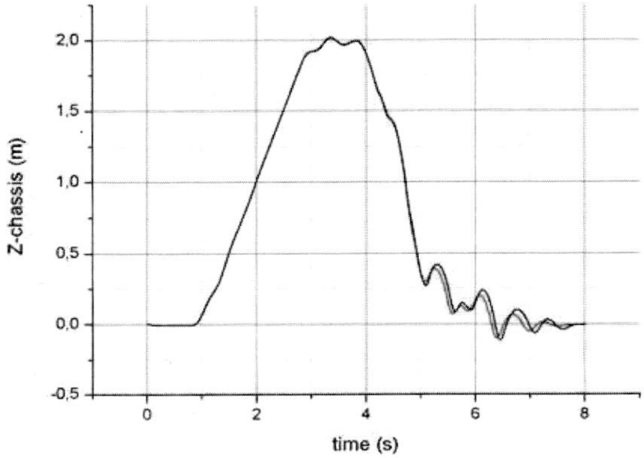

Figure 14. The fastest simulation with an error under $4 \cdot 10^{-2}$ m vs. the reference simulation.

2.8 Application Criteria

At the view of the presented results, and taking into account that more factors than just the size of the system must be considered when deciding the more suitable method for a certain problem, the following simplified practical recommendations can be given for real-time applications:

a) For small- and moderate-size systems (under 50 absolute variables), the preferred selection is the method in absolute coordinates combined with the trapezoidal rule.

b) For large systems (over 100 absolute variables), the best results are obtained through the method in relative coordinates in combination with a Newmark dissipative scheme.

3 Experimental Validation

To further confirm the good behavior of the method developed by the authors, the dynamic analysis of a complex and realistic case is addressed in this section. The objective consists of obtaining the stresses suffered by the chassis of a prototype car during its motion by using the proposed formulation, and validating the corresponding results through comparison with those obtained from experimental measurements.

In what follows, the prototype will be described, along with the maneuver chosen to carry out the experiment. Then, the instrumentation installed onboard the prototype to get the measurements will be shown, and details about the computational model of the vehicle will be provided. Finally, the experimental and numerical results will be compared.

3.1 The Prototype

The prototype car, which can be seen in Fig. 15, has been manufactured in steel tubing and parts from old cars have been cannibalized and conveniently adapted (i.e. engine, suspension and steering systems, wheels, etc.).

Figure 15. The actual prototype car.

Inertial data of the different components are listed in Table 9, giving a total mass of the car of 597 kg. Front shock-absorbers are characterized by a stiffness constant of 12,360 N/m and a free length of 0.750 m, while their rear counterparts have values of 10,595 N/m and 0.916 m for the same magnitudes. Damping of these front and rear elements has been set to 1,000 Ns/m. Radii of the tires are 0.27 and 0.29 m for the front and rear tires respectively, which show a vertical stiffness of 60,000 N/m.

3.2 The Maneuver

A maneuver has been defined in order to compare the resulting stresses at the chassis coming from measurement and simulation: starting from rest, the car accelerates and travels direct until reaching an approximate speed of 5 m/s, then drops down a curb of 8 cm height, and finally brakes until complete stop is attained. In this way, the bending modes of the chassis are

Table 9. Mass of each component.

Body description	Mass (kg)
Chassis	107.00
Engine	180.00
Driver + Seat + Experimental boxes	154.00
PC + accessories	12.00
Front wheel	14.00
Rear wheel	17.50
Front shock-absorber	7.00
Rear shock-absorber	6.00
Short spindle	4.00
Long spindle	6.00
Front hub	8.00
Front A-arm	2.00
Rear hub	9.50
Rear suspension coupler	3.00

excited. The described maneuver is illustrated in Fig. 16.

Figure 16. The actual maneuver.

3.3 Instrumentation

To obtain the motion of the car, four tri-axial accelerometers have been attached to the chassis at the four corners of its base. Accelerations provided by the sensors in the motion direction have been integrated twice in order to obtain the position and velocity of the chassis at all times. On the other hand, four Wheatstone bridges have been installed at four points of the chassis to measure bending stresses. They are half-bridges, with an upper and a lower band each, to measure the normal bending stresses on the steel tubes surfaces. The location of the sensors (accelerometers and strain gauges) on the chassis is shown in Fig. 17.

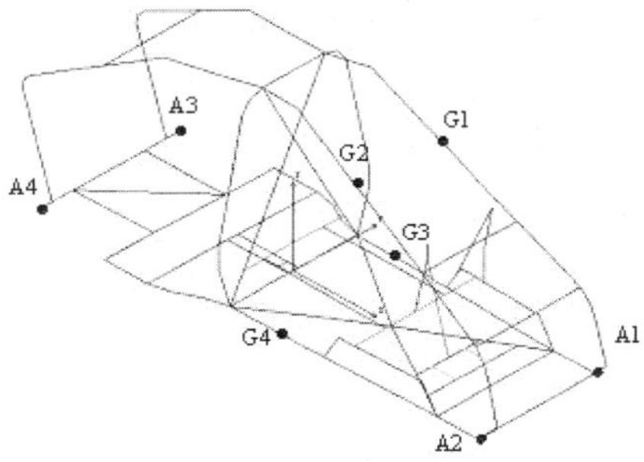

Figure 17. Location of the sensors.

The sixteen captured signals -twelve accelerations and four strains- are recorded by a laptop PC after passing through a data-adquisition board. Figure 18 offers two views of the experimental setup, while Fig. 19 shows the connection scheme.

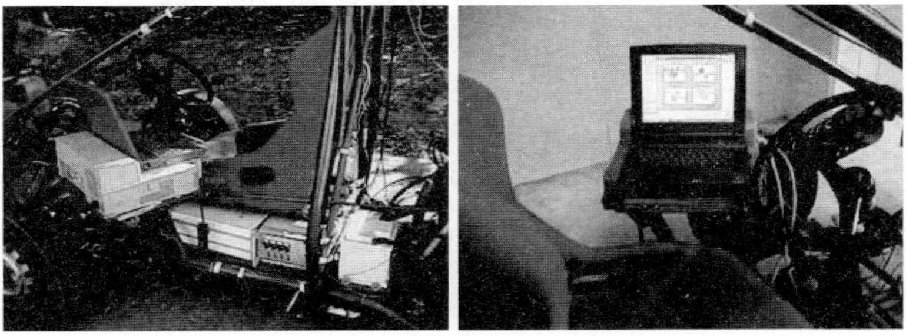

Figure 18. Experimental setup.

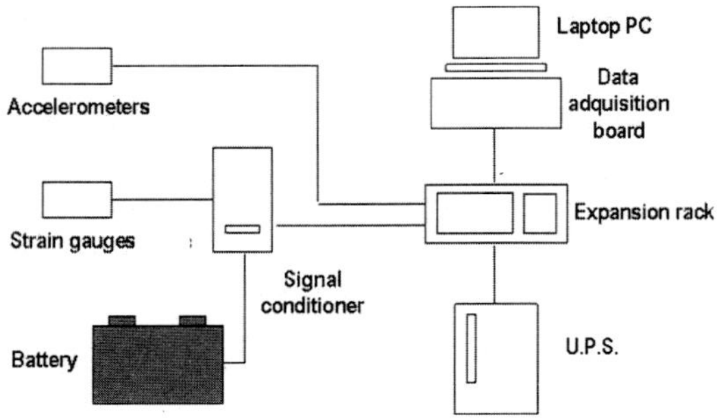

Figure 19. Connection scheme of the data-adquisition system.

3.4 The Virtual Prototype

The corresponding virtual system has been built according to the method in absolute coordinates. Of course, the method in relative coordinates could have been used as well, but the former was chosen due to its easiness of implementation and the fact that real-time performance was not required for this test.

For this example, consideration of flexible bodies is required, since this is the correct way to obtain stresses of the system components during the motion. Due to space limitations, the extension of the two variants of the proposed method to flexible bodies will not be presented here. However, the interested readers can resort to Cuadrado et al. (2004b), for the method in absolute coordinates, and to Lugris et al. (2007), for the method in relative coordinates. In both cases, flexibility is considered by the floating frame of reference approach along with static and dynamic deformation modes.

The modeling in natural coordinates can be seen in Fig. 20. The chassis in steel tubing has been considered as flexible, while all the other bodies of the car have been considered as rigid. The driver has not been explicitly modeled, his inertial properties having been included in those of the seat. The whole model consists of 58 points, 16 unit vectors, 5 distances, 1 angle, 60 static modes and 5 dynamic modes, which gives a total number of variables of 293 for the whole multibody system.

The forces applied in the model are: the weight of each component, acting at the mass center of it; the forces provided by the shock-absorbers, considered as linear springs and dampers; the contact forces at the tires, which in this case have been limited to the compressive forces, modeled as linear springs; the engine torque, acting on both the engine and the rear wheels; the braking torques, acting on the four wheels and hubs.

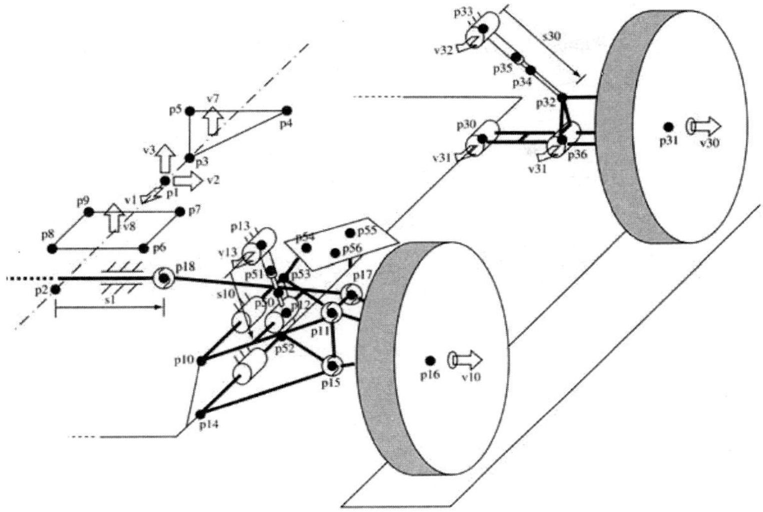

Figure 20. Modeling of the prototype car in natural coordinates.

To obtain the static and dynamic modes of the chassis, as well as its finite element mass and stiffness matrices (all of them needed for the proposed formulation), a finite element model of this structure has been prepared, illustrated in Fig. 21. The commercial code COSMOS/M has been used for this purpose. The model consists of 247 beam elements and 211 nodes with 6 degrees of freedom each, so that the problem size is 1,266. A structural damping matrix condensed to the modes of 1/100 times the corresponding stiffness matrix has been adopted, which provided an adequate level of dissipation.

The number of static modes rises to 60, as said before. This is due to the high number of boundaries the chassis has (points and unit vectors which connect the chassis to neighbor bodies), which, in addition to the variables forming the body local base (point **p1** and unit vectors **v1**, **v2** and **v3**), are represented in Fig. 22. In fact, not all the possible static modes have been taken into account (theoretically, three for each boundary point and two for each boundary unit vector), as long as the contribution of some of them to the chassis deformation can be neglected (e.g. static modes producing axial deformation of the steel tubes).

The number of dynamic modes should be decided by the analyst. In this occasion, 5 dynamic modes have been considered, covering the natural frequencies up to 100 Hz.

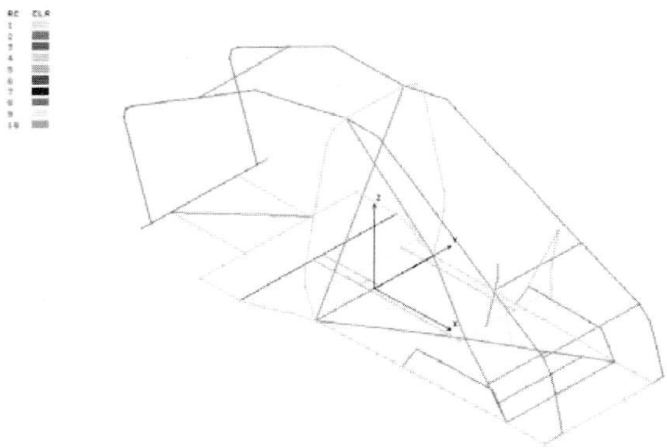

Figure 21. Finite element model of the chassis.

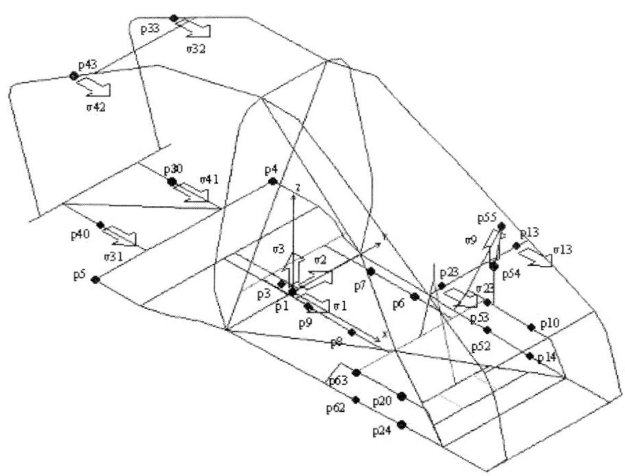

Figure 22. Boundary points and unit vectors of the chassis.

In order to reproduce at the virtual analysis the maneuver performed with the real car, x-positions, x-velocities and x-accelerations of the chassis, previously obtained from the experimental test and later integration as already commented, have been used to kinematically guide the forward motion of the origin of the chassis local reference frame (point **p1** in the model shown in Fig. 20). However, the actual prototype is not dragged or stopped by a powerful hand acting at point **p1**, but powered by the engine and slowed down by the brakes. Therefore, forces caused by engine and brakes must be considered, as long as their effect on the stresses suffered by the chassis is crucial.

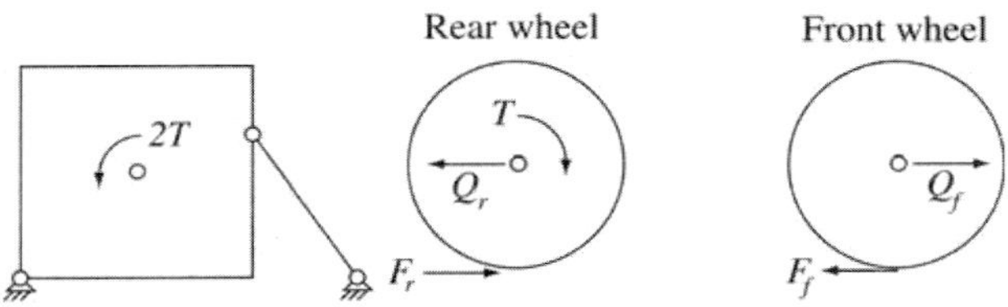

Figure 23. Equilibrium of engine and wheels during acceleration.

Figure 23 shows the engine and wheels equilibrium when the car is accelerating. The dynamic equilibrium equations lead to the following results,

$$T = \left(\frac{mR}{2} + \frac{2I_R}{R} \right) a \tag{79}$$

$$F_r = \left(\frac{m}{2} + \frac{I_R}{R^2} \right) a \tag{80}$$

$$Q_r = \left(\frac{m}{2} + \frac{I_R}{R^2} - m_R \right) a \tag{81}$$

$$F_f = \left(\frac{I_R}{R^2} \right) a \tag{82}$$

$$Q_f = \left(\frac{I_R}{R^2} + m_R \right) a \tag{83}$$

where T is the engine torque at each rear wheel, F_f and F_r are the front and rear friction reactions at the ground rolling contacts, Q_f and Q_r are the front and rear hub reactions, m is the total mass of the car, R the wheel radius, m_R and I_R are the wheel mass and inertia moment, and a is the forward acceleration of the car. Therefore, the engine torque T and the friction forces F_f and F_r can be obtained as functions of the forward acceleration a at each time by means of Eq. (79), (80) and (82), and introduced into the force vector of the model.

The braking case is treated analogously. Figure 24 shows the equilibrium of any of the wheels when the car is braking. Opposite torque T and force Q are applied to the hub. The dynamic equilibrium equations yield,

$$T = \left(\frac{mR}{4} + \frac{I_R}{R} \right) a \tag{84}$$

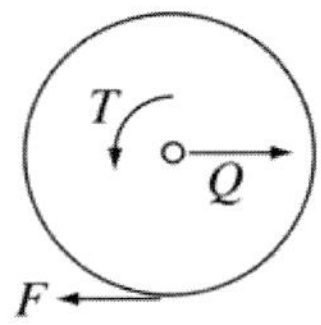

Figure 24. Wheel equilibrium during braking.

$$F = \left(\frac{m}{4}\right) a \tag{85}$$

$$Q = \left(\frac{m}{4} - m_R\right) a \tag{86}$$

where T is the braking torque at each wheel, F is the friction reaction at the ground rolling contact, and Q is the hub reaction. Hence, the braking torque T and the friction force F can be obtained as functions of the forward acceleration a at each time by means of Eq. (84) and (85), and introduced into the force vector of the model.

3.5 Results and Comparison

The actual maneuver of 15 s has been simulated by implementation of the proposed formulation, with a fixed time-step of 0.01 s, the same sample period used for data adquisition during the experimental test, where a low-pass filter was applied with a cutoff frequency of 10 Hz to avoid noise from engine vibration. The penalty factor was adjusted to 10^9, due to the presence of high stiffness values. The history of the forward acceleration of the car is shown in Fig. 25, obtained from both test and simulation. It can be seen that an excellent correlation has been achieved.

The comparison between measured and calculated bending moments at points G2 and G4 (see Fig. 17) is shown in Figure 26. The corresponding plots for points G1 and G3 have been omitted as long as the problem is almost symmetric. At the view of the results, three time intervals can be distinguished.

The first one is the [0,5] interval. During this period of time, the car is at rest, but some peaks can be detected at about 2 s. They are explained because, at that time, the gear was set on *drive* position. Obviously, this kind of engine loads are not considered by the model since, as previously explained, the engine torque is derived from acceleration information. Therefore, if the car is at rest, no engine torque is introduced in the model. This fact justifies the discrepancies shown between measurements and calculations at G4 during this time interval.

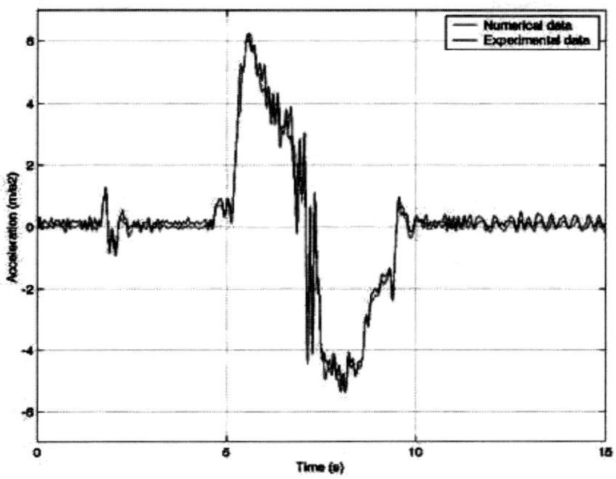

Figure 25. Forward acceleration of the car.

The second time interval is [5,10]. During these five seconds, the car accelerates, goes over the curb, and brakes until reaching the rest again. Good correlation is found during the accelerating and braking phases. The high peaks due to the falling action are not perfectly reproduced, although their intensities are reasonably captured. An explanation for such peak discrepancy can be the appearance of percussion forces in the actual prototype, due to the presence of clearances in some joints which have not been modeled.

The third and final time interval is [10,15], when the vehicle is again at rest. Some amount of error in the measurements can be detected in the figures, since the bending moments should return to zero given that the initial and final positions of the car are exactly the same. However, the calculations come back correctly to zero.

4 Real-Time Application

In order to show the aptitude of the proposed method for addressing real-time applications, this section describes the development of a low-cost and efficient computational framework for the design and evaluation of automobile motion controllers. The core elements of the tool are the proposed formalism for the dynamics of multibody systems, a virtual reality (VR) interface for human-in-the-loop simulation, and Matlab for the controllers.

The starting point of the work has been the prototype car described

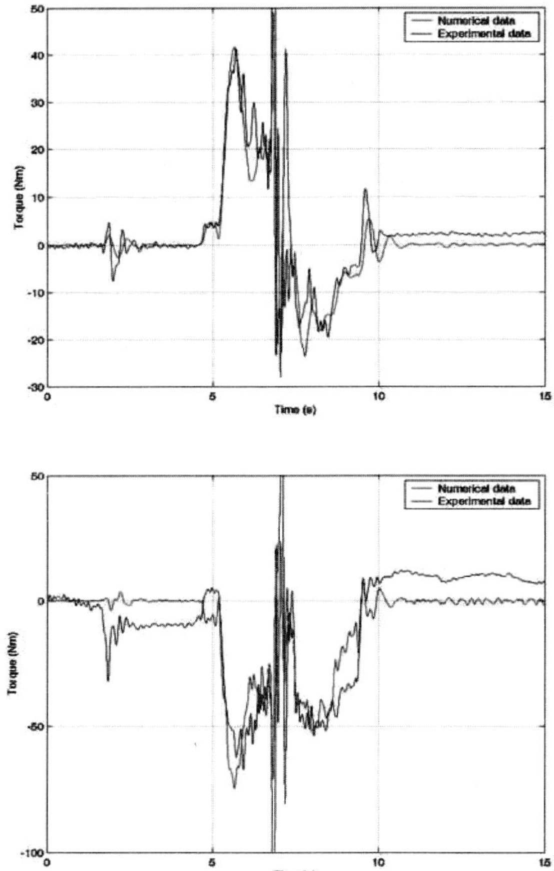

Figure 26. Comparison of bending moments: a) At point G2; b) At point G4.

in the previous section, whose virtual counterpart has been implemented in Fortran language through the proposed dynamic formulation but, this time, all the bodies, the chassis included, have been considered as rigid. Virtual reality capabilities have been given to the program by means of a realistic graphical output (either on screen or HMD) and game-type driving peripherals (steering wheel and pedals), in order to enable participation of human drivers in the controller design and evaluation processes. Although the framework is independent on the control method selected, fuzzy logic has been chosen for the design of the control algorithms, which have been implemented on the Matlab environment. Figure 27 illustrates the general structure of the computational tool developed.

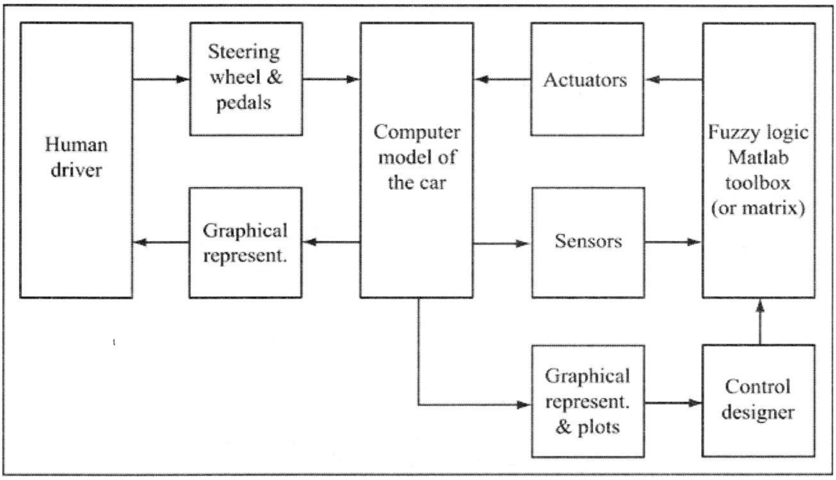

Figure 27. Computational framework for controller design and evaluation.

This section is organized as follows: subsection 1 describes the computational model of the car; subsection 2 focuses on the virtual reality features of the tool, and their role in the controller design and evaluation processes; subsection 3 explains the structure of the control module, justifies the use of fuzzy logic in the work, and shows the different alternatives available to connect the Fortran code containing the dynamics of the car with the Matlab functions implementing the control algorithms, pointing out their preferred contexts of application; subsection 4 details the process followed to design the controllers for two examples, and illustrates the advantages of human-in-the-loop simulation for both the design and evaluation of the controllers; finally, subsection 5 makes some efficiency remarks.

4.1 The Computational Model of the Car

Although the prototype car is the same, its mathematical description is different from the one presented in the previous section, since this time the chassis has been considered as rigid, and possible maneuvers are not limited to straight line maneuvers which must replicate a previously measured motion, but can be any. Again, the method in absolute coordinates has been applied due to its simpler implementation. It will be seen that, although this variant of the formalism is less efficient than the method in relative coordinates, it is capable of providing the necessary performance to address highly challenging applications.

The modeling of the chassis, the steering and suspension systems, and the wheels, carried out in natural coordinates, is illustrated in Fig. 28: 44 points, 7 unit vectors, 5 distances and 5 angles have been used as problem variables, leading to a total problem size of 163.

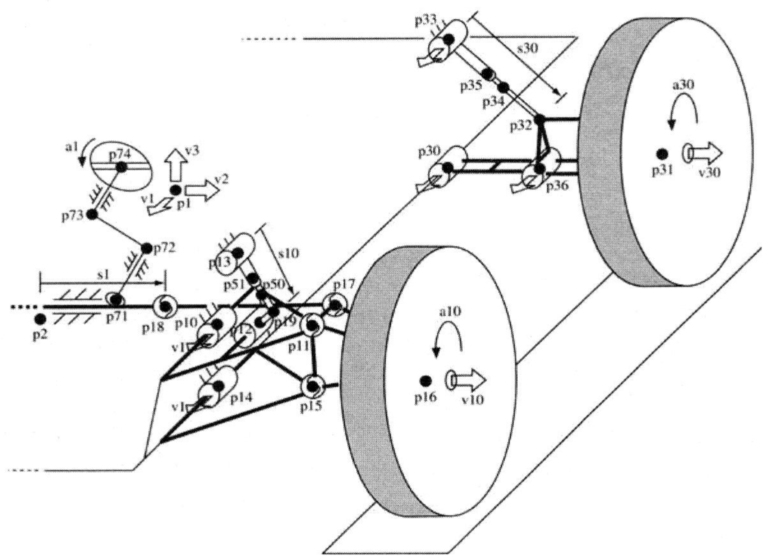

Figure 28. Prototype model in natural coordinates.

The steering wheel is kinematically guided. Forces which deserve to be described are the following:

a) The chassis inertial parameters have been obtained from a model developed with the CAD commercial package I-deas, in which the actual structural elements have been replicated. The engine inertial parameters have been experimentally estimated.

b) Suspension forces: linear models of springs and dampers have been considered.

c) Tire forces: lateral and longitudinal forces, as well as self-aligning torques,

have been introduced through the *Magic Formula* (Bakker and Pacejka (1991)) with coefficients provided by the tire manufacturer. Moreover, rolling resistance has been taken into account.
d) Power transmission forces: from the torque-speed engine relationships and the gear ratios, both provided by the engine maker, the automatic gearing has been modeled. Then, for a certain value of the car velocity, the engine speed and, consequently, the engine torque, can be easily derived. The torque is applied on the rear wheels. The model of power transmission includes engine braking torque at closed throttle position.
e) Brake forces: the braking torque has been estimated from disk geometry, being applied to the four wheels.
f) Aerodynamic forces: they have been considered as proportional to the square of the car speed, with an estimated aerodynamic coefficient.
g) Impact forces: the volume of the car has been approximated by several spheres, and the indentation method has been applied in order to model the impacts of the car against other objects of the environment.

4.2 The Virtual Reality Interface

The computational model of the car has been used to create a driving simulator, shown in Fig. 29, by combining the already mentioned vehicle dynamics code, along with a realistic graphical output (received by the user either through the computer screen or through a HMD) and game-type driving peripherals (steering wheel and pedals). Both the gearing and the driver's point of view are operated from buttons on the wheel. For the graphical output, the application programming interface (API) selected has been OpenGL, due to its portability and intuitive programming structure. Communication between the program and the driving peripherals has been managed with the library DirectInput from the API DirectX.

Such a virtual reality interface serves to several purposes. Prior to the control design stage, a human driver is expected to carry out the targeted maneuver on the simulator, in order to provide an initial assessment on maneuver feasibility and realistic values of the involved parameters: time required to perform the maneuver, maximum speed and acceleration for stability, more challenging parts of the maneuver, etc. Moreover, the performance achieved by the human driver may be used as a reference to measure the quality of the successive controllers obtained during the design stage, until the desired behavior is attained. Additionally, when the aim is to partially control the motion of the car, like in driver assistance applications (as opposed to cases in which the total automatic control of the car is pursued), both the human driver and the controller must simultaneously operate (see

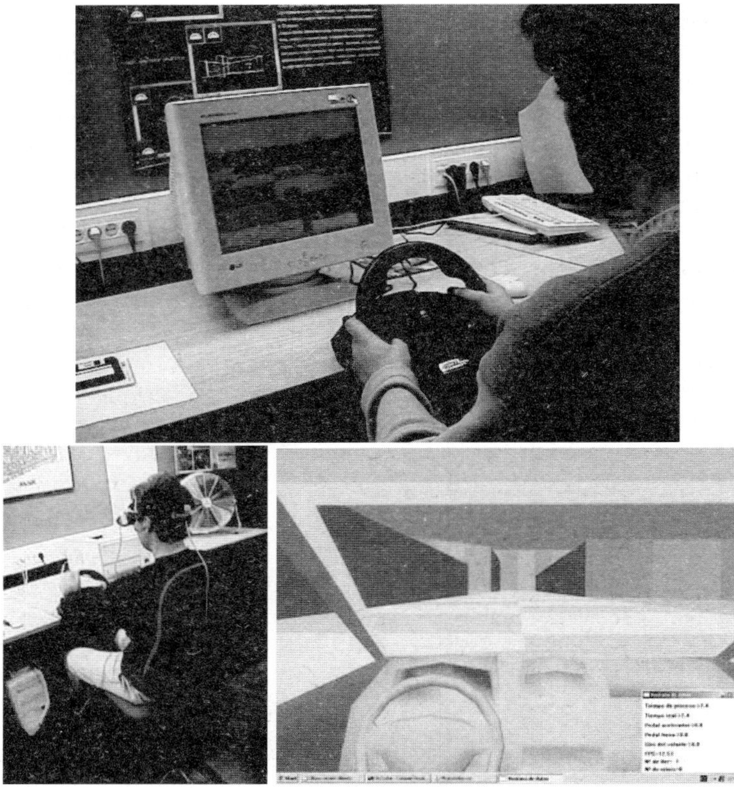

Figure 29. Car driving simulator: a) Screen; b) HMD.

Fig. 27): this is also enabled by the virtual reality interface. On the other hand, it must be pointed out that the graphical representation of the car motion, along with the plots provided by the program, are essential elements which allow the designer to tune the controller parameters until a satisfying result is achieved, as illustrated in Fig. 27.

4.3 The Control Module

As said at the beginning of the section, the controllers have been based on fuzzy logic, implemented by using the corresponding toolbox of Matlab. The controllers receive information from virtual sensors defined in the computational model of the car, and act upon actuators, also defined in the virtual model, as illustrated in Fig. 27. For sensor and actuator definition, the objective has been to replicate as much as possible their counterparts in the actual prototype, so that the designed controllers may be tested in the future onboard the real car for their experimental validation. In the case of the actuators, this has been completely achieved: the controller acts upon the steering wheel and pedals (throttle and brake); since the engine possesses automatic gearing, no actuation is needed upon the gear-lever. In what respects to the sensors, it will be seen later that all the magnitudes required as inputs by the controllers can be derived from the information about position, velocity and acceleration of the car; while such information is directly provided by the computational model of the vehicle, it is not so straightforwardly obtained in the real case: an intermediate module for motion recovery from sensor data will be required for the experimental task. On the other hand, the controller design process, also included in Fig. 27, is iteratively carried out by the designer, based on the results achieved by each new design, which may be appraised by means of the graphical representation and plots provided by the tool.

Reasons for the use of fuzzy logic. Designing a conventional controller, such as a proportional, integral and derivative (PID) controller, normally follows a standard procedure of modeling the plant, constructing a controller and evaluating the performance. A complete ground vehicle is a highly nonlinear system. Developing a model which preserves the nonlinear characteristics of the system, and is simple enough to represent the plant of the complete system, is not easy at all. Due to its capability of handling systems nonlinearity, this technique seems a good choice to control either a ground vehicle (Kageyama and Pacejka (1991)), or partial aspects of vehicles (Zhang (2003)).

When addressing the path tracking problem, it is common to resort to a simplified model, as the bicycle model, in order to design a controller. Using fuzzy logic, such a simplified model is not required, thus streamlining the process.

Fuzzy control is knowledge-based control technology that can mimic human strategies to control complex systems. In order to mimic the driver behavior, his physical capabilities should be considered by the model, such as preview, adaptation to changing dynamic characteristics of the controlled vehicle, learning, anticipation, and planning abilities. In fact, the driver employs a certain *internal*, intuitive model, or input-output understanding of the vehicle, that allows him to compare the expectation of the vehicle state at some future time with the directly observed preview input requirements (Macadam (2003)). The boundaries of this *internal model* are not very accurate, and this fact fits very well on the fuzzy logic concept of membership of a fuzzy set.

Finally, another valuable reason is that a reliable toolbox for fuzzy logic programming is provided by Matlab environment. This fact highly simplifies both the programming and tuning of the controller.

The Fortran-Matlab connection. Efficiency is a great concern in many applications dealing with vehicle control systems, either for speeding-up the design cycle, or for real-time performance when required. Both situations appear in this work. On the one hand, during the design stage, fast simulation of the car governed by the controller is desirable, since it allows the designer to carry out more design iterations in shorter times, thus reaching earlier the final tuning of the controller (this demand would be even stronger if the designer was substituted by an optimization algorithm); however, if the pursued objective is not the total control of the vehicle but partial only, like in driver assistance applications, the simulation requires human-in-the-loop participation too, and then real-time performance is needed. On the other hand, once the controller has been designed, it may be used inside other real-time applications, provided its implementation is fast enough for that purpose.

Once justified the relevance of the efficiency issue, it must be said that the bottleneck from this point of view is the Fortran-Matlab connection, and the heavy evaluation of the controllers in Matlab. Here, some considerations are to be done on the different alternatives to connect both programming languages. The second aspect will be addressed in subsection 4.5.

In order to introduce the control in the prototype model, the Fortran code containing the dynamics of the car and the Matlab functions implementing the fuzzy logic control algorithms must be combined. Matlab always employs double-precision variables, so that the different types of Fortran variables must be converted to double-precision for Matlab compatibility. To connect Fortran and Matlab, two alternatives have been investigated: Matlab Engine and MEX files. For both of them, compatibility between Matlab and the Fortran compiler is required.

Fortran-Matlab communication through Matlab Engine requires opening a communication channel between Fortran and Matlab. Matlab features several compilation functions which enable data transfer between both languages, as well as executing functions in Matlab (Matlab (2000)). Then, when the Fortran program is running, a Matlab session is started, which implies some delay. Likewise, data transmission through the communication channel and the execution of Matlab instructions both slow down the program. In fact, to execute a Matlab function, it must be written and executed on Matlab command window through the communication channel. Despite this fact, the simulation CPU-times obtained are kept moderate and, therefore, Matlab Engine can be considered as a suitable tool to design new control algorithms.

Matlab allows the user to write new functions by means of the so-called MEX files. Through this method, the user can write the whole program in Matlab language with the exception of bottleneck functions, which could be written in more efficient languages, like Fortran or C, and directly executed from Matlab as its own functions. A command library is available in order to communicate Matlab with the other programming language. The compilation is carried out in Matlab. Consequently, this option is just opposite to the previous one. In the present work, the whole Fortran program containing the dynamics of the car has been converted into a MEX file, which can be executed in Matlab, thus enabling the access to the functions of the fuzzy logic toolbox. The efficiency has shown to be a little lower than that obtained with the first option.

There exists a third possible option of Fortran-Matlab connection: the conversion of Matlab functions into DLL files, in order to compile them along with the functions written in the other language. However, this option has been considered impractical, since it would imply to encapsulate the controller, which is the part of the program that must be changed (at each iteration) during the design process. Moreover, it seems that Matlab only provides this option for C compilers (Vaculin et al. (2004)).

4.4 Examples

To illustrate the use of the proposed tool, controllers have been designed for two maneuvers of the car. The first maneuver, a 20 m straight path, is a very simple one, only requiring throttle-brake control. The second maneuver may be referred to as obstacle avoidance, and needs both throttle-brake and steering control. The procedure for development of the controllers starts with a human driver carrying out the desired maneuver. Then, the designer creates an initial controller based on the obtained information and on his perceptions about the maneuver. This controller is tuned manually in an iterative process, repeating simulations until the desired performance is achieved. Fuzzy logic controllers have been developed employing Mamdani-type inference and its typical defuzzification process (Mamdani and Assilian (1975)), finding the centroid of a two-dimensional function to determine the value of the output variable from its membership to the output fuzzy set.

First example: straight path. In the first maneuver, shown in Fig. 30, the car starts from rest, covers a distance of 20 m following a straight line, and stops. The objective is to stop the car as close to the target point as possible. First, the maneuver was carried out by a human driver, through the already described virtual reality interface. Based on the behavior of the car under human guidance, a maximum speed value of 5 m/s has been allowed, and an acceleration limit of 3 m/s^2 has been imposed. The time spent in the maneuver has not been limited.

Figure 30. First example: straight path.

The control scheme for automatic actuation of the pedals has been represented in Fig. 31. It receives position, velocity and acceleration of the car (sensors or inputs), and acts upon throttle and brake (actuators or outputs). The steering of the car should be considered in a real maneuver, due to irregularities of the terrain, initial misalignment, etc. However, in a theoretical maneuver on a completely flat floor, this control is not relevant, and hence has not been taken into account. Fuzzy sets have been chosen in an intuitive way to reflect a simplified linguistic description of the car during the maneuver, and the logic followed by the driver.

The first input is the position of the car. The total traveling distance has been divided into five intervals: start (from 0 to 4.5 m), taxi (covering the largest space, between 2.5 to 17.5 m), brake (16.5 to 19.5 m), stop (18.75 to 20 m), and over (beyond 20 m). The second and third inputs are the velocity and acceleration of the car, respectively. In order to avoid a great number of combinations in the formulation of rules, only three sets have been employed for each of them: low, medium and high values of the variables, as illustrated in Fig. 32. The case of null velocity has not been considered, since such state may only occur at the start or the end of the maneuver and, in both situations, it is possible to formulate the rules without taking into account the velocity (e.g. "if position is stop, pedal is full brake"). Trapezoidal fuzzy sets have been employed to model the input

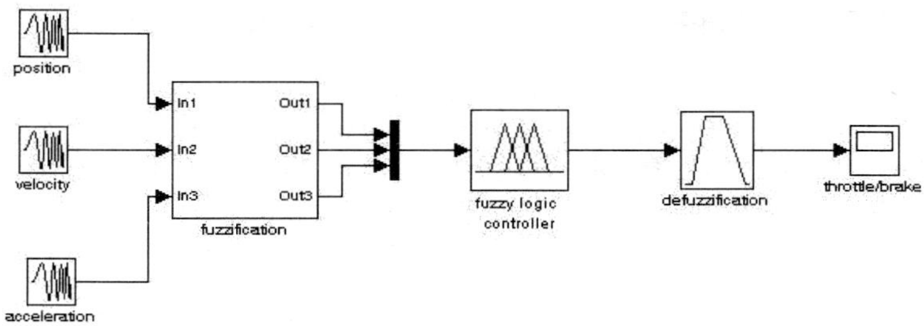

Figure 31. Control diagram for throttle-brake actuation.

variables, and overlapping at the medium of the slopes of adjacent sets has been sought. In this way, a smooth transition in the actuation is ensured for input values between intervals.

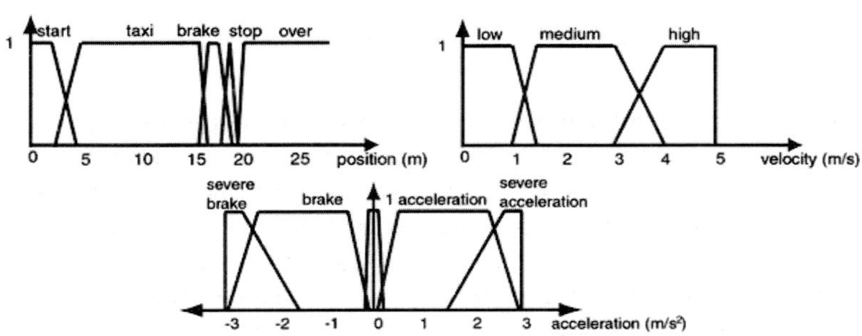

Figure 32. Input fuzzy sets for position, velocity and acceleration.

Regarding the output, the actuation on throttle and brake has been reduced to only one output. The domain of the actuation variable is [-1, 1], corresponding respectively to a fully pressed position of the brake pedal (-1), and to the same position of the throttle (1). Values sent to both actuators range from 0 to 1, corresponding to null and maximum displacement of the respective devices. Such a procedure avoids a simultaneous operation on throttle and brake. Another advantage derived from the use of a unique variable is that, if once designed and tuned, the controller is mapped and stored in a matrix for faster evaluation (see subsection 4.5), the memory space required will be lower. Again, trapezoidal fuzzy sets have been employed to describe the displacement of the pedal, except for fully or null pressed situations, where triangular sets have been selected, as depicted in Fig. 33.

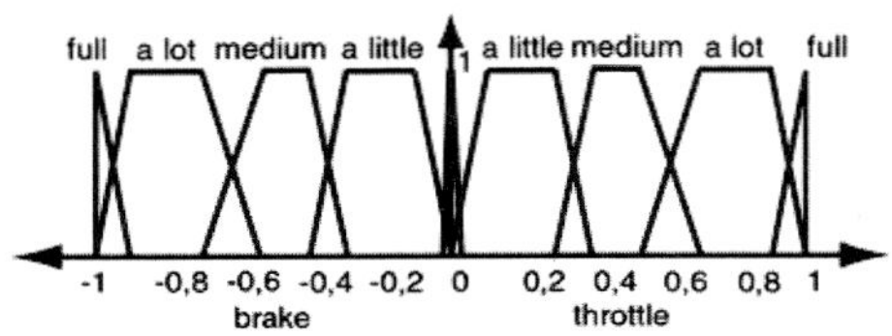

Figure 33. Output fuzzy sets for throttle-brake.

A total number of 21 rules have been imposed to relate inputs and output. Position and acceleration have been more weighted than speed in the controller logic, especially at the start and stop intervals. Characteristic values of the boundaries of the trapezoidal sets have been achieved after an iterative process, consisting of manually adjusting the values, and then looking at the resulting motion obtained from car simulation (see Fig. 27). In Fig. 34, the surface relating two of the inputs (position and velocity), and the output (position of throttle-brake) is represented. The nonlinear character of the control surface can be clearly observed in the figure.

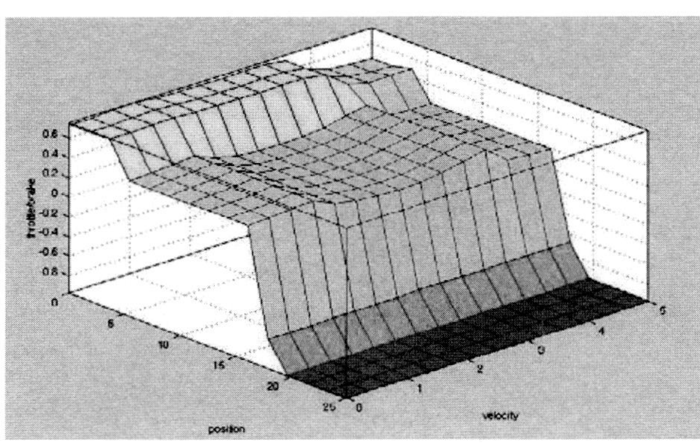

Figure 34. Control surface generated by rules for throttle-brake control.

The performance of the obtained controller has been evaluated by comparing it with that of a human driver. Figure 35 shows the comparison of the position history of the car when driven by either the human or the controller, as well as throttle and brake actuation histories in both cases. As it can be observed, the maneuver carried out by the controller is cleaner.

The controller does not need to make several approximations, as the human driver does. Actuation on both throttle and brake is more efficient when the car is automatically controlled, and less time is required to complete the maneuver.

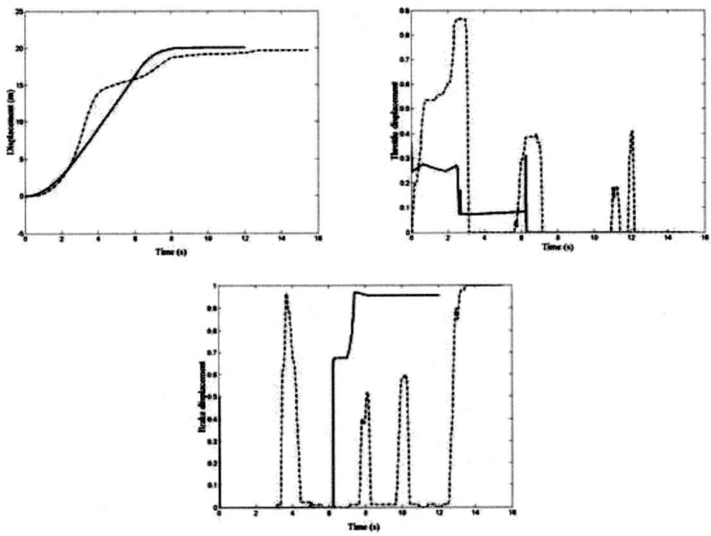

Figure 35. Controller vs. human driver (dashed): a) position history; b) throttle actuation; c) brake actuation.

Table 10 provides numerical values of the characteristic magnitudes of the maneuver. Less time is employed by the controller, although velocities are kept smaller than in the human driven case, and acceleration and brake are not so severe. This result is due to a more constant actuation that avoids abruptness. The error attained by the controller is one order of magnitude lower than that achieved by the human driver. The negative error value obtained by the latter means that the vehicle stops before reaching the target position.

Second example: obstacle avoidance. The second example is an obstacle avoidance maneuver, illustrated in Fig. 36. Starting from rest, the car covers an initial straight path of 20 m, then follows a full period (from peak to peak) of a sinoidal path of amplitude 1.75 m and, finally, must return to the straight line. First, a human driver carried out the maneuver

Table 10. Characteristic values obtained for the first maneuver.

	Maneuver Time (s)	Error (m)	Max Vel (m/s)	Max Accel (m/s^2)	Max Brake (m/s^2)
Driver	13.1	-0.314	8.11	5.8	-9.6
Controller	8.6	+0.081	3.7	1.7	-4.1

so as to determine realistic performance values for the maneuver. In such a way, a region of feasible velocities was established, which served to establish the fuzzy sets for such magnitude. As a result, the speed has been kept below 8 m/s, and the acceleration has been restricted to the interval ± 3 m/s^2, in order to avoid severe actions.

Figure 36. Second example: obstacle avoidance.

Controls of both, the steering and the throttle-brake couple, have been addressed separately. The throttle-brake control follows the same strategy explained in the first example, excepting that the position, used there as first input, has been substituted here by the path tracking error. Such error is calculated as the average of the two errors used for the steering control that are described below, and ranges from 0 to 1. The pursued objective is that the controller slows down the car when the path tracking error increases.

The steering control scheme is illustrated in Fig. 37. Two error functions have been defined. The first one is the position error of the vehicle at the current time, ε_{pos}. This error, shown in Fig. 38, is obtained as the distance from the car to the intended path, measured along the normal to the actual path. This error takes values between -5 and 5 m, but values over 1 m are severely penalized. Errors greater than 5 m are reduced to this value.

Preview is an essential part of human driver control behavior (Sharp et al. (2000)) and, therefore, the second error employed, the attitude error, ε_{att}, aims to preview the car attitude in the next instants. For this purpose, the

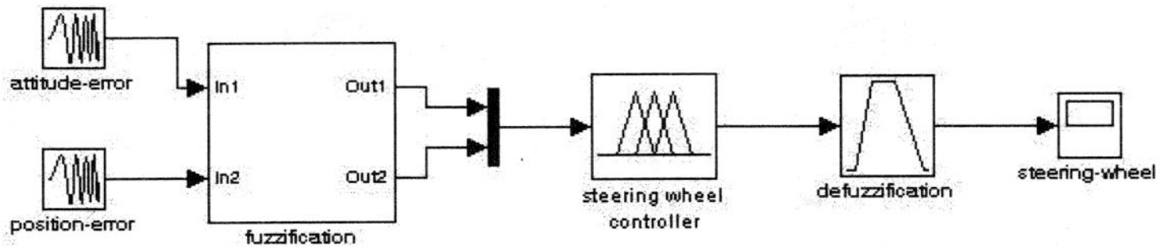

Figure 37. Control diagram for steering actuation.

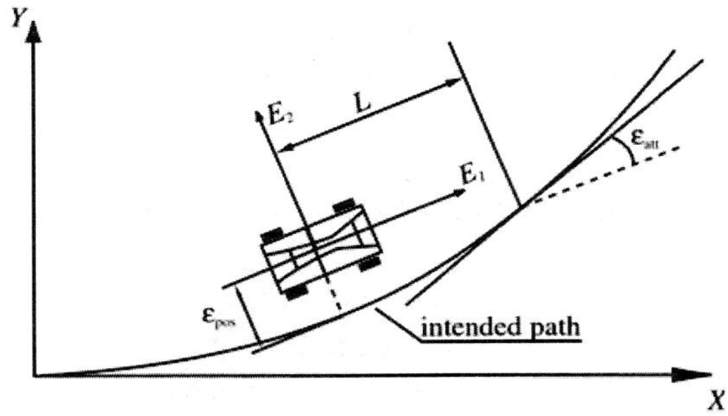

Figure 38. Error functions for steering actuation.

tangent to the trajectory at the current point is compared with the tangent to the desired trajectory at a more advanced point. The driver monitors the path ahead by projecting forward an optical lever L. The distance of the lever represents the extent of the preview available for a preview time T_{P}. In this way, the distance becomes a function of the vehicle longitudinal velocity:

$$L = v_x \cdot T_{\mathrm{P}} \tag{87}$$

A T_{P} value of 0.425 s has experimentally shown to be the most appropriate for the maneuver. A maximum deviation of 50° is considered, and a value of 50° is assigned to exceeding values.

Of course, different definitions of the errors are possible too. For example, it might be convenient to consider more than one preview point (Sharp et al. (2000)) for the attitude error, but the present work focuses on developing a tool for designing controllers rather than on obtaining the most accurate controller.

Fuzzy sets that describe the errors are depicted in Fig. 39. For the attitude error, sets are narrower near the null error, so that a deviation from the intended path can be quickly corrected. For the position error, the figure zooms the central part of the whole [-5, 5] interval. Again, fuzzy sets are not so wide in the surroundings of null error, in order to more quickly react in the presence of a slight deviation.

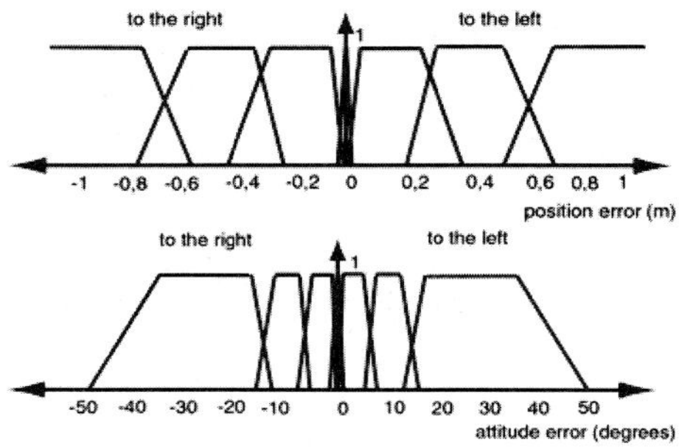

Figure 39. Input fuzzy sets for position and attitude errors.

The output sets are represented in Fig. 40. The output represents the absolute position of the steering wheel, whose extreme position is 450° both to the right and to the left. A maximum displacement of 15° between consecutive time-steps of the numerical integration has been allowed. This value has been chosen having in mind the future implementation of the controller in the prototype. In that context, if the steering wheel is actuated by a step motor, such a displacement is equivalent to an excitation frequency of 833 Hz (time-step 0.01 s, motor step 1.8°).

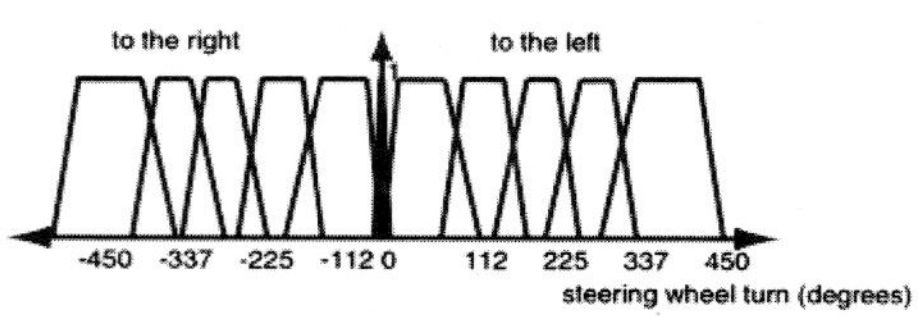

Figure 40. Output fuzzy sets for steering.

A total number of 49 rules have been established to manage the steering wheel, the resulting surface being shown in Fig. 41. The surface provides the absolute angular position of the steering wheel for any pair of values of the errors.

Coming back now to the throttle-brake control for this second maneuver, the fuzzy sets for velocity and acceleration inputs are like those described for the first maneuver. Only the range of the velocity has been raised from a maximum of 5 m/s to 12 m/s, and sets have been consequently amplified.

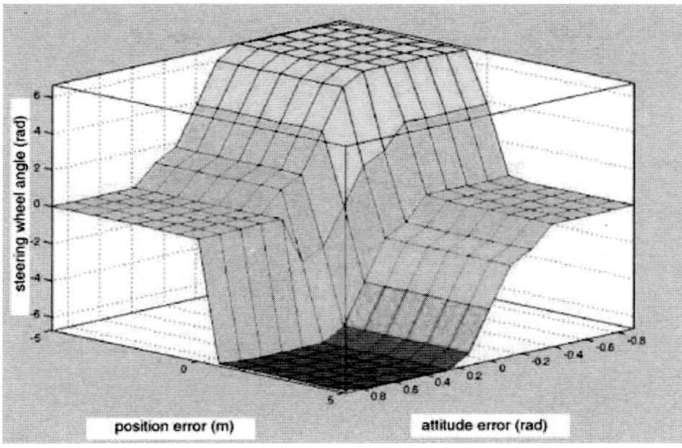

Figure 41. Control surface generated by rules for steering control.

However, as pointed out at the beginning of this subsection, the position error of the first maneuver has been substituted in this case for the path tracking error, calculated as the average of the position and attitude errors used for the steering control. The fuzzy sets employed for this error are shown in Fig. 42.

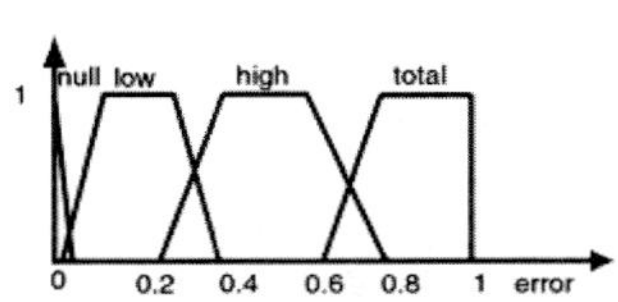

Figure 42. Input fuzzy sets for path tracking error.

As for the first maneuver, the performance of the obtained controller has been evaluated by comparing it with that of a human driver. Fig. 43 shows the comparison of the trajectory, and the histories of both the velocity and the steering wheel angle of the car when driven by either the human or the controller.

It can be seen that the human driver tends to smooth the trajectory at the turns. Due to the preview, this trend is less acute for the controller. In fact, the controller could be tuned to better fit the desired path by reducing the maximum attitude error allowed, and by considering the attitude error without preview. However, the controller tuned in that way is too sensitive to small disturbances and, consequently, is rather unstable. Therefore, a

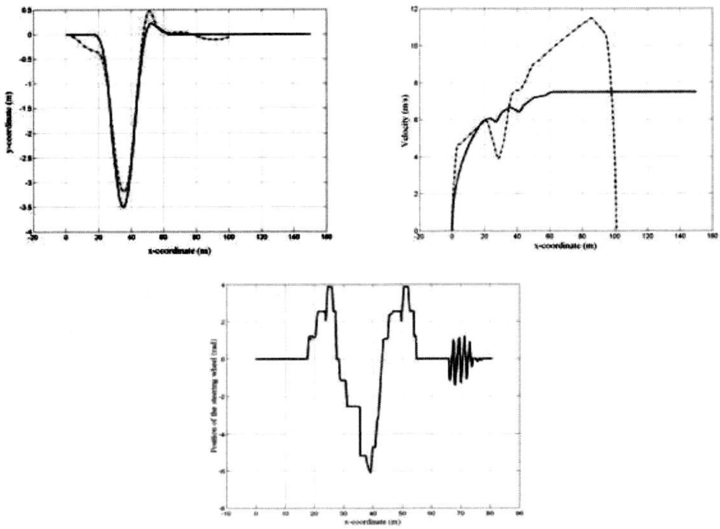

Figure 43. Controller vs. human driver (dashed): a) trajectory; b) velocity history; c) steering wheel angle history.

compromise between these two factors must be sought (note the instability of the steering wheel actuation at the end of the maneuver). It is surprising that the driver does not anticipate the last turn, likely due to an excessive speed. During the curved part of the trajectory, the error incurred by the controller is negligible, and the speed reduction is less acute than in the human-driven case. The maximum error committed by both the human driver and the controller occurs at the time of returning to the straight path. For lower speeds, such error can be made almost zero for the controller, but the maneuver has been managed at quasi-critical speed from the stability point of view.

Table 11 provides numerical values of the characteristic magnitudes of the maneuver. The maximum velocity appearing in the table corresponds to the curved part of the trajectory only. The mean position error has been defined as the algebraic mean of the position error along the maneuver,

$$\bar{\varepsilon}_{pos} = \frac{\sum_{i=1}^{n} \left| \varepsilon_{pos}\left(i\right) \right|}{n} \tag{88}$$

where n is the number of time-steps of numerical integration required for the maneuver. This value may be considered as an evaluation of the accuracy on trajectory tracking.

Table 11. Characteristic values obtained for the second maneuver.

	Maneuver Time (s)	Max Pos Error (m)	Mean Pos Error (m)	Max Vel (m/s)
Driver	14.05	0.455	0.1294	7.55
Controller	14.91	0.205	0.0656	6.55

It can be observed that, although the human driver reaches higher velocities, the time spent for the maneuver is similar to that required by the controller. Furthermore, the human driver commits higher errors too: note that both the maximum and mean errors of the controller are half of the human driver's errors. Therefore, it can be affirmed that the vehicle follows the track more accurately when guided by the controller.

4.5 Efficiency Remarks

The CPU-times obtained for the two maneuvers reported when using the Matlab Engine option on a Pentium IV @ 2 GHz are listed in Table 12. CPU-times* include the time spent in opening a Matlab session, while CPU-times** are obtained when the running program is attached to an already opened Matlab session. It must be said that, if no control is considered, the Fortran program comfortably achieves real-time performance when solving for the dynamics of the vehicle. It can be seen that the second maneuver needs a greater computational effort, due to the evaluation of two controllers, while only one is evaluated for the first maneuver. Evaluation of a fuzzy controller implies carrying out a *fuzzification* (i.e. determining the membership degree in the input fuzzy sets), applying the inference rules and, finally, performing a *defuzzification* (i.e. mapping an output value to its appropriate membership value in the output fuzzy sets).

Table 12. Efficiency during the design stage of the controller.

Maneuver	Time (s)	CPU-time* (s)	CPU-time** (s)
1	12	18.00	11.99
2	24	36.14	30.45

The CPU-times needed when using the MEX file option are slightly higher. The Matlab Engine option seems to be preferable, since the computational effort required is lower, and since less code must be added to the original program. Indeed, only commands for the opening and closure of the communication channel along with those relative to data transfer must be added (when the MEX file alternative is chosen, a new heading file must be added too, in order to enable the program to be called from Matlab). Therefore, Matlab Engine represents a good solution when, as in the present case, only some parts of the program need to be executed in Matlab. Furthermore, it is adequate for the stage of controller tuning, since real-time performance is not required.

However, if controllers are to be employed in stand alone applications that require real-time, the previous procedure is not a good option. Time delay is due to Fortran-Matlab connection and the heavy evaluation of the controllers in Matlab. In order to find a solution, it must be taken into account that the fuzzy logic generates, by means of rules of membership and actuation, a hyper-surface which relates the input (error, velocity, etc.) and output variables (throttle-brake, steering). An evaluation of a matrix containing all the possible combinations of inputs of the controller can be carried out. Matlab provides function *evalfis* to this end. A matrix may be used as first input of the function, in such a way that each row of the matrix contains one of the possible inputs required by the controller. *Evalfis* function will return another matrix whose rows are the corresponding outputs. If the controller is stored in this way, its later evaluation just consists of an interpolation of the output matrix.

Following this procedure, arbitrary values of a variable cannot be employed and, therefore, its domain must be divided into a discrete set of values (discretization). This step plays a very important role. Fuzzy logic establishes degrees of membership to fuzzy sets. Discretization must reflect the membership of an input to a fuzzy set. Membership varies between 0 and 1 over an interval. The shape of membership functions is not any: in the present work, trapezoidal and triangular membership functions have been used. The intervals of variation of membership are different for different sets, as illustrated in Fig. 44. The step of discretization, s, employed, is one tenth of the minimum interval of variation of membership, b in Fig. 44. With these values of the step, the behavior of the controller is perfectly reproduced.

Table 13 shows the CPU-times obtained with controllers stored in matrices and loaded in the program as it starts. CPU-time* is the time obtained with Matlab Engine attached to a Matlab previous session (the faster case presented above), and CPU-time** is the time consumed when the controllers are stored into matrices. As it can be seen, now real-time is comfortably achieved.

To give an idea about the practical capabilities of the described framework, the two following applications have been developed:

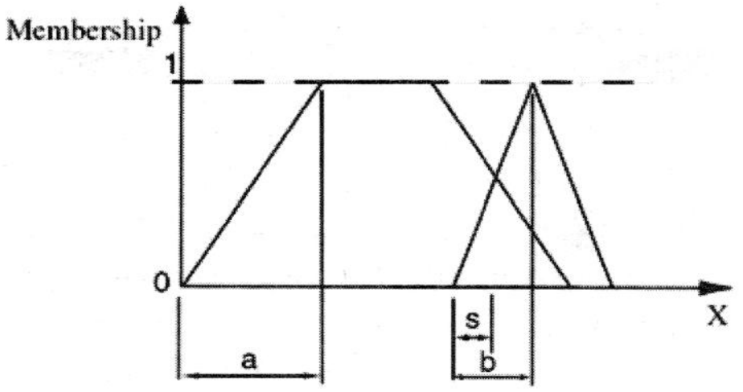

Figure 44. Fuzzy sets and step of discretization.

Table 13. Efficiency for real-time applications.

Maneuver	Time (s)	CPU-time* (s)	CPU-time** (s)
1	12	11.99	4.87
2	24	30.45	11.75

a) The environment full of obstacles illustrated in Fig. 45a. The aim of this application was to highlight the robustness of the dynamic formulation when dealing with a full model of a car (compared to that provided by the usual real-time explicit integrators). The result was that the model can be driven in real-time by a human driver through the different obstacles, i.e. undergoing highly violent motion, during very long simulation times. This means that the proposed framework can be used not only for smooth and easy maneuvers, but also for demanding and critical situations.

b) The simultaneous simulation of three cars in real-time, one driven by a human driver and the two others automatically driven by controllers developed within the proposed framework for the replica of Catalua circuit shown in Fig. 45b. A module for collisions was included, so that the cars could impact each other, thus forcing the controllers to react in order to keep the vehicles inside the road and to follow the correct trajectories. This application served to demonstrate the efficiency of the whole approach, since it is capable of dealing with three detailed car models (163 variables each) at the same time, two of them operated by controllers, and to show that robust controllers can be designed with the proposed tool.

5 Conclusions

In this contribution, two variants of an efficient augmented Lagrangian formulation with structural integrators for the simulation of the dynamic of multibody systems have been presented. The first variant describes the sys-

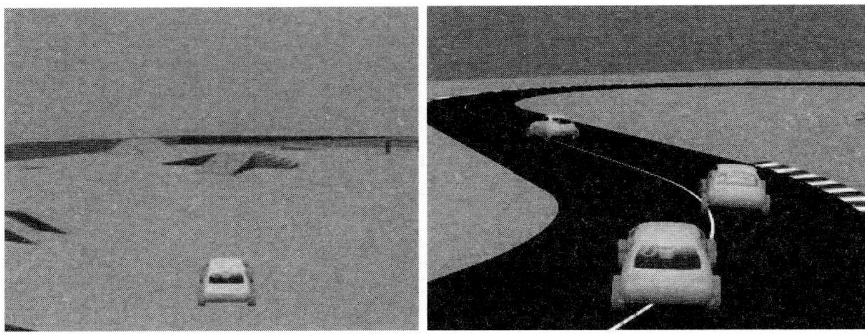

Figure 45. Demo applications: a) robustness; b) efficiency.

tem in absolute (natural) coordinates, while the second variant defines the system in relative coordinates. The two variants have been tested through several examples, showing excellent efficiency and robustness, along with acceptable accuracy. The comparison between both alternatives reveals that the method in relative coordinates is more efficient but also more involved to implement than the method in absolute coordinates. Therefore, both may have chances to be used, depending on the pursued application. The experimental validation of the method has been addressed too, by comparison between measured and calculated stresses of the chassis of a prototype car. Finally, the method has been used to develop a tool for the design and evaluation of automobile controllers, which benefits from the efficiency of the algorithm at the design stage, and demands real-time performance at the human-in-the-loop evaluation stage. In this way, it has been demonstrated that the method can be the base of practical real-time applications.

Bibliography

ADAMS 11.0.0 *Product Guides.* 2000.

J. Angeles and O. Ma, Dynamic simulation of n-axis serial robotic manipulators using a natural orthogonal complement. *Int. Journal of Robotic Research*, 7:32-47, 1988.

E. Bakker and H.B. Pacejka, The magic formula tyre model. In *Proceedings of the First International Colloquium on Tyre Models for Vehicle Dynamics Analysis*, pages 1-18, 1991.

E. Bayo and R. Ledesma, Augmented Lagrangian and mass-orthogonal projection methods for constrained multibody dynamics. *Nonlinear Dynamics*, 9:113-130, 1996.

K.E. Brenan, S.L. Campbell and L.R. Petzold, *The Numerical Solution of Initial Value Problems in Differential-Algebraic Equations.* Elsevier, 1989.

J. Cuadrado, J. Cardenal, P. Morer and E. Bayo, Intelligent simulation of multibody dynamics: space-state and descriptor methods in sequential and parallel computing environments. *Multibody System Dynamics*, 4:55-73, 2000.

J. Cuadrado, D. Dopico, M. Gonzalez and M.A. Naya, A combined penalty and recursive real-time formulation for multibody dynamics. *Journal of Mechanical Design*, 126:602-608, 2004a.

J. Cuadrado, R. Gutierrez, M.A. Naya and M. Gonzalez, Experimental validation of a flexible MBS dynamic formulation through comparison between measured and calculated stresses on a prototype car. *Multibody System Dynamics*, 11:147-166, 2004b.

S. Frik, G. Leister and W. Schwartz, Simulation of the IAVSD road vehicle benchmark bombardier Iltis with FASIM, MEDYNA, NEWEUL and SIMPACK. In W. Kortum and R.S. Sharp, editors, *Multibody Computer Codes in Vehicle System Dynamics.* Swets and Zeitlinger, 1993.

J. Garcia de Jalon and E. Bayo, *Kinematic and Dynamic Simulation of Multibody Systems -The Real-Time Challenge-.* Springer-Verlag, 1994.

M. Geradin and A. Cardona, *Flexible Multibody Dynamics.* John Wiley and Sons, 2001.

I. Kageyama and H.B. Pacejka, On a new driver model with fuzzy control. *Vehicle System Dynamics*, 20:314-324, 1991.

U. Lugris, M.A. Naya, F. Gonzalez and J. Cuadrado, Performance and application criteria of two fast formulations for flexible multibody dynamics. *Mechanics Based Design of Structures and Machines*, 35:381-404, 2007.

C.C. Macadam, Understanding and modelling the human driver. *Vehicle System Dynamics*, 40:101-134, 2003.

E.H. Mamdani and S. Assilian, An experiment in linguistic synthesis with fuzzy logic controller. *International Journal of Man-Machine Studies*, 7:1-13, 1975.

Matlab 6.1 Documentation, *User's Guide Version 2.1.* 2000.

R.S. Sharp, D. Casanova and P. Symonds, A mathematical model for driver steering control, with design, tuning and performance results. *Vehicle System Dynamics*, 33:289-326, 2000.

O. Vaculin, W.R. Kruger and M. Valasek, Overview of coupling of multibody and control engineering tools. *Vehicle System Dynamics*, 41:415-429, 2004.

Q. Zhang, A generic fuzzy electrohydraulic steering controller for off-road vehicles. *Proc. Instn. Mech. Engrs. Part D: J. Automobile Engineering*, 217:791-799, 2003.

Advanced Applications

P. Eberhard, A. Ast, C. Henninger and P. Ziegler
Institute of Engineering and Computational Mechanics, University of Stuttgart, Stuttgart, Germany

1 Introduction

In this chapter, advanced applications* of the theory of rigid and flexible multibody systems introduced in the previous chapters are presented. The first two examples show how multibody system simulation can be used to improve design and control of parallel kinematic machines. The third example demonstrates the analysis of the dynamics of gearwheels subject to impact-like contacts based on flexible multibody systems.

Because only comparatively small masses are required, machines with parallel kinematics feature low inertia forces, while simultaneously a high accuracy and stiffness can be achieved. Such machines are currently under investigation in various fields of engineering like robotics, measurement systems or manufacturing technologies, see for example Merlet (2000), Boër, Molinari-Tosatti and Smith (1999), or Riebe and Ulbrich (2003). The growing demand for highly productive machine tools which are able to operate rapidly while moving accurately along a given trajectory has led in the last decade to the development of new machine types such as parallel and hybrid kinematic machines, Winkler et al. (2003).

The dynamics of parallel kinematic machine tools is inherently nonlinear and the motion of the drives is strongly coupled. Therefore, the usual assumption of independently controllable drives, compare Sciavicco and Siciliano (1996), becomes questionable and conventional linear control approaches as described by Weck (1995) lead to a significant loss of accuracy.

*This chapter contains in a modified and unified way material which was published before by the authors in the Archive of Applied Mechanics (Ast and Eberhard (2006), Ziegler, Eberhard and Schweizer (2006)) and Multibody System Dynamics (Kübler, Henninger and Eberhard (2005)). In the CISM course on "Simulation Techniques in Applied Dynamics" which is documented in this book also two lectures on optimization have been held that are not covered here. Some details on these topics can be found in Sedlaczek and Eberhard (2006) and Sedlaczek and Eberhard (2007).

Thus, new approaches in the control of such machines are needed like the nonlinear flatness-based control presented in Section 2, see Ast and Eberhard (2006).

A crucial point concerning the design of parallel kinematic machines is the resulting stiffness of the endeffector as well as the location of kinematic singular configurations in the workspace Merlet (2000). Here, significant improvements of the performance can be achieved by using a systematic simulation-based design and optimization approach, see Section 3 and Kübler, Henninger and Eberhard (2005).

Impact-like contacts on gearwheels very often appear in gear trains. One characteristic of gear trains is that the transmitted average load is small compared to dynamic forces, like gas forces. One consequence is that teeth frequently lift-off and the re-establishment of contact is typically impact-like.

Besides the exhibited noise due to the impacts, the very high and short peaks in the contact forces also imply a design problem. Particularly because the gearwheels in gear trains are typically light-weight designs, classical rigid body models cannot be used anymore. Instead, elastic models are necessary to simulate contact forces precisely. Here, a modal model is used that allows an elastic description of structures and is numerically very efficient, see Ziegler, Eberhard and Schweizer (2006).

2 Flatness-based Control of Parallel Kinematics Using Multibody Systems

In this section, the design of a flatness-based position control algorithm as described for example by Rothfuß, Rudolph and Zeitz (1997) is proposed based on a multibody system model of the machine tool with parallel kinematics, Schiehlen and Eberhard (2004). Such a nonlinear and multi-variable control concept is able to take into account the nonlinear dynamics as well as the coupling of the drives. Simulation results presented for a high performance machine tool called Lambda Kinematics already showed the high potential of the flatness-based position control concept in comparison to standard linear control algorithms as they are implemented in the actual control system of the machine tool, Ratering and Eberhard (2005a) and Ratering and Eberhard (2005b). The machine tool has a parallel kinematic structure for the motion in the horizontal plane driven by linear direct drives and is used for the processing of plate-like workpieces, e.g. milling of large wooden plates, see Figure 1. It has been build at the Institute of Machine Tools headed by Prof. Heisel at the University of Stuttgart, Germany, see for example Heisel and Maier (2002) and Winkler et al. (2003), and is in-

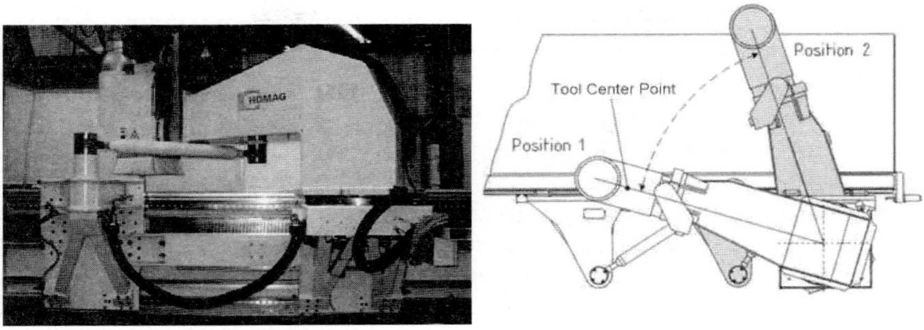

Figure 1. Photograph and extremal positions and pivoting of the cantilever of the Lambda Kinematics as seen from above [source: www.ifw.uni-stuttgart.de].

vestigated in a joint project, see Ast et al. (2007).

Experimental and simulation results are presented here for an equivalent but smaller test machine with the same kinematic structure but a flexible programmable position control system, see Figure 2. The test machine has been build by Altenburger at the ZFS Stuttgart, Germany, see Berkemer et al. (2004). The linear position control approach is compared to the nonlinear flatness-based controller design. The experiments demonstrate too the superior performance of the nonlinear approach by reducing the trajectory error significantly.

2.1 Experimental Machine with Scissors-Like Kinematics

Figure 2 shows a photograph and a schemativ view for the multibody system model of the experimental machine. It has a scissors-like, parallel kinematic structure for the motion in the xy-plane. The supports are driven by two linear direct drives, which are mounted opposite to each other. Connected to them via revolute joints A and B around the z-axis are two rods, which are again connected to each other via a revolute joint at point C closing a kinematic loop. The point C corresponds to the Tool Center point (TCP) where in a machine tool the processing tool is located. Moving both supports in the same direction results in a movement along the x-axis, moving in opposite directions causes a movement along the y-axis of the machine.

Mechanical Model and Equations of Motion For the design of the nonlinear position controller and the simulations, the experimental machine

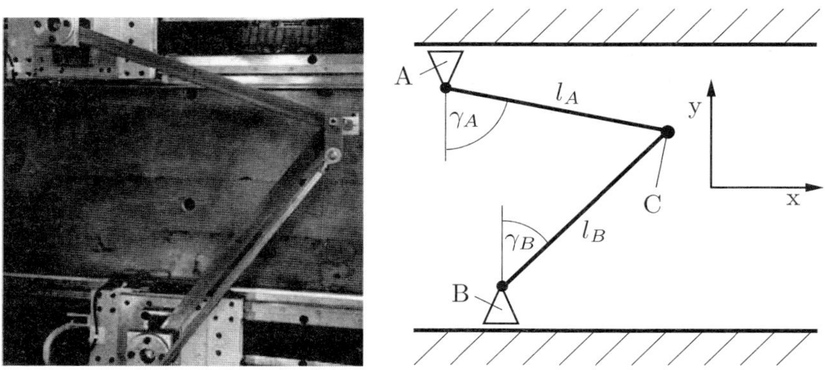

Figure 2. Photograph and schematic view of the experimental machine.

is modeled as a rigid multibody system, see Schiehlen and Eberhard (2004). It consists of four rigid bodies: the two supports and the two rods, see Figure 2. Hence, the number of degrees of freedom of the multibody system is two. The x-positions (x_A, x_B) of the two supports, where also the revolute joints are attached, are chosen as the minimal coordinates

$$\boldsymbol{q}_m = \begin{bmatrix} x_A & x_B \end{bmatrix} \tag{1}$$

determining the motion of the machine. They are available as measurements at the actual machine tool as well.

The inverse kinematics as defined by Craig (1989) describes the relation between the position (x_C, y_C) of point C and the position of the two supports

$$x_A = x_C - \sqrt{l_A^2 - (y_A - y_C)^2}, \quad x_B = x_C - \sqrt{l_B^2 - (y_C - y_B)^2} \tag{2}$$

with l_A and l_B being the distances between the revolute joints, see Figure 2, and y_A and y_B being the constant y-positions of the supports. There are two possible singular positions, where all three revolute joints lie on a straight line in the xy-plane. Here, the solution of the kinematic equations is no longer well-defined. Due to the design of the experimental machine, it is not possible to reach those positions in reality. Therefore, they are of no further concern in the following sections.

In general, there are two possibilities to formulate the equations of motion for a multibody system with kinematic loops, Woernle (2001):

- equations of motion in joint coordinates $\boldsymbol{q}_t$ of the open-loop or tree structure with the loop closing conditions as algebraic constraints resulting in a system of differential-algebraic equations (DAEs) or
- equations of motion in minimal coordinates $\boldsymbol{q}_m$ resulting in a system of ordinary differential equations (ODEs).

By opening the kinematic loop at the revolute joint C connecting the rods, the equations of motion for each branch of the open tree structure can be formulated using D'Alembert's Principle, see Schiehlen and Eberhard (2004), in joint coordinates

$$\boldsymbol{q}_t = [x_A \;\; x_B \;\; \gamma_A \;\; \gamma_B], \tag{3}$$

where γ_A and γ_B are the angles between the y-axis of the inertial coordinate system and the rods, see Figure 2. In doing so, the generalized constraint or reaction forces $\boldsymbol{Q}^r$ resulting from the opening of the kinematic loop have to be taken into account. The equations of motion of the open tree structure have the form

$$\boldsymbol{M}(\boldsymbol{q}_t)\cdot\ddot{\boldsymbol{q}}_t + \boldsymbol{k}(\boldsymbol{q}_t,\dot{\boldsymbol{q}}_t) = \boldsymbol{Q}^a(\boldsymbol{q}_t,\dot{\boldsymbol{q}}_t) + \boldsymbol{H}\cdot\boldsymbol{u}(t) + \boldsymbol{Q}^r, \tag{4}$$

where $\boldsymbol{M}$ is the symmetric mass matrix, $\boldsymbol{k}$ the vector of the generalized gyroscopic and centrifugal forces, $\boldsymbol{Q}^a$ the vector of the generalized applied forces - e.g. process forces - and $\boldsymbol{H}\cdot\boldsymbol{u}$ the vector of the motor forces with $\boldsymbol{H}$ being an appropriate Boolean matrix. The vector $\boldsymbol{u}$ consists of the forces exerted by the linear drives on the two supports in x-direction

$$\boldsymbol{u} = [u_A(t) \;\; u_B(t)]. \tag{5}$$

The joint coordinates $\boldsymbol{q}_t$ are subject to the constraint equations, which define implicitly the connection between the independent, minimal coordinates $\boldsymbol{q}_m$ and the dependent coordinates

$$\boldsymbol{q}_h = [\gamma_A \;\; \gamma_B]. \tag{6}$$

For the given system the point C on both branches of the open tree structure must always coincide resulting in two constraint equations

$$\boldsymbol{g}(\boldsymbol{q}_t) = \begin{bmatrix} x_A - x_B + l_A\sin(\gamma_A) - l_B\sin(\gamma_B) \\ y_A - y_B - l_A\cos(\gamma_A) - l_B\cos(\gamma_B) \end{bmatrix} = \boldsymbol{0}. \tag{7}$$

With the help of the constraint equations (7) on velocity-level

$$\dot{\boldsymbol{g}} \equiv \frac{\partial \boldsymbol{g}}{\partial \boldsymbol{q}_t}\cdot\dot{\boldsymbol{q}}_t = \boldsymbol{G}(\boldsymbol{q}_t)\cdot\dot{\boldsymbol{q}}_t = \begin{bmatrix} 1 & -1 & l_A\cos(\gamma_A) & -l_B\cos(\gamma_B) \\ 0 & 0 & l_A\sin(\gamma_A) & l_B\sin(\gamma_B) \end{bmatrix}\cdot\dot{\boldsymbol{q}}_t = \boldsymbol{0} \tag{8}$$

the vector of the generalized constraint forces $\boldsymbol{Q}^r$ can be written as

$$\boldsymbol{Q}^r = \boldsymbol{G}^T \cdot \boldsymbol{\lambda}, \tag{9}$$

where $\boldsymbol{\lambda}$ are the Lagrangian Multipliers, see Woernle (2001), corresponding to the forces needed to close the kinematic loop.

The Equations (4) and (7) with (9) form the equations of motion of the multibody system in DAE-form. To convert the system of DAEs into a system of ODEs, it is necessary to formulate the constraint equations explicitly as functions of the minimal coordinates

$$\boldsymbol{q}_t = \boldsymbol{q}_t(\boldsymbol{q}_m) \tag{10}$$

and to eliminate the dependent coordinates γ_A and γ_B, their derivatives and the constraint forces in Equation (4). To find an explicit formulation of the constraint equations is often very complicated and sometimes only possible numerically. Here, it can be done analytically, for example assisted by a computer-algebra system, resulting in six equations of the form

$$\gamma_i = \gamma_i(\boldsymbol{q}_m), \tag{11}$$

$$\dot{\gamma}_i = \dot{\gamma}_i(\boldsymbol{q}_m, \dot{\boldsymbol{q}}_m), \tag{12}$$

$$\ddot{\gamma}_i = \ddot{\gamma}_i(\boldsymbol{q}_m, \dot{\boldsymbol{q}}_m, \ddot{\boldsymbol{q}}_m) \tag{13}$$

with $i = A, B$. The equations are always well-defined in the given workspace. The resulting equations of motion in ODE-form are quite complicated and, as it turns out, not necessary for the nonlinear controller design. They are therefore omitted here.

Alternatively, if the constraint equations (7) cannot be solved analytically, a Newton algorithm as defined in Bronstein et al. (1997) can, for example, be used to solve the constraint equations for $\boldsymbol{q}_h$. The angular velocities and accelerations can then be determined successively from the first and second time derivatives of Equation (7) resulting in

$$\dot{\boldsymbol{q}}_h = -\left(\frac{\partial \boldsymbol{g}}{\partial \boldsymbol{q}_h}\right)^{-1} \left(\frac{\partial \boldsymbol{g}}{\partial \boldsymbol{q}_m} \cdot \dot{\boldsymbol{q}}_m\right), \tag{14}$$

$$\ddot{\boldsymbol{q}}_h = -\left(\frac{\partial \boldsymbol{g}}{\partial \boldsymbol{q}_h}\right)^{-1} \left(\frac{\partial \boldsymbol{g}}{\partial \boldsymbol{q}_m} \cdot \ddot{\boldsymbol{q}}_m + \dot{\boldsymbol{G}} \cdot \dot{\boldsymbol{q}}_t\right). \tag{15}$$

2.2 Control System

Figure 3 shows a block diagram of the complete structure of an electromechanical system with trajectory planning and cascaded position and current control.

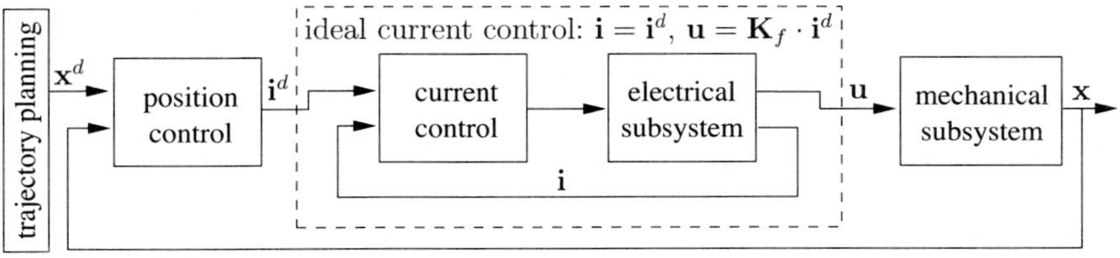

Figure 3. Block diagram of the complete electro-mechanical system structure with position and current control.

The trajectory planning allows the user to specify the desired trajectory in TCP-coordinates (x_C^d, y_C^d) and generates the corresponding support trajectories (x_A^d, x_B^d) via the inverse kinematics, see Equation (2). For machine tools, the desired trajectory in TCP-coordinates is usually programmed as NC-code, see Weck (1995), from which the desired trajectory is generated in real-time during processing. Alternatively, it would also be possible to generate the trajectory off-line once before processing, avoiding the problem of calculating the desired support positions, velocities etc. in real-time.

The position control is implemented to ensure that the system tracks the desired path with the specified velocity and acceleration in the face of disturbances, e.g. process forces. Linear as well as nonlinear control concepts can be used here. The standard linear control concept with independent position controllers for each drive typically used in machine tools is described first. Then, a nonlinear, flatness-based control concept is proposed as an alternative to the standard control concept. As the equations of motion (4), (7) with (9) are nonlinear and coupled, it is obvious that a nonlinear, multivariable controller should improve the system performance significantly.

The underlying current control ensures that the electrical drives provide the system with the forces required by the position control. Each drive has its own independent current controller. The controlled electrical subsystems are usually much faster than the mechanical subsystem. Therefore, a valid assumption for the position controller design is that the current control is ideal and the dynamics of the electrical subsystems can be neglected, Brahms (1998). The relation between the generated motor forces $\boldsymbol{u}$ and the desired current $\boldsymbol{i}^d$ being the output of the position control system hence reduces to

$$\boldsymbol{u} = \boldsymbol{K}_f \cdot \boldsymbol{i}^d \tag{16}$$

with $\boldsymbol{K}_f$ being a diagonal matrix with the elements being the force constants

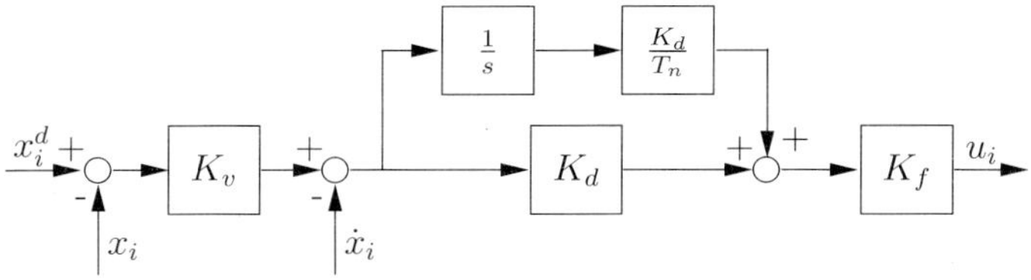

Figure 4. Block diagram of the linear, cascaded P-PI position control implemented independently for each drive.

of the corresponding drives. The assumption of ideal current control is used throughout the whole paper.

Linear Position Control The traditionally implemented control system of the real machine tool consists of two independent, standard, cascaded linear controllers for each support as stated by Weck (1995), with a PI control loop for velocity with the proportional gain K_d and time constant T_n and in cascade a P control loop for position with the proportional gain K_v, see Figure 4, resulting in the control law

$$\begin{aligned} u_i &= K_f\left[K_d\left(K_v\left(x_i^d - x_i\right) - \dot{x}_i\right) + \frac{K_d}{T_n}\int\left(K_v\left(x_i^d - x_i\right) - \dot{x}_i\right)\mathrm{dt}\right] \\ &= K_f K_d\left[K_v\left(x_i^d - x_i\right) - \frac{1}{T_n}x_i + \frac{K_v}{T_n}\int\left(x_i^d - x_i\right)\mathrm{dt} - \dot{x}_i\right] \end{aligned} \tag{17}$$

with $i = A, B$. This controller design is equivalent to an incomplete PID controller for position in the sense that no desired velocities are used for comparison. Moreover, no feedforward control, see Föllinger (1992), was implemented. As such a design is the established and widely-used method for the control of machine tools in industry, compare to Weck (1995), it is used in the following for comparison with the nonlinear, flatness-based controller design, even though using the desired velocities and accelerations in the control law would certainly improve the tracking behavior of the system.

The linear control design does not take into account that the movement of the TCP is determined by the coordinated motion of the two supports and neglects that the dynamic behavior of the machine tool changes nonlinearly over the workspace depending on the relative position $\Delta x = x_A - x_B$ of the two supports. Hence, the machine tool performance varies over the workspace and can only be optimized for a single relative position Δx of the machine. This is an undesirable property for a processing tool, because the machining quality varies as well over the workspace. In the worst case the two position controllers may even work against each other and the machine tool can become unstable. As a result, it is not possible to use the full potential of the linear drives in such a machine tool with parallel kinematics, since the control parameters have to be reduced as well as the maximal admissible operation speed compared to a single axis with a linear drive alone. Table 1 lists the control parameters used for simulation and experiments. The parameter K_f is already included as a factor in K_d. Higher K_v factors are not possible as the test machine quickly becomes unstable. The velocity loop was tuned by the designer of the test machine.

Table 1. Parameters of the linear controller in the experimental machine in Figure 4.

parameter	K_v	K_d	T_n
value	500 1/s	7000 Ns/m	6.7 ms

Nonlinear, Flatness-based Position Control In the following it is proposed to use a flatness-based position controller design for several reasons, Rothfuß (1997):

- The desired trajectory of the control variables can be calculated off-line under consideration of the limitations of input variables and states.
- The dynamics of the trajectory error is exactly linearized resulting in a "natural" control behavior and small input expenses.
- Simple and standard controller design, e.g. using pole placement, under consideration of the nonlinear dynamics and stabilization against disturbances and model errors is possible.
- With a (quasi-)static controller design it is possible to implement a controller without the use of differential equations, which can save computational time in real-time applications.

A nonlinear system in state space representation

$$\dot{\boldsymbol{x}} = \boldsymbol{f}(\boldsymbol{x}, \boldsymbol{u}), \quad \boldsymbol{x}(0) = \boldsymbol{x}_0 \in I\!R^n, \quad \boldsymbol{u} \in I\!R^m \tag{18}$$

with the state $\boldsymbol{x}$ and multiple inputs $\boldsymbol{u}$ ($\boldsymbol{x}$, $\boldsymbol{u}$ sufficiently often differentiable) is called (differentially) flat, if there exists a (fictitious) output

$$\boldsymbol{y} = \boldsymbol{\phi}\left(\boldsymbol{x}, \boldsymbol{u}, \dot{\boldsymbol{u}}, \ldots, \boldsymbol{u}^{(\alpha)}\right) \quad \text{with} \quad \dim \boldsymbol{y} = \dim \boldsymbol{u}, \tag{19}$$

which meets the conditions that all states $\boldsymbol{x}$ and inputs $\boldsymbol{u}$ can be expressed as algebraic functions $\boldsymbol{\Psi}_1$ and $\boldsymbol{\Psi}_2$ of $\boldsymbol{y}$ and a finite number of its time derivatives $\boldsymbol{y}^{(k)}$

$$\boldsymbol{x} = \boldsymbol{\Psi}_1\left(\boldsymbol{y}, \dot{\boldsymbol{y}}, \ldots, \boldsymbol{y}^{(\beta_m)}\right), \tag{20}$$

$$\boldsymbol{u} = \boldsymbol{\Psi}_2\left(\boldsymbol{y}, \dot{\boldsymbol{y}}, \ldots, \boldsymbol{y}^{(\beta_m+1)}\right). \tag{21}$$

In other words, the nonlinear system admits a complete, free and finite differential parameterization with $\boldsymbol{y}$ as the parameter. If these conditions are fulfilled at least locally, the output $\boldsymbol{y}$ is called flat output, see for example Rothfuß (1997), Rothfuß, Rudolph and Zeitz (1997) and Rudolph (2003).

A possible flat output for the model of the experimental machine (4), (7) with (9) is given by the minimal coordinates

$$\boldsymbol{y} = \boldsymbol{q}_m \tag{22}$$

as it is in general the case for fully actuated multibody systems. Hence, it is obvious that the system state

$$\boldsymbol{x} = \begin{bmatrix} x_A & x_B & \dot{x}_A & \dot{x}_B \end{bmatrix} = \begin{bmatrix} \boldsymbol{y} & \dot{\boldsymbol{y}} \end{bmatrix} = \boldsymbol{\Psi}_1\left(\boldsymbol{y}, \dot{\boldsymbol{y}}\right) \tag{23}$$

is an algebraic function of the flat output $\boldsymbol{y}$ and its first time derivative $\dot{\boldsymbol{y}}$. The dependent coordinates γ_A, γ_B and their first and second time derivatives can be calculated with Equations (10) as functions of $\boldsymbol{y}$, $\dot{\boldsymbol{y}}$ and $\ddot{\boldsymbol{y}}$. They are used in the following calculations as auxiliary variables in order to keep the calculations simple and to avoid the more complicated ODE-form of the nonlinear equations of motion. Equation (4) with (9) is now a system of linear equations for the unknown constraints $\boldsymbol{\lambda}$ and the motor forces $\boldsymbol{u}$

$$\boldsymbol{M} \cdot \ddot{\boldsymbol{q}}_t + \boldsymbol{k} - \boldsymbol{Q}^a = \begin{bmatrix} \boldsymbol{H} & \boldsymbol{G}^T \end{bmatrix} \cdot \begin{bmatrix} \boldsymbol{u} \\ \boldsymbol{\lambda} \end{bmatrix}. \tag{24}$$

Solving Equation (24) and using the auxiliary variables, the motor forces can be calculated as algebraic functions of the flat output only

$$\boldsymbol{u} = \begin{bmatrix} u_A & u_B \end{bmatrix} = \boldsymbol{\Psi}_2(\boldsymbol{y}, \dot{\boldsymbol{y}}, \ddot{\boldsymbol{y}}). \tag{25}$$

The function $\boldsymbol{\Psi}_2$ is often called the inverse system or the inverse dynamics in the case of mechanical systems, Rothfuß, Rudolph and Zeitz (1997). Figure 5 shows a schematic view of the flatness analysis of the multibody system with x_i, u_i and y_i meaning the ith component of the corresponding vector. As it can be seen, all states and inputs can be calculated by $\boldsymbol{y}$ and a

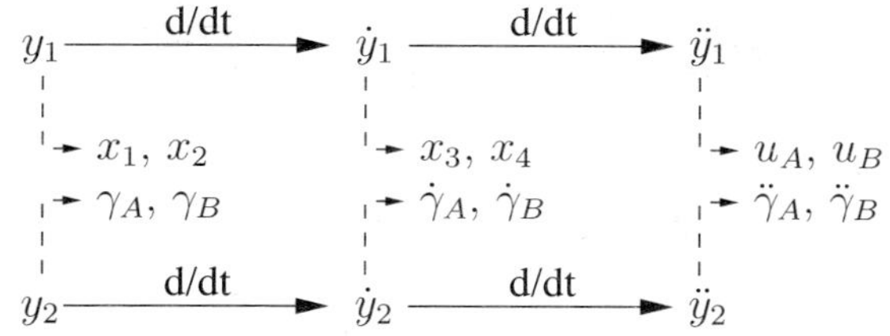

Figure 5. Schematic view of the flatness analysis.

finite number of its time derivatives and the conditions of the definition of flatness are met. Hence, the system is flat and $\boldsymbol{y}$ characterizes completely the dynamic behavior of the system. Due to the flatness property of the system, it is also controllable and observable, since the flat output is also measured, Rothfuß, Rudolph and Zeitz (1997).

Using $\boldsymbol{\Psi}_2$ as state feedback and introducing $\boldsymbol{v}$ as new input

$$\boldsymbol{u} = \boldsymbol{\Psi}_2(\boldsymbol{y}, \dot{\boldsymbol{y}}, \boldsymbol{v}) \tag{26}$$

results in an exactly linearized system of two uncoupled linear ODEs of second order

$$\ddot{\boldsymbol{y}} = \boldsymbol{v}. \tag{27}$$

Hence, this procedure can be interpreted as a general state transformation resulting in a system in Brunovsky normal form as defined in Isidori (1995)

$$\dot{\boldsymbol{z}} = \begin{bmatrix} 0 & 1 & 0 & 0 \\ 0 & 0 & 0 & 0 \\ 0 & 0 & 0 & 1 \\ 0 & 0 & 0 & 0 \end{bmatrix} \cdot \boldsymbol{z} + \begin{bmatrix} 0 & 0 \\ 1 & 0 \\ 0 & 0 \\ 0 & 1 \end{bmatrix} \cdot \boldsymbol{v} \tag{28}$$

with the new state

$$\boldsymbol{z} = \begin{bmatrix} y_1 & \dot{y}_1 & y_2 & \dot{y}_2 \end{bmatrix}. \tag{29}$$

Here, $\boldsymbol{z}$ has the same state variables as the original state vector $\boldsymbol{x}$. Hence, the flatness-based controller design coincides here with the control via a (static) exact linearization and nonlinear decoupling through state feedback, Isidori (1995). A standard linear PID controller with feedforward

$$\boldsymbol{v} = \ddot{\boldsymbol{y}}^d + \boldsymbol{K}_1 \cdot (\dot{\boldsymbol{y}}^d - \dot{\boldsymbol{y}}) + \boldsymbol{K}_2 \cdot (\boldsymbol{y}^d - \boldsymbol{y}) + \boldsymbol{K}_3 \cdot \int (\boldsymbol{y}^d - \boldsymbol{y})\,\mathrm{dt} \tag{30}$$

can be used for the now linear system with diagonal matrices $\boldsymbol{K}_1$, $\boldsymbol{K}_2$ and $\boldsymbol{K}_3$. In order to guarantee asymptotic stability, the matrices $\boldsymbol{K}_1$, $\boldsymbol{K}_2$ and

Table 2. Parameters of the nonlinear controller in the test machine.

parameter	K_1	K_2	K_3
value	$1000\,1/\mathrm{s}$	$220000\,1/\mathrm{s}^2$	$4000000\,1/\mathrm{s}^3$

$\boldsymbol{K}_3$ have to be chosen appropriately. Using Equation (30) with (27) results in exact linear dynamics

$$\ddot{\boldsymbol{e}} + \boldsymbol{K}_1 \cdot \dot{\boldsymbol{e}} + \boldsymbol{K}_2 \cdot \boldsymbol{e} + \boldsymbol{K}_3 \cdot \int \boldsymbol{e}\,\mathrm{dt} = \boldsymbol{0} \tag{31}$$

for the trajectory error

$$\boldsymbol{e} = \boldsymbol{y}^d - \boldsymbol{y}. \tag{32}$$

In order to achieve a desired dynamic behavior of the trajectory error, standard methods of linear control systems such as pole placement can be easily used now. If the control parameters are chosen such that $\boldsymbol{K}_i = K_i\boldsymbol{E}, i = 1, 2, 3$ the decoupled trajectory errors exhibit the same dynamic behavior, which is usually desirable for machine tools. Table 2 lists the experimentally tuned control parameters, the chosen factors correspond to the poles of error dynamics ≈ (-20, -291, -690).

Figure 6 shows a block diagram of the nonlinear, flatness-based position control. The computation of the state feedback follows exactly the procedure outlined before using γ_A and γ_B and their time derivatives as auxiliary variables. Such a controller design results in a homogeneous machine tool behavior over the whole workspace as the dynamics of the trajectory error (31) are now independent of the relative position Δx of the supports.

An alternative to the exact feedback linearization described above is the feedforward linearization as described by Hagenmeyer and Delaleau (2003). Instead of the measured state $\boldsymbol{x}$ in the state feedback the desired variables are used

$$\boldsymbol{u} = \boldsymbol{\Psi}_2 \left(\boldsymbol{y}^d, \dot{\boldsymbol{y}}^d, \boldsymbol{v} \right). \tag{33}$$

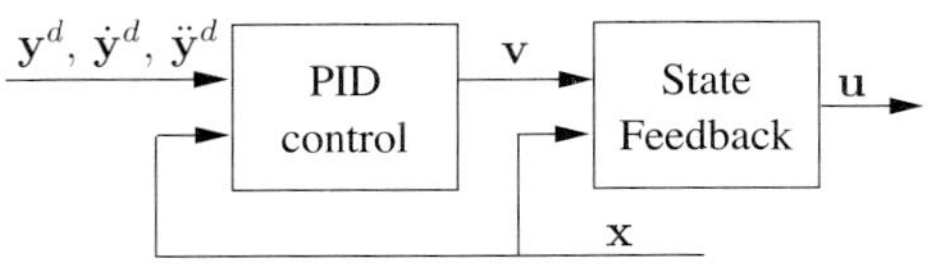

Figure 6. Block diagram of the nonlinear flatness-based position control.

The PID controller (30) remains unchanged. Hence, the system is exactly linearized, when it is on the desired trajectory and stabilized in the vicinity. Such an approach can lead to a better robustness of the control system especially if the measurements are strongly corrupted with noise.

Optimization of the control parameters to further improve the performance of the control system is of course possible, but was not pursued any further here, since already the nonlinear control system with this simple basic tuning shows superior results compared to the linear control concept.

Solving the constraint equations (7) numerically - if no analytical solution is possible - presents no further difficulty to the controller implementation in real-time. As only comparably small changes from one time step to next occur in the machine tool position, the Newton algorithm converges rapidly, if $\boldsymbol{q}_h$ from the previous time step is used as initial value for the algorithm. For the here presented examples the algorithm converges within 3-4 iterations to machine precision for the described trajectories.

2.3 Simulation and Experimental Results

In the following, experimental results for the test machine with the linear and the nonlinear flatness-based position control with feedback linearization are shown and compared to simulation results. Furthermore, the effect of the acceleration feedforward on the control error is demonstrated. Tables 1 and 2 list the control parameters used in the experiments and simulation.

One can clearly see that the flatness-based control concept taking into account the nonlinearities and the coupling in the system is able to improve the system performance significantly reducing the control error by several orders of magnitude. The highly superior results demonstrate unambiguously that the potential of parallel kinematics can only be fully judged and exploited if not only the design but also the control concept is adapted to the parallel machine tool setup.

Experimental Setup The digital control system of the test machine has a master-slave architecture with RTX as real-time operating system. The trajectory planning is done on the master PC with a standard NC-program in C coordinates (x_C^d, y_C^d). The program generates the desired variables for the support positions (x_A^d, x_B^d) online with a cycle time of $\Delta t_m = 6\,\text{ms}$. The master PC coordinates also the overall process communication and initiates the test runs. The desired variables are passed on to the slave PC, where the actual position control algorithm runs with a cycle time of $\Delta t_s = 0.6\,\text{ms}$. A precision interpolation generates the desired variables for the intermediate steps in between the 6 ms communication time of master and slave. The

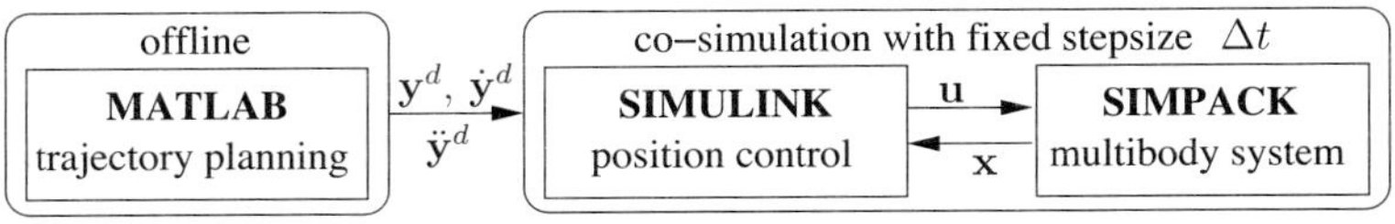

Figure 7. Structure of the simulation environment.

position control algorithm is freely programmable in C++ code, allowing to use independent, standard linear as well as multi-variable and nonlinear position control concepts. The position control output is then passed on to the individual drive units.

The drive units can be used in two ways: either in current control mode, in which only the current control is active and the slave PC transmits force values u_i $(i = A, B)$ directly to it, or in velocity control mode, in which an additional PI control loop for velocity in cascade to the current control is active and the slave PC transmits the output of the P control loop for position to it, compare Figure 4. The PI control loop for velocity on the drive unit then runs with a higher sampling rate $\Delta t_v = 0.1\,\text{ms}$. Measurements are taken with the slave PC at the sample rate Δt_s. The desired variables x_A^d and x_B^d can be stored simultaneously as well. The position of point C could not be measured directly in the experiments. Corresponding results were calculated with the system kinematics from the measured support positions.

Simulation Environment The multibody system model is implemented in SIMPACK, see INTEC GmbH, a commercial software tool for multibody systems able to deal with kinematic loops as well as with elastic bodies. The different control algorithms are implemented in Simulink. During a co-simulation between SIMPACK and Simulink the motor forces $\boldsymbol{u}$ generated by the position control respectively the system state $\boldsymbol{x}$ are exchanged with a fixed step size Δt, see Figure 7. This simulation setup is comparable to a digital control system operating with the same sample time Δt as the co-simulation. All simulations are done with a sample time equal to the cycle rate of the actual position control system $\Delta t = \Delta t_s = 0.6\,\text{ms}$. The faster sub-sampling of the PI velocity control loop in the experiment is neglected.

Trajectories The circular trajectory in Figure 8, which for example occurs in milling, starts at $t_{start} = 0\,\text{s}$ from the marked position, accelerates to a constant track speed of $0.5\,\text{m/s}$ following first clockwise the right circle,

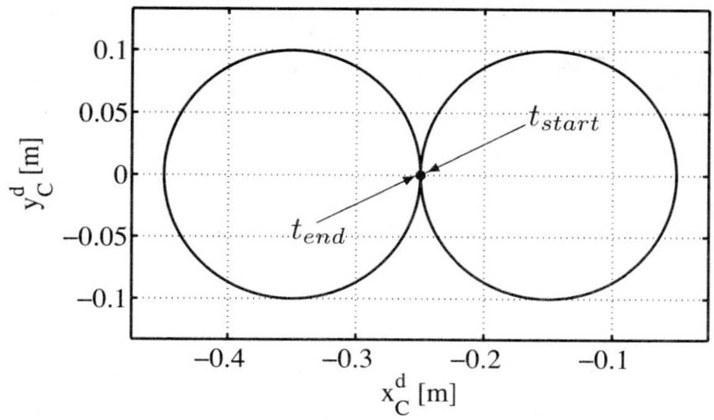

Figure 8. Definition of the circular trajectory, dots mark rest positions.

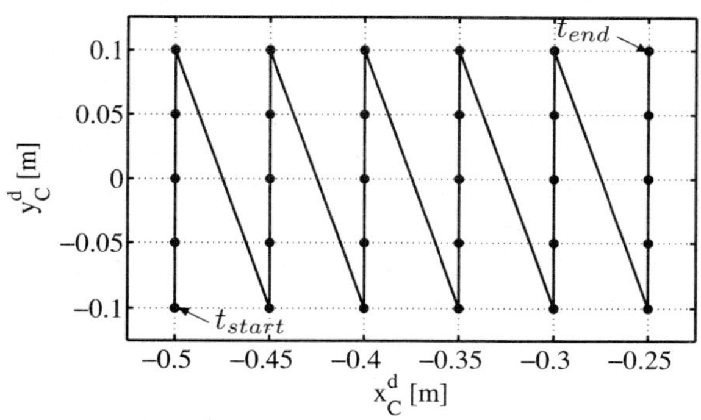

Figure 9. Definition of the grid-like trajectory, dots mark rest positions.

then counterclockwise the left circle describing a lying eight in the xy-plane. After following the eight three more times keeping the constant track velocity it comes to a stop at the start position at time $t_{end} = 11.1984\,\mathrm{s}$.

The grid-like trajectory in Figure 9 for example occurs in pick-and-place operations or drilling. Each dot marks a rest position. The maximum track velocity, to which the system is accelerated in between the rest positions, varies with every second column and the time at the rest position is shortened as well. For the first two columns the maximal track velocity is equal to $1/6\,\mathrm{m/s}$ with a rest time of $0.1\,\mathrm{s}$ at each point. The maximal velocity for the next two columns is equal to $0.5\,\mathrm{m/s}$ with a rest time of $0.05\,\mathrm{s}$. For the last two rows the maximal track velocity is equal to $5/6\,\mathrm{m/s}$ with a very

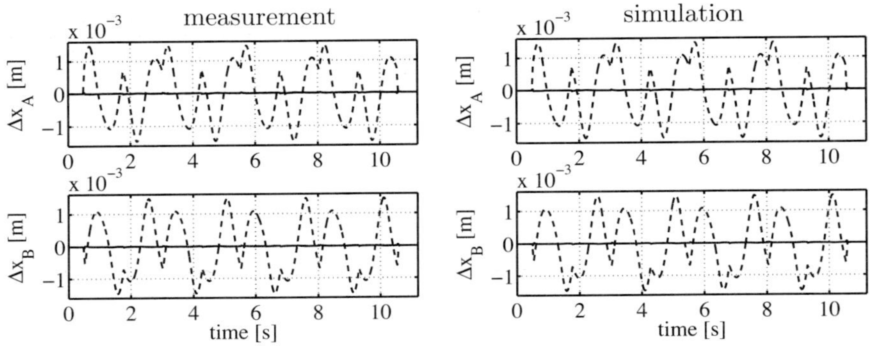

Figure 10. Position error of the supports A and B with linear (dashed) and nonlinear (solid) position control, circular trajectory.

short rest time of 0.006 s. The trajectory ends after $t_{end} = 11.385$ s at the last dot.

Comparison of measurements and simulations In Figures 10 - 15 a comparison of measurement and simulation results for the independent, linear cascaded position control and the nonlinear, flatness-based position control with feedback linearization is shown for both trajectories. The simulation results fit very well to the measurements apart from some small mea-

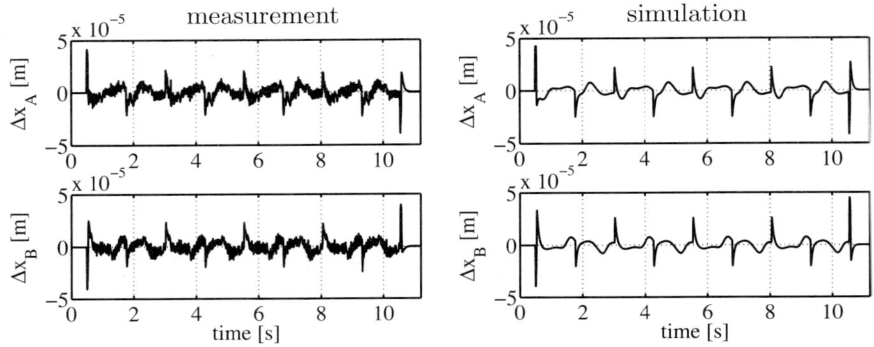

Figure 11. Position error of the supports A and B with nonlinear position control, circular trajectory.

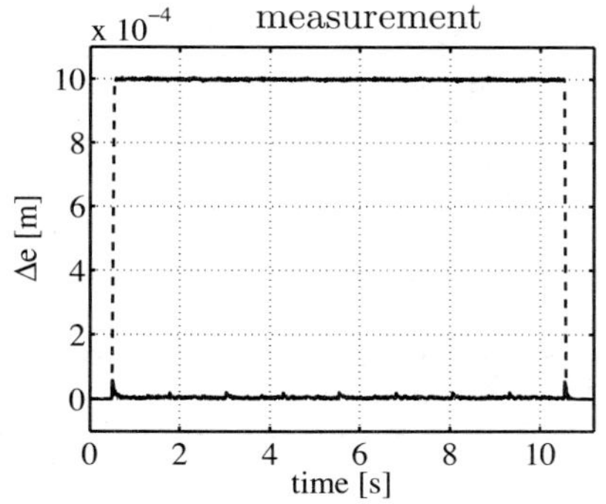

Figure 12. Absolute position error of point C with linear (dashed) and nonlinear (solid) position control, circular trajectory.

surement noise present in the experimental data. Even the peaks in Figures 11, 13 and 15 due to jumps in the desired acceleration coincide quite well. This also demonstrates that the assumptions made, e.g. concerning current control or modeling, are valid and that the multibody system simulation is a useful tool for the controller design for mechanical systems.

A comparison of the results with linear and with the flatness-based position control proves in the experiments too the great potential of the nonlinear control concept to reduce the control error significantly. In Figures 10, 12 and 14 the position error for the nonlinear controller appears in the di-

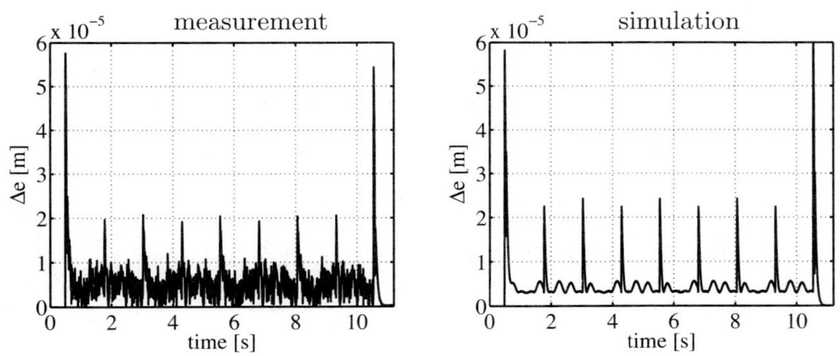

Figure 13. Absolute position error of point C with nonlinear position control, circular trajectory.

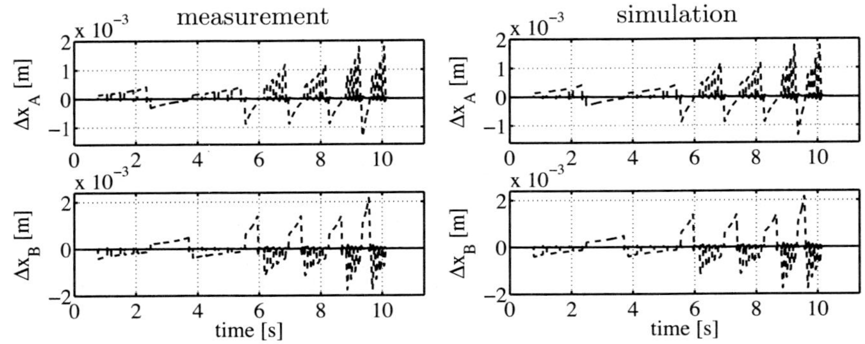

Figure 14. Position error of the supports A and B with linear (dashed) and nonlinear (solid) position control, grid-like trajectory.

rect comparison to the linear position controller as a nearly flat line being very close to zero. Figures 11, 13 and 15 show that the control error can be reduced by nearly two orders of magnitude. It also becomes clear that the main cause for relatively large error amplitudes are the jumps in the desired accelerations resulting from the trajectory planning. As mentioned before, a smoother trajectory planning or an optimized controller tuning can reduce these peaks. Furthermore, as the following paragraph shows, acceleration feedforward can reduce the height of the peaks significantly.

For the grid-like trajectory the effect of the jumps in the desired accel-

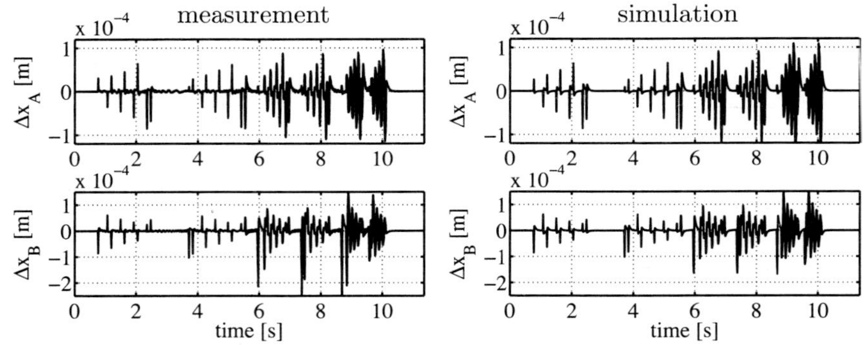

Figure 15. Position error of the supports A and B with nonlinear position control, grid-like trajectory.

eration is particularly pronounced, see Figures 14 and 15. The especially jerky motion in between the last two columns of rest positions, see Figure 8 and Figure 9, brings the linear as well as the nonlinear controller to the limit of their performance, as for example the actuator forces nearly reach the limit of 400 N. The system has hence only a very small capacity left to eliminate the influence of disturbances and model uncertainties and acts close to a bang-bang behavior. Such problems primarily resulting from insufficient trajectory planning could be avoided, if the desired trajectories were calculated off-line and limitations of the input variables and the state were checked with the help of the inverse system in Equation (25) beforehand.

Effect of Acceleration Feedforward Acceleration feedforward $\ddot{\boldsymbol{y}}^d$, see Equation (30) was not used in the experiments in order to avoid numeric quantization effects, since already the desired velocity variables have to be calculated via numeric differentiation. Figure 16 shows simulation results for the nonlinear position controller with acceleration feedforward for both trajectories. The acceleration feedforward $\ddot{\boldsymbol{y}}^d$ was calculated by differentiating the desired position variables $\boldsymbol{y}^d$ twice numerically. One can clearly see, that the height of the peaks due to jumps in the acceleration as well as the overall position error can be reduced further in comparison to the performance of the nonlinear position control without acceleration feedforward. Including acceleration feedforward into the control algorithm is also a simple and cheap additional measure to further improve the control sys-

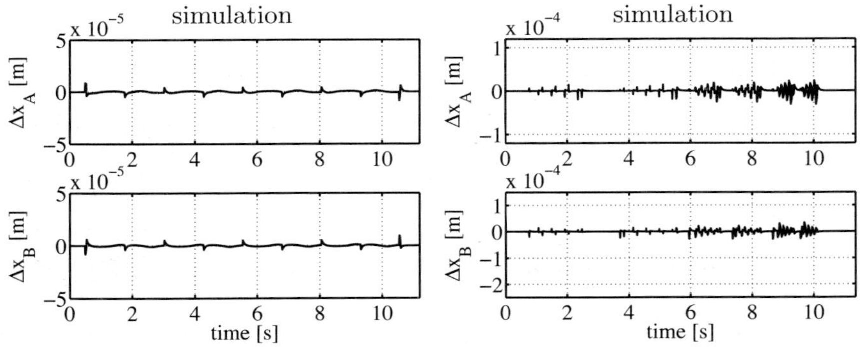

Figure 16. Position error of the supports A and B with nonlinear position control and acceleration feedforward, circular trajectory (left) and grid-like trajectory (right).

tem performance, as the values of the desired acceleration or at least of the desired velocities are known after trajectory planning and only have to be passed on to the control algorithm without any additional computational costs.

2.4 Summary

The experimental and simulation results prove that the flatness-based position control leads to highly superior results in performance and accuracy. They also show that it is possible to implement the nonlinear control algorithm in a regular control system for machine tools operating under real-time conditions. The trajectory error can be reduced by several orders of magnitude in comparison to conventional linear control concepts. Several characteristics of the flatness-based concept such as feasibility tests of trajectories, relatively simple controller tuning and physical insight make it a promising concept.

3 Multi-Criteria Optimization of a Hexapod Machine

In this section the hexapod robot HEXACT is investigated. HEXACT is a research machine tool with parallel kinematics, developed by Prof. Heisel and his coworkers at the Institute for Machine Tools at the University of Stuttgart, Germany, see Figure 17. Hexapod machines feature the advantages of parallel kinematics and are characterized by a simple, cost-effective

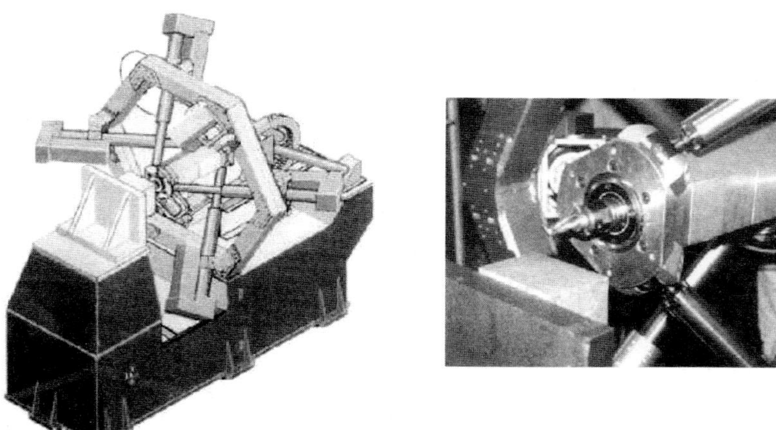

Figure 17. Hexapod machine HEXACT.

design, Heisel, Maier and Lunz (1998). The tool carrier has six degrees of freedom in space and is positioned by actuators integrated in the six telescope struts. Furthermore, the specific construction of this machine allows for compensation of joint clearance by pre-stressing of the symmetrically arranged struts with an initial tension. However, pre-stressing leads to loss of one degree of freedom, as the rotation about the tool axis is not controllable anymore. Nevertheless 5-axis processing of cubic parts is possible in the pre-stressed case with high static and dynamic accuracy.

A general drawback of parallel kinematic robots is the appearance of kinematic singular configurations that have to be avoided during operation and, hence, reduce the usable workspace. For the current design of the investigated hexapod machine an especially unpleasant kinematic singularity is located along the tool axis in the central position of the machine, where the rotational stiffness about the tool axis decreases to zero, see Liu (2001). One way to overcome this problem and to reduce the manifold of singularities is to use a higher number of actuators than the degree of freedom of the end-effector, so-called redundant actuation, see e. g. Dasgupta and Mruthyunjaya (1998), O'Brien and Wen (1999), or Valášek et al. (2001). However, this approach leads to additional difficulties. For example special techniques for the control of the over-constrained systems are required. In this article, redundant actuation is not considered. The idea followed here is to remove unpleasant singularities in the workspace by changing the design of the machine.

A brief description of the modeling of the hexapod is given in the first part of this contribution. The dynamic behavior of the system is described by nonlinear differential equations of motion obtained from the multibody system approach. In order to define tool trajectories, the articular coordinates have to be derived by inverse kinematics. Both topics are described in Dignath (2004), where the hexapod robot was modeled and optimizations were carried out. We utilize a re-engineered, extended model in order to do further optimizations and to evaluate alternative designs, proposed here, with less or even no singular configurations in the interesting workspace.

The description then continues with the design process optimization. After definition of performance criteria and design variables a multi-criteria optimization approach is applied in order to find an 'optimal' system, see Bestle (1994), Eberhard (1996), Eberhard and Bestle (1994). The design goals considered here are to reduce flexibility and improve accuracy under consideration of a sufficiently large workspace of the machine utilizing our alternative designs. This implies the decrease of the manifold of singular configurations of the machine.

Further, a general approach to evaluate the stiffness/flexibility behavior of the machine is introduced. Here, the tangential stiffness or flexibility matrices, respectively, are determined for the entire workspace of the machine. The results found with this method are compared with those from the previous optimization.

3.1 Modeling of the Hexapod Machine

A crucial point within the modeling of the hexapod machine is the description of the telescope struts. A main question is whether elastic deformations of the relatively slender struts have to be considered or not. In Dignath (2004) flexible struts were investigated with the conclusion that in the low frequency range the strut elasticity has a negligible influence on the vibration behavior of the machine. In contrast, the elasticity of the joints turned out to be important to get sound results corresponding closely to real measured data. Hence, a rigid multibody system (MBS) model with elastic joints for the connection of the struts is utilized. In order to eliminate singular configurations and reduce flexibility, alternative design variants are proposed.

Multibody Systems In the common multibody system approach, deformations are neglected and the bodies are considered as rigid. The bodies are interconnected with ideal joints and coupling elements like springs, dampers and elements for active control. The dynamics of multibody systems is described by the differential equations of motion, see Schiehlen and Eberhard (2004). Using f generalized coordinates $\boldsymbol{y}$ they can be written in general form as differential equations

$$\boldsymbol{M}(t,\boldsymbol{y}) \cdot \ddot{\boldsymbol{y}} + \boldsymbol{k}(t,\boldsymbol{y},\dot{\boldsymbol{y}}) = \boldsymbol{q}(t,\boldsymbol{y},\dot{\boldsymbol{y}}) \,, \tag{34}$$

where $\boldsymbol{M} \in \mathbb{R}^{f\times f}$ is the symmetric, positive definite mass matrix, $\boldsymbol{k} \in \mathbb{R}^f$ the vector of generalized centrifugal and Coriolis forces and $\boldsymbol{q} \in \mathbb{R}^f$ the vector of generalized applied forces. Equation (34) can be derived from Newton's and Euler's laws and d'Alembert's principle.

Starting from initial conditions for position and velocity $\boldsymbol{y}^0$, $\dot{\boldsymbol{y}}^0$ the equations of motion have to be solved by numerical time integration, where various algorithms exist, e.g. Runge-Kutta methods or the Shampine-Gordon method that is also applied to the presented example.

MBS Model of the Hexapod The rigid MBS model of the hexapod used here is similar to the one in Dignath (2004), except that it was re-engineered for computational reasons and in order to extend the model to further analysis and optimization tasks. The nonlinear equations of mo-

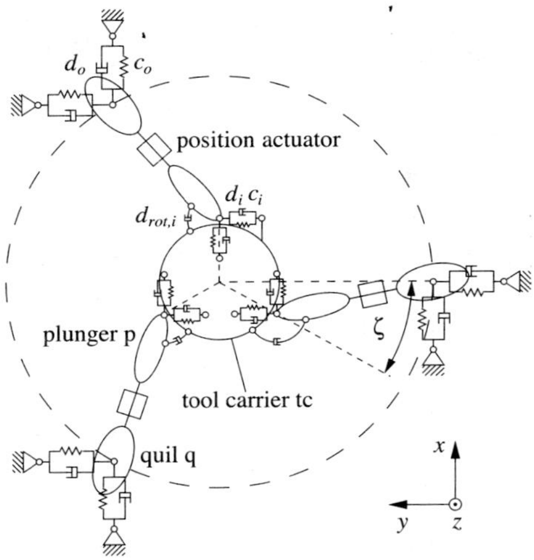

Figure 18. MBS model of the hexapod machine (front struts only).

tion Equation (34) are created symbolically with NEWEUL, see Kreuzer and Leister (1991). In Figure 18 a section of the spatial rigid multibody model of the hexapod machine in central position is illustrated, compare also the real machine in Figure 17. A front view of the model in the xy-plane is shown and for lucidity only three of the six telescope struts are drawn.

The model has $f = 36$ degrees of freedom. The generalized coordinates

$$\boldsymbol{y} = \begin{bmatrix} \boldsymbol{y}_{\mathrm{tc}} & \boldsymbol{y}_{\mathrm{p1}} & \cdots & \boldsymbol{y}_{\mathrm{p6}} \end{bmatrix} \in \mathbb{R}^f \tag{35}$$

describe the motion of the tool carrier $\boldsymbol{y}_{\mathrm{tc}} = [x_{\mathrm{tc}}\ y_{\mathrm{tc}}\ z_{\mathrm{tc}}\ \alpha_{\mathrm{tc}}\ \beta_{\mathrm{tc}}\ \gamma_{\mathrm{tc}}]$ and the six strut plungers $\boldsymbol{y}_{\mathrm{p}i} = [x_{\mathrm{p}i}\ y_{\mathrm{p}i}\ z_{\mathrm{p}i}\ \eta_{\mathrm{p}i}\ \vartheta_{\mathrm{p}i}]$. Since the telescope struts cannot rotate about the strut axis only five coordinates are required for each plunger. The strut quils are connected to the plungers via position actuators, which realize the positioning travel of the telescope struts.

All parameters of the model like geometry, masses, spring and damper coefficients, etc. are taken from Dignath (2004) or from the Institute of Machine Tools (IfW) at the University of Stuttgart. Details on the measured elasticities of the bearings are given in Dignath and Hempelmann (2002).

The articular coordinates, i.e. the amplitudes of the actuators in the struts, are treated as time dependent variables, or rather rheonomic constraints. For desired tool trajectories they are computed by solving the inverse kinematic problem, which is to determine the articular coordinates belonging to a defined position and orientation of the tool carrier in the workspace. For parallel kinematic structures like the hexapod, the inverse kinematic problem can be solved directly via geometric consideration whereas the direct kinematic problem, i.e. to find the tool carrier's position and orientation belonging to a given set of articular coordinates, must be solved numerically, in general. This is a remarkable difference to serial kinematic structures where the direct kinematic problem can be solved analytically and the inverse problem becomes very difficult, see e.g. Tsai (1999).

Alternative Designs In the current design of the hexapod machine the joints are placed equidistantly on the perimeter of the tool carrier. Therefore, the struts are placed radial at the tool carrier in its central position. This leads to a kinematic singularity where the flexibility around the tool axis approaches infinity and is very high in the nearby surrounding. For machining processes this is very undesirable as high flexibility leads to loss of accuracy and low surface quality of the produced parts.

In order to improve the flexibility behavior of the hexapod and to eliminate singular configurations two options for alternative designs are proposed and a third variant, described in Dignath (2004), is discussed. All variants, including the current design, are illustrated in Figure 19.

The angle between the struts and the tool can be varied in different ways. As a first possibility one may rotate the joint positions at the tool carrier by an angle ζ, see Figure 19 (variant I). Thus, the singular point in the center vanishes and a lower flexibility is expected. However, the mentioned pre-stressing is no more possible. A second option is to rotate the front and back joints in counterwise directions, respectively, Figure 19 (variant II). It is the advantage of this design that a pre-stressing of the struts for eliminating joint clearance is still possible. However, this modification leads to a kinematic coupling of the tool rotation around the z-axis (tool axis) and its translation in z-direction. A third possibility, proposed in Dignath (2004), is to change the positions of the two upper and lower joints each counterwise while the left and right joints remain in their original positions, as illustrated in Figure 19 (variant III).

All alternatives involve a reduction of the workspace in the xy-plane as the pivoting angle of the inner joints is restricted. This has to be considered when modifying the design of the hexapod.

3.2 Optimization of MBS

It is the purpose of optimizing multibody systems to find a mechanical design that leads to improved dynamic behavior of the technical system,

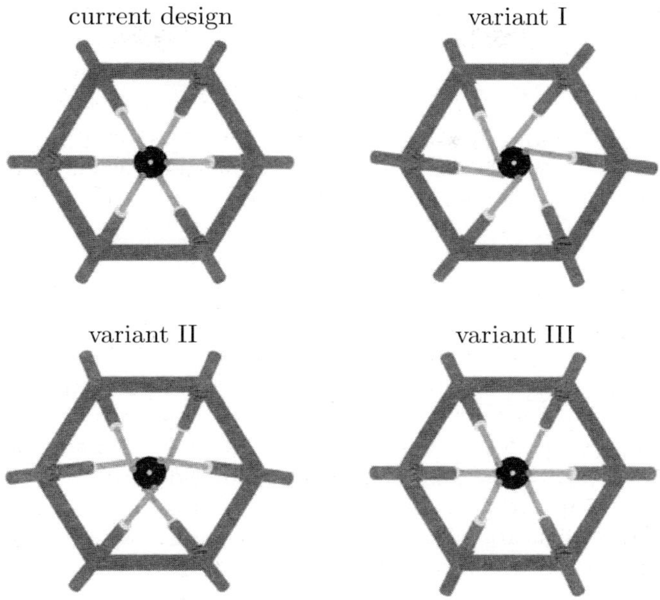

Figure 19. Alternative designs: current design with radial struts; variant I with equally rotated joint positions; variant II with counter-rotated joint positions; variant III with counterwise rotated joint positions for four struts.

e.g., increased damping of undesired vibrations. In order to evaluate the dynamic behavior, the simulation of the motion by numerical integration of the equations of motion is required. During the optimization process these time integrations are repeatedly executed using different sets of design variables $\boldsymbol{p}$. These design variables are the parameters of the system that can be changed within given ranges for optimizing the dynamic behavior, for example, parameters of the structure, geometric quantities, stiffnesses or active controls of the technical system.

More information concerning the procedure and available solution methods is given in the textbooks Bestle (1994) and Fletcher (1987) or the article Bestle and Eberhard (1992).

Optimization Criteria Use of numerical optimization methods requires that technical restrictions and design goals are formulated as mathematical functions. From the motion of the system, which is determined by the time trajectories of the generalized coordinates, the optimization criteria can be evaluated. For the evaluation of vibrations, energies, etc. these criteria are usually of the form

$$\xi_i = G_i(t_{\mathrm{end}}, \boldsymbol{y}_{\mathrm{end}}, \dot{\boldsymbol{y}}_{\mathrm{end}}, \boldsymbol{p}) + \int_{t_0}^{t_{\mathrm{end}}} F_i(t, \boldsymbol{y}, \dot{\boldsymbol{y}}, \ddot{\boldsymbol{y}}, \boldsymbol{p}) \mathrm{d}t \tag{36}$$

which is closely related to optimal control. In Equation (36) the dynamic behavior during the complete time interval $[t_0, t_{\mathrm{end}}]$ is taken into account by F_i, while G_i only considers the final state and, hence, accounts for cases where special values for the final state or a minimum time or energy must be achieved. The final time t_{end} may be fixed or given implicitly by the final state

$$t_{\mathrm{end}} : \; H_{\mathrm{end}}(t_{\mathrm{end}}, \boldsymbol{y}_{\mathrm{end}}, \dot{\boldsymbol{y}}_{\mathrm{end}}, \boldsymbol{p}) = 0. \tag{37}$$

Simple explicit functions $\xi_i = \xi_i(\boldsymbol{p})$ or user-defined functions are also utilized. For example, the bounds of the design variable space can be defined as

$$p_{i\mathrm{l}} \leq p_i \leq p_{i\mathrm{u}} \tag{38}$$

or implicitly

$$\xi_1 = p_i - p_{i\mathrm{l}} \geq 0, \;\; \xi_2 = p_{i\mathrm{u}} - p_i \geq 0. \tag{39}$$

All criteria ξ_i may be used either as equality constraints g_i, inequality constraints h_i, or optimization criteria ψ_i.

Multi-Criteria Optimization The optimization of technical systems usually cannot be performed by the optimization of only one criterion. Instead, several conflicting criteria must often be considered for optimization. The theory of multi-criteria optimization helps to avoid such discrepancies, Eberhard (1996). Several criteria can be minimized simultaneously, and optimal compromise solutions can be derived if conflicts arise. In general, these solutions are not unique but a whole set of solutions exists which are not comparable to each other, so-called Edgeworth-Pareto-optimal (EP-optimal) solutions. All of these solutions have to be considered as optimal since no strict order exists in vector optimization. In general, the multi-criteria optimization task can be written as follows

$$\text{find} \; \underset{\boldsymbol{p} \in \mathbb{P}}{\mathrm{opt}} \; \boldsymbol{\psi}(\boldsymbol{p}) \quad \text{where} \quad \mathbb{P} := \left\{ \boldsymbol{p} \in \mathbb{R}^h \,|\, \boldsymbol{g}(\boldsymbol{p}) = \boldsymbol{0}, \boldsymbol{h}(\boldsymbol{p}) \geq \boldsymbol{0} \right\} \tag{40}$$

with several optimization criteria ψ_i summarized in the vector $\boldsymbol{\psi}$ and analogous the vectors $\boldsymbol{g}$ and $\boldsymbol{h}$ of equality and inequality constraints. Goal of the optimization of the vector criterion $\boldsymbol{\psi}(\boldsymbol{p})$ is to minimize all criteria ψ_i.

A simultaneous minimization of all criteria is rarely possible as generally no point is feasible in the design space where all criteria have their minimal values simultaneously. Such a point in the criteria space is called *utopian solution*. Instead, multi-criteria approaches try to find a compromise that favorably lies on the set of EP-optimal solutions. Which solution to choose from the set of EP-optimal solutions then depends on the preferences of the designer.

Common to most of the strategies is the reduction of the vector optimization problem to nonlinear programming problems. This reduction is based on two fundamental principles: scalarization and hierarchization.

In the case of scalarization the optimization criteria are combined to a scalar utility function $u(\boldsymbol{\psi})$ which is then optimized instead of the original vector criterion. Often the weighted criteria method is applied which can be problematic for dynamic problems, as the weighting coefficients are difficult to define physically and the dependency of the result on this choice is highly nonlinear.

Hierarchical methods have in common, that the designer has to assign a level of importance to each optimization criterion ψ_i. Optimization starts with a scalar optimization of the first, most important, criterion. Then the next important criterion can be minimized, where the results reached in the previous levels are considered as inequality constraints and so on. An often useful modification introduces worsening factors that allow a defined worsening of results reached in previous levels in order to give the optimization of the criterion in the actual level some margin.

Optimization Algorithms During the multi-criteria optimization every scalar optimization task reads

$$\text{find} \min_{\boldsymbol{p} \in \mathbb{P}} \psi(\boldsymbol{p}) \quad \text{where} \quad \mathbb{P} := \left\{ \boldsymbol{p} \in \mathbb{R}^h \,|\, \boldsymbol{g}(\boldsymbol{p}) = \boldsymbol{0}, \boldsymbol{h}(\boldsymbol{p}) \geq \boldsymbol{0} \right\} . \tag{41}$$

In order to solve this nonlinear optimization problem with equality and inequality constraints, deterministic or stochastic optimization algorithms can be applied. Deterministic algorithms usually use gradient information and have the advantage of locating a minimum within a few iteration steps. On the other hand, they usually only find local minima and require, at least in theory, a smooth optimization criterion. Stochastic methods have the advantage that they may find a global minimum without posing severe restrictions on the differentiability or convexity of the optimization criterion, however, they often need many criteria evaluations. Therefore, they are not well suited for systems which require a high computational effort for evaluation. This drawback can be reduced by using parallel criteria evaluation in combination with a stochastic evolution strategy, see e.g. Eberhard, Dignath and Kübler (2003).

For the use of gradient-based optimization algorithms, like sequential quadratic programming (SQP) algorithms that are implemented in NEWOPT/AIMS, Bestle and Eberhard (1994), gradients of the optimization criteria with respect to the design variables $\boldsymbol{p}$ are required. Since the state variables $\boldsymbol{y}_0$, $\dot{\boldsymbol{y}}_0$ can not generally be given as explicit functions of $\boldsymbol{p}$, direct differentiation methods cannot be applied. In NEWOPT/AIMS gradients can be derived by purely numerical methods like finite differences or more sophisticated methods like the adjoint variables method, Eberhard (1996), the direct differentiation approach, Bestle (1994), and automatic differentiation using e.g. ADIFOR, Bischof et al. (1996). In the symbolic Matlab-based MBS code SYMBS, see Kurz, Henninger and Eberhard (2008), also adjoint variables and direct differentiation approaches are implemented.

3.3 Optimization of the Hexapod Machine

In this section the spatial MBS model of the hexapod machine described before is optimized with respect to flexibility and workspace. The optimization is carried out under consideration of the alternative designs as proposed earlier. In a first step, the flexibility of the system is investigated without considering geometric restrictions in order to obtain the unconstrained theoretical minimum. This is advisable as a premature restriction of the design space often leads to loss of innovative solutions of the problem. Afterwards, the geometric boundary conditions are taken into account and the influence of the new configuration on the workspace is discussed, where a possible compromise is proposed. All optimizations are carried out using an SQP algorithm with gradients computed by finite differences or automatic differentiation.

Optimization Criteria and Design Variables It is the goal of the optimization to increase the stiffness, i.e. to reduce the flexibility of the tool for machining purposes. Therefore, we define an exemplary tool trajectory with relative large, but still realistic machining forces and moments. When running the trajectory, the actual translational and rotational position is computed. The flexibility can be evaluated by comparing the actual trajectory with the target values. This can be formulated applying an integral optimization criterion, see Equation (36). Regarding the rotational flexibility about the tool axis, the average deflection of the tool rotation can be used

$$\psi_1 = \frac{1}{t_{\text{tot}}} \int_{t_0}^{t_{\text{end}}} |\gamma_{\text{tc}} - \gamma_{\text{tc,target}}| \, \mathrm{d}t. \tag{42}$$

The translational flexibility can be evaluated analogously

$$\psi_2 = \frac{1}{t_{\text{tot}}} \int_{t_0}^{t_{\text{end}}} |\Delta \boldsymbol{x}| \, \mathrm{d}t \tag{43}$$

with

$$\Delta \boldsymbol{x} = [x_{\text{tc}} - x_{\text{tc,target}} \;\; y_{\text{tc}} - y_{\text{tc,target}} \;\; z_{\text{tc}} - z_{\text{tc,target}}] \,. \tag{44}$$

As design variable p the joint rotation angle ζ is chosen, see Figure 18. Due to geometric restrictions the admissible range of values for the design variable is bounded by $0 \leq p \leq \pi/2$.

Optimizing Design Variant I During the first optimization run, only criterion ψ_1 evaluating the rotational flexibility is taken into account. In order to achieve an optimal angle ζ^* in the central position, the struts have to be fixed tangentially at the tool, here $\zeta^* = 80.2°$. The question is whether this is also the optimum for a typical machining trajectory. In order to answer this question, the following trajectory is defined, see Figure 20.

- 0 – 1 sec: adjust joint rotation angle ζ, see Figure 18,
- 1 – 2 sec: point to point motion to $\boldsymbol{y}_{tc}$=[-0.1 m 0 m 0 m 0° 0° 0°],

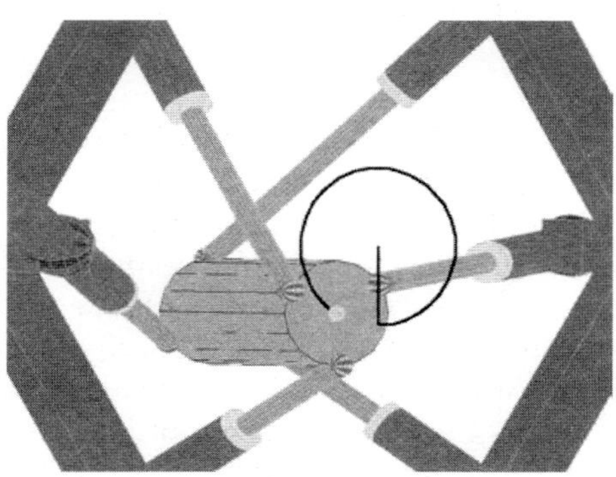

Figure 20. Exemplary machining trajectory for evaluating flexibility behavior.

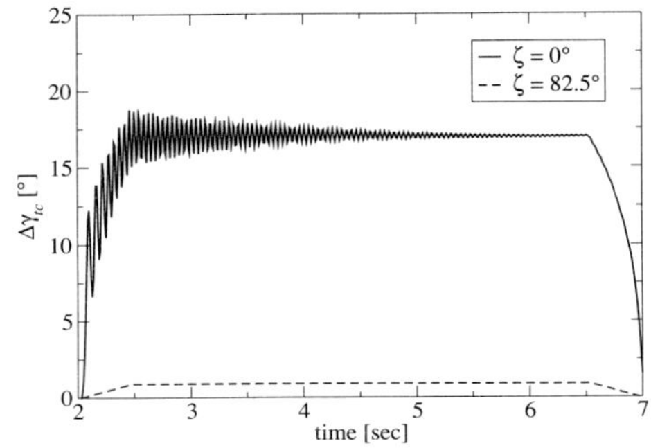

Figure 21. Deviation of tool rotation γ_{tc} due to an applied torque for the original design ($\zeta = 0°$) and the optimized design ($\zeta = 82.5°$).

- 2 – 7 sec: circle in the xy-plane with radius $r = 0.1\,\mathrm{m}$ and midpoint at $[0\ 0\ 0]\,\mathrm{m}$ with an applied torque of $l_{\mathrm{z}} = 1000\,\mathrm{Nm}$ around the tool axis.

The time integration of this trajectory takes approximately 50 sec on a 3 GHz Linux PC. The optimum is found for approximately the same value for ζ as stated for the zero position, $\zeta^* = 82.5°$. Figure 21 shows the deviation of the tool rotation γ_{tc} both for the original design ($\zeta = 0°$) and the optimized design ($\zeta^* = 82.5°$). For the "stationary" state ($3 < t < 6$ sec), the average rotational flexibility around the tool axis calculates as $c_{\gamma\gamma} = \Delta\gamma_{\mathrm{tc}}/l_{\mathrm{z}} = 1.77 \cdot 10^{-2}\,°/\mathrm{Nm}$ for the original design. The average rotational flexibility decreases to $c_{\gamma\gamma} = 9.00 \cdot 10^{-4}\,°/\mathrm{Nm}$ which is an improvement by a factor of approximately 20. This factor can be seen in Figure 21 from the ratio $\Delta\gamma_{\mathrm{tc}}(\zeta = 0°)$ to $\Delta\gamma_{\mathrm{tc}}(\zeta = 82.5°)$.

At real machining processes, also force loads are applied to the tool, of course. Therefore, the influence of the joint position angle ζ on the translational flexibility, evaluated by ψ_2, must also be considered. For the original design, the resulting translational flexibility $c_{\mathrm{trans}} = |\Delta\boldsymbol{x}|/|\boldsymbol{f}|$ has a relative high value of $3.39 \cdot 10^{-6}\,\mathrm{m/N}$. For the optimization, the above described trajectory is chosen again with an additional force loading of $\boldsymbol{f} = [1\ 1\ 1]\,\mathrm{kN}$. Now a vector optimization problem with criteria ψ_1 and ψ_2 has to be solved. For this purpose a hierarchical optimization is performed. In the first level, criterion ψ_1 is optimized. The solution is identical with the result already derived for the first optimization ($\zeta = 82.5°$). The resulting translational flexibility, which is not considered in ψ_1, also improves by a factor of about 7 to a value of $4.96 \cdot 10^{-7}\,\mathrm{m/N}$, see Figure 22.

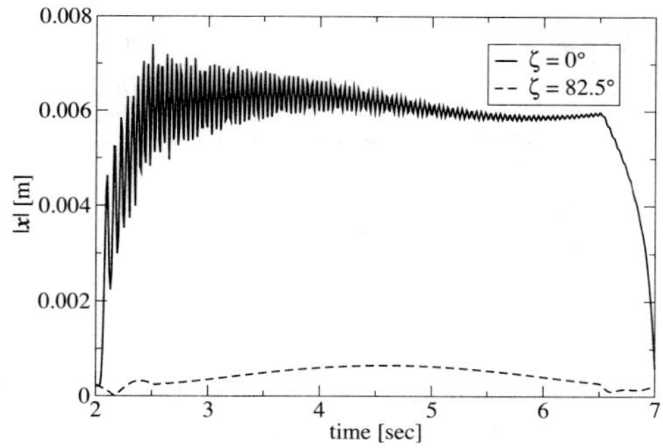

Figure 22. Deviation of the absolute tool position $\boldsymbol{x}$ due to applied torque and force for the original design ($\zeta = 0°$) and the optimized design at level 1 ($\zeta = 82.5°$).

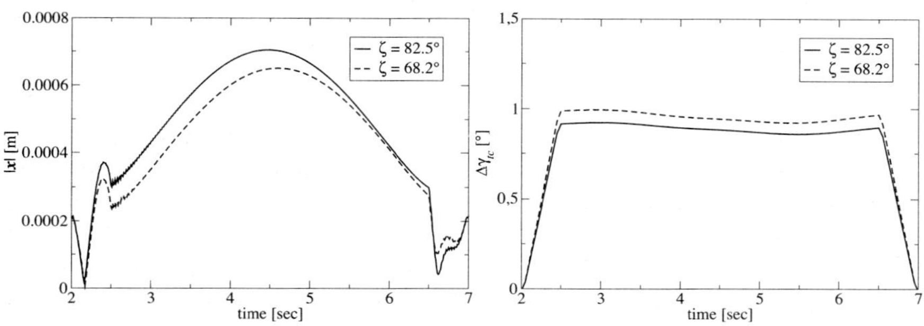

Figure 23. Deviation of resulting tool position $\boldsymbol{x}$ (left) and tool rotation γ_{tc} (right) due to applied torque and force load for the optimal design of variant I at level 1 ($\zeta = 82.5°$) and the final design ($\zeta = 68.2°$).

In the second level of the hierarchical optimization, criterion ψ_2 evaluating the translational flexibility is optimized. Now the result for criterion ψ_1 is considered as an inequality constraint. In order to give the optimization of ψ_2 some margin in the design space, the optimal value for ψ_1 may increase, i.e. may be worsened by 50 %. The final optimum is stated for $\zeta = 68.2°$ where the translational flexibility c_{trans} decreases by only 4.3 % to $4.76 \cdot 10^{-7}$ m/N. Figure 23 (left) shows the decrease of deviation of the translational position for the final optimum ($\zeta = 68.2°$) compared to the result of the first optimization level ($\zeta = 82.5°$).

The rotational flexibility increases to $c_{\gamma\gamma} = 9.68 \cdot 10^{-4}\,{}^\circ/\mathrm{Nm}$, see Figure 23 (right). As this is still far away from the maximum allowed worsening, i.e. the constraints given by the worsening factor are not active, the optimum of ψ_2 corresponds to the unconstrained case.

Optimizing Design Variant II Contrary to the expected behavior, a counterwise rotation of the strut joints at the tool by an angle ζ does not improve the rotational stiffness of the tool, but even worsens it, see Figure 24. For the desired trajectory, the average rotational flexibility increases to a value of $3.3 \cdot 10^{-2}\,{}^\circ/\mathrm{Nm}$ for $\zeta = 82.5^\circ$ which is an increase by a factor of about 2. In fact, from an optimization of ψ_1 the optimal solution $\zeta = 0^\circ$ is found.

Due to the coupling of the translation and the rotation with respect to the z-axis, the applied torque also causes a non-negligible deviation of the tool position in z-direction, see Figure 25.

This behavior is not acceptable when considering machining processes with the hexapod. Hence, the alternative design II is discarded.

Results from an Optimization of Design Variant III Design variant III has already been optimized for a spiral trajectory with a combined force and torque load, Dignath (2004). As additional design variable also the pre-stressing force was taken into account. The optimum for the joint position angle was found at $\zeta = 34.1^\circ$. However, for the different loadcase and trajectory observed here, this design does not show a better behavior regarding rotational flexibility than the original design, compare Figure 26.

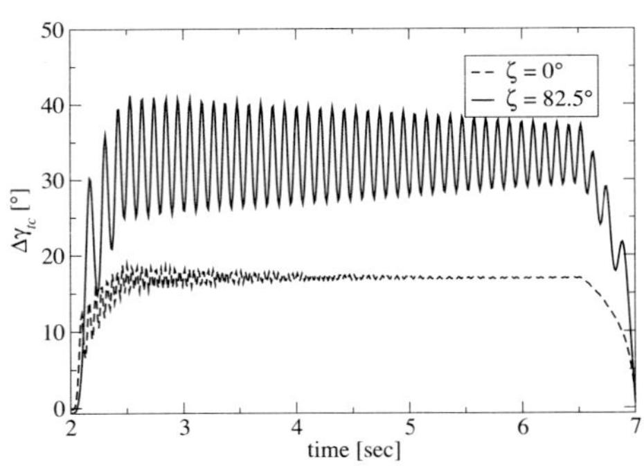

Figure 24. Increase of the rotational flexibility around the z-axis when rotating the struts counterwise (design variant II).

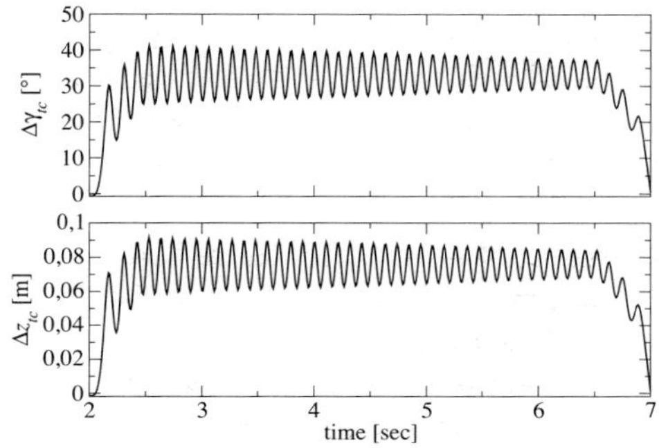

Figure 25. Deviation of the tool's z-position due to the coupling of the tool's z-rotation and z-translation for design variant II.

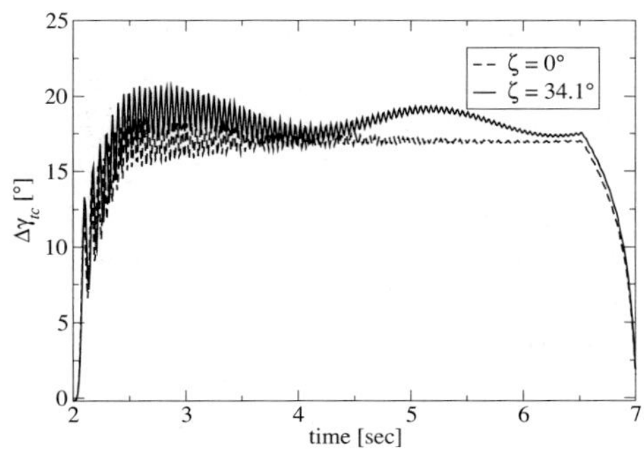

Figure 26. Rotational flexibility around the z-axis for design variant III.

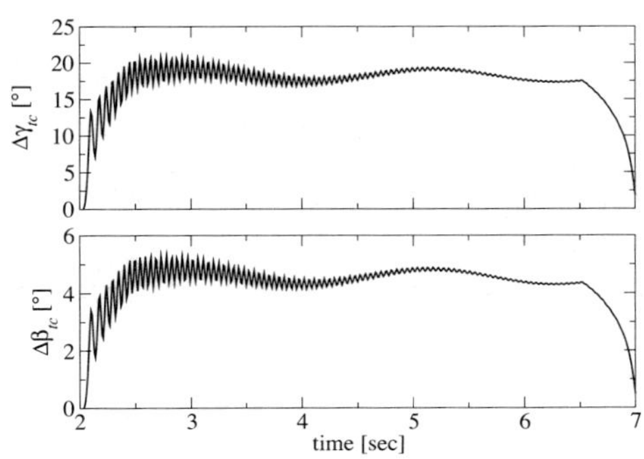

Figure 27. Deviation of the tool's y-rotation due to the coupling of the y- and z-rotations for design variant III.

The average rotational flexibility here is $c_{\gamma\gamma} = 1.81 \cdot 10^{-2}\,{}^\circ/\mathrm{Nm}$. In addition, a very strong coupling between the z- and y-rotation can be observed, as illustrated in Figure 27.

3.4 Investigation of Flexibility – General Approach

So far the flexibility behavior of the hexapod robot was investigated by observing the deviation of the tool position for a certain trajectory within a numerical simulation. Therefore, the results are only valid for this specific trajectory. Now, an approach is introduced that allows investigation of the flexibility behavior all over the workspace.

In robotics, besides flexibility another very important criterion for the evaluation of different design variants is the location of singular points in the workspace. At these points, at least one direction exists where a small variation of the tool carrier position is associated with an arbitrary change in the articular coordinates. Or vice versa, a small variation of one or several articular coordinates would cause a very intensive reaction of the tool carrier. These positions are disadvantageous when located in the workspace as there does not exist a definite relationship between the tool carrier's position and the articular coordinates.

For the stiffness behavior this means that there exists one combination of rotation and translation for which the stiffness vanishes or the flexibility approaches infinity, respectively. Mathematically, these positions can be determined by regarding the eigenvalues of the tool carrier's Jacobian matrix $\boldsymbol{J}$ which associates the articular velocities of the driven joints $\dot{\boldsymbol{\Theta}} = [\dot{\Theta}_1, \dot{\Theta}_2, \ldots, \dot{\Theta}_6]$ with the tool carrier's translational and rotational velocity $\boldsymbol{\mathcal{V}} = [\dot{\boldsymbol{x}}, \boldsymbol{\omega}]$,

$$\boldsymbol{\mathcal{V}} = \boldsymbol{J}(\boldsymbol{y}_{\mathrm{tc}}) \cdot \dot{\boldsymbol{\Theta}} \,. \tag{45}$$

It should be mentioned that different definitions of the Jacobian matrix exist in literature and one has to clearly define which one is used. Positions where at least one eigenvalue approaches infinity are kinematic singular positions. As infinite values are numerically difficult to handle and the Jacobian matrix itself is difficult to compute, we regard the inverse Jacobian matrix instead which can be obtained directly via geometric consideration. Here, at kinematic singular positions at least one eigenvalue is zero. The associated eigenvector describes the motion of the tool carrier for which stiffness disappears. As there also exist very small eigenvalues in the neighborhood of singular positions, not only the exact singular position but also its neighborhood is critical. For parallel kinematic mechanisms much effort is devoted to the investigation of singular constellations and their neighborhood, see Takeda and Funabashi (1996), Takeda and Funabashi (1999),

Merlet (1989), Gosselin and Angeles (1990).

For the determination of kinematic singularities, the evaluation of the inverse Jacobian matrix is useful. However, when investigating the neighborhood of kinematic singularities, the point of interest is the flexibilitity behavior of the machine. The direct usage of the inverse Jacobian matrix for investigations is disadvantageous here. The results are not significant as neither resulting flexibilities nor the corresponding axes are available. Therefore, it is better to evaluate the flexibility behavior directly.

Matrices of Tangential Stiffnesses and Flexibilities Stiffness and flexibility of the end-effector depend on its translational and rotational position in the workspace, the so-called pose. The relation between applied forces and torques $\boldsymbol{\mathcal{F}} = [\boldsymbol{f}, \boldsymbol{l}]$ on the one hand and the evasive displacements $\Delta \boldsymbol{y}_{\mathrm{tc}}$ of the tool carrier on the other hand is highly nonlinear. For the evaluation of the stiffness and flexibility behavior, it is sufficient to regard the linearized relations

$$\mathrm{d}\boldsymbol{\mathcal{F}} = \boldsymbol{K}_{\mathrm{t}} \cdot \mathrm{d}\boldsymbol{y}_{\mathrm{tc}} \qquad \text{and} \qquad \mathrm{d}\boldsymbol{y}_{\mathrm{tc}} = \boldsymbol{C}_{\mathrm{t}} \cdot \mathrm{d}\boldsymbol{\mathcal{F}} \,, \tag{46}$$

where the tangential stiffness matrix $\boldsymbol{K}_{\mathrm{t}}$ describes the forces caused by a given rotational and translational displacement and the tangential flexibility matrix $\boldsymbol{C}_{\mathrm{t}}$ (often also called compliance matrix) gives the relationship between a given force and torque load and the resulting tool carrier's displacement.

In robotics, the elastic behavior is usually described in a global coordinate system using the vector of infinitesimal rotation $\mathrm{d}\boldsymbol{\varphi} = \boldsymbol{\omega}\mathrm{d}t$ in order to describe the deflection of orientation. A simplified description of the compliance behavior can be achieved by assuming that all driven joints have a linear elasticity k, which is sufficient to show the basic behavior of the machine. In our case the stiffnesses of the inner and outer joints can be summarized to a resulting strut elasticity. The relation between the infinitesimal applied load $\mathrm{d}\boldsymbol{\mathcal{F}}$ and the displacement of the tool carrier $[\mathrm{d}\boldsymbol{x}, \mathrm{d}\boldsymbol{\varphi}]$ can be derived by means of the principle of virtual work as

$$\begin{bmatrix} \mathrm{d}\boldsymbol{x} \\ \mathrm{d}\boldsymbol{\varphi} \end{bmatrix} = \underbrace{-\frac{1}{k}\,\boldsymbol{J} \cdot \boldsymbol{J}^{\mathrm{T}}}_{=:\,\boldsymbol{C}_{\mathrm{t}\varphi}} \cdot \mathrm{d}\boldsymbol{\mathcal{F}} \,, \tag{47}$$

see e.g. Tsai (1999). In order to describe the deflection of orientation by the change of elementary rotation angles like Kardan angles, which are applied here with $\mathrm{d}\boldsymbol{\phi} = [\mathrm{d}\alpha, \mathrm{d}\beta, \mathrm{d}\gamma]$, we can write

$$\begin{aligned} \mathrm{d}\boldsymbol{y}_{\mathrm{tc}} &= \begin{bmatrix} \mathrm{d}\boldsymbol{x} \\ \mathrm{d}\boldsymbol{\phi} \end{bmatrix} \cdot \begin{bmatrix} \boldsymbol{Id} & \mathbf{0} \\ \mathbf{0} & \boldsymbol{H}_{\mathrm{R}} \end{bmatrix} \cdot \begin{bmatrix} \mathrm{d}\boldsymbol{x} \\ \mathrm{d}\boldsymbol{\varphi} \end{bmatrix} = \\ &= \underbrace{\begin{bmatrix} \boldsymbol{Id} & \mathbf{0} \\ \mathbf{0} & \boldsymbol{H}_{\mathrm{R}} \end{bmatrix} \cdot \boldsymbol{C}_{\mathrm{t}\varphi}}_{=: \boldsymbol{C}_{\mathrm{t}}} \cdot \mathrm{d}\boldsymbol{\mathcal{F}} = \boldsymbol{C}_{\mathrm{t}} \cdot \mathrm{d}\boldsymbol{\mathcal{F}} \end{aligned} \tag{48}$$

with the identity matrix $\boldsymbol{Id}$ and the Jacobian matrix of rotation $\boldsymbol{H}_{\mathrm{R}}$, see e.g. Schiehlen and Eberhard (2004). The inverse relation is then given by

$$\mathrm{d}\boldsymbol{\mathcal{F}} = \boldsymbol{K}_{\mathrm{t}} \cdot \mathrm{d}\boldsymbol{y}_{\mathrm{tc}} \tag{49}$$

with the tangential stiffness matrix

$$\boldsymbol{K}_{\mathrm{t}} = -k\, \boldsymbol{J}^{-\mathrm{T}} \cdot \boldsymbol{J}^{-1} \cdot \begin{bmatrix} \boldsymbol{Id} & \mathbf{0} \\ \mathbf{0} & \boldsymbol{H}_{\mathrm{R}}^{-1} \end{bmatrix} . \tag{50}$$

In the general case both matrices are symmetric and fully occupied. This means that a load in one direction causes rotational and translational evasive motions in every direction. Furthermore, they consist of four blocks having different units. For some of the further considerations, it is useful to normalize the coefficients using characteristic forces, torques, lengths and angles. These values should be chosen with care as they influence the results of the following optimization. For our example, we choose the normalization coefficients $F_0 = 1\,\mathrm{N}$, $l_0 = 1\,\mathrm{Nm}$, $r_0 = 10^{-4}\,\mathrm{m}$, and $\varphi_0 = 10^{-4\,\circ}$. The resulting dimensionless matrices are denoted with $\hat{\boldsymbol{K}}_{\mathrm{t}}$ and $\hat{\boldsymbol{C}}_{\mathrm{t}}$.

Principal Stiffnesses and Flexibilities For parallel kinematics, the inverse Jacobian matrix $\boldsymbol{J}^{-1}$ is directly available. The flexibility matrix $\boldsymbol{C}_{\mathrm{t}}$ is then computed by inversion of $\boldsymbol{J}^{-1}$ under consideration of Equations (47) and (48) or inversion of $\boldsymbol{K}_{\mathrm{t}}$ from Equation (50). However, direct evaluation of the flexibility matrix $\boldsymbol{C}_{\mathrm{t}}$ is often not possible as the stiffness matrix $\boldsymbol{K}_{\mathrm{t}}$ or rather $\boldsymbol{J}^{-1}$ are badly conditioned in the neighborhood of kinematic singular positions and, of course, singular at singular positions. Therefore, it is useful to transform the stiffness matrix into its system of principal axes. Here the eigenvalues are the resulting stiffnesses for the motions defined by the corresponding eigenvectors. The corresponding principal flexibilities are then the reciprocals of the principal stiffness constants as the transformed stiffness matrix is diagonal.

In order to achieve more insight into the global flexibility behavior of the machine, it is necessary to evaluate either the principal stiffnesses or the flexibilities at many points all over the workspace. Here exemplary points on regular grids lying in planes $x =$ const., $y =$ const. and $z =$ const. are regarded.

As criterion for the stiffness, it is both possible to regard the lowest normalized principal stiffnesses $\min(\hat{k}_i^*)$ or the mean value of the normalized principal flexibilities $\text{mean}(\hat{c}_i^*)$ at each point, see also Kirchner (2000), Zhengyi, Fengfeng and Mechefske (2002). However, the numerical evaluation of flexibilities at kinematic singular points is problematic because they approach infinity. Therefore, it is often recommendable to regard the principal stiffnesses even if the flexibilities are the point of interest.

In the following section, the flexibility/ stiffness behavior of the alternative design variants is evaluated with the described approach. In order to be able to explain the results from Section 3.3 variants I and II are discussed for the theoretically optimal joint position angle $\zeta = 82.5°$ and variant III for $\zeta = 34.1°$. In the following, several pictures are given, some showing the flexibility constants c_{ij} along the vertical axes, some the minimum principal stiffnesses $\min(\hat{k}_i^*)$. It is desirable to have very low flexibilities, i.e. low values of c_{ij} or high stiffnesses, i.e. high values of $\min(\hat{k}_i^*)$.

Current Design For the original design the kinematic singularity at the center position is obvious, compare Sections 3.1 and 3.3. The flexibility matrix at this point also confirms this observation

$$\hat{\boldsymbol{C}}_{\text{t}} = \begin{bmatrix} 0.1772 & 0 & 0 & 0 & 0 & 0 \\ 0 & 0.1772 & 0 & 0 & 0 & 0 \\ 0 & 0 & 0.1330 & 0 & 0 & 0 \\ 0 & 0 & 0 & 1.9477 & 0 & 0 \\ 0 & 0 & 0 & 0 & 1.9477 & 0 \\ 0 & 0 & 0 & 0 & 0 & \infty \end{bmatrix}. \tag{51}$$

At all other points on the described planes, always one principal stiffness vanishes. This means that every point in the workspace is kinematic singular with one principal flexibility approaching infinity. The axis of maximum flexibility (which is the eigenvector corresponding to the lowest eigenvalue of $\hat{\boldsymbol{K}}_{\text{t}}$ and, therefore, the highest one of $\hat{\boldsymbol{C}}_{\text{t}}$) depends on the position in workspace. In the central position the respective eigenvector is given by [0 0 0 0 0 1] corresponding to the kinematic singularity observed in Section 3.3. The fact that at all other points also a singular direction is existent accounts for the bad behavior of this design variant observed for the circular sample trajectory.

Design Variant I During the optimization and evaluation of the designs in Section 3.3, the alternative design variant I achieved a strongly superior flexibility behavior in comparison to all other designs discussed here. A look at the tangential stiffnesses and flexibilities for $\zeta = 82.5°$ explains the advantage of this design. Here, no kinematic singularities appear in the workspace. This further allows the general conclusion that the favorable behavior can be achieved not only for the sample trajectory but also for arbitrary motions in the overall workspace. However, due to space limitations it is impossible to show the reader enough plots to sufficiently mirror the encouraging stiffness behavior for the entire workspace. Therefore, only some exemplary planes in the workspace are discussed. In Figure 28 (left) the minimum normalized principal stiffnesses $\min(\hat{k}_i^*)$ in xy- and yz-planes with $z = 0$ and $x = 0\,\mathrm{m}$, respectively, are illustrated (xz-planes show similar results to yz-planes). These are critical planes during the following

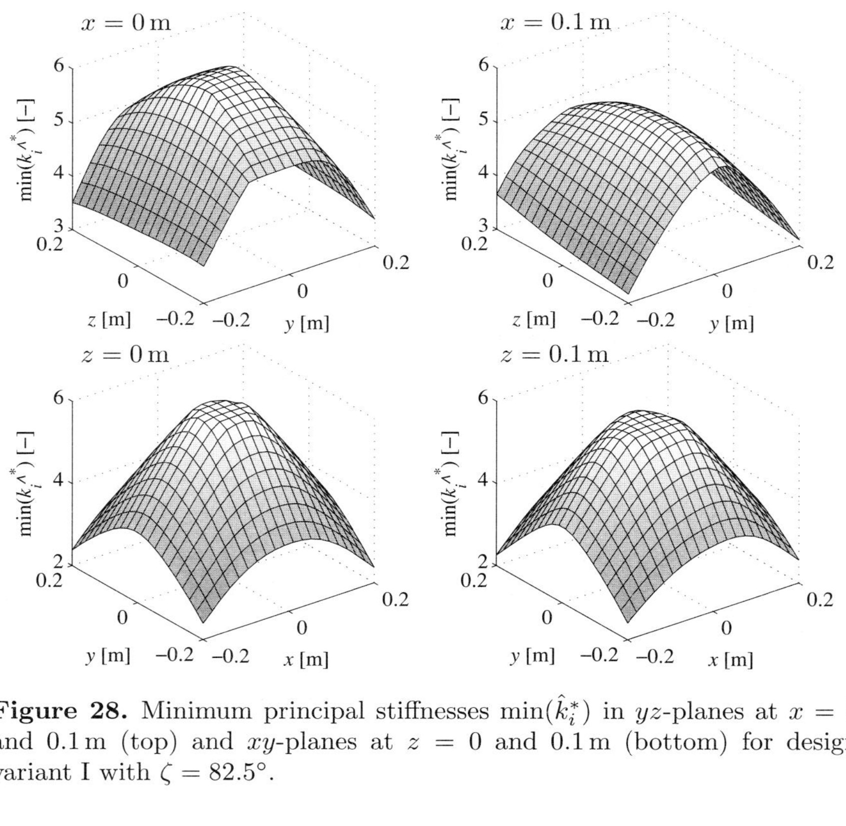

Figure 28. Minimum principal stiffnesses $\min(\hat{k}_i^*)$ in yz-planes at $x = 0$ and $0.1\,\mathrm{m}$ (top) and xy-planes at $z = 0$ and $0.1\,\mathrm{m}$ (bottom) for design variant I with $\zeta = 82.5°$.

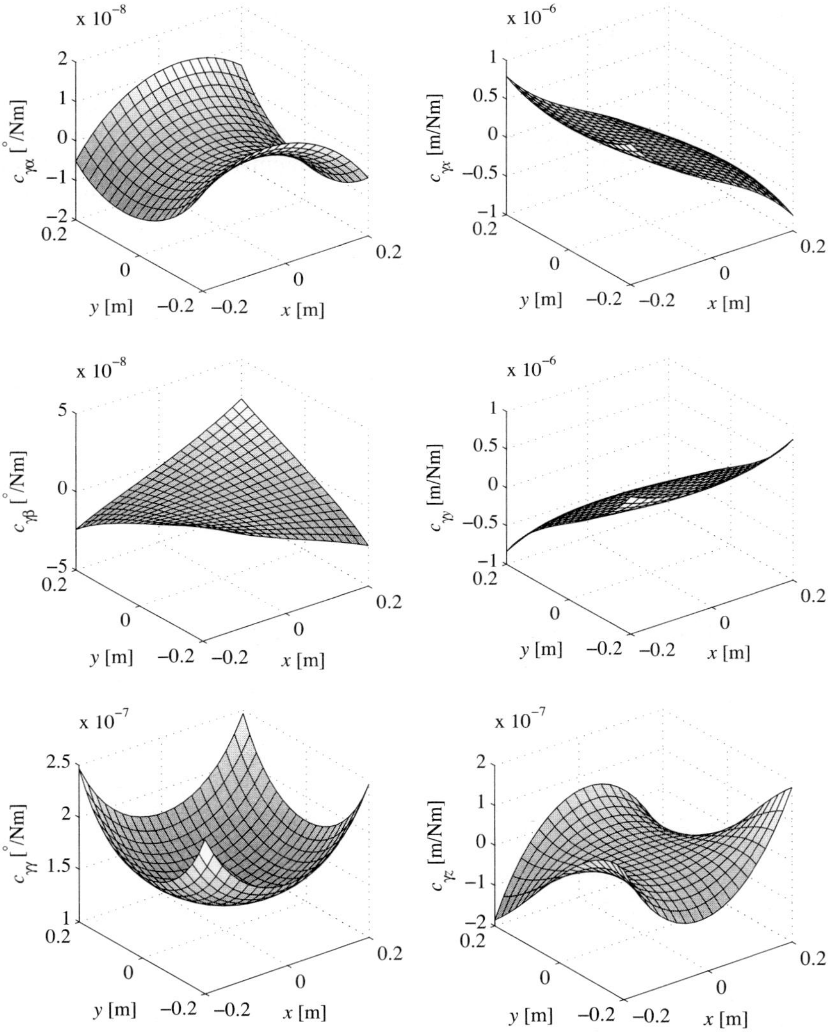

Figure 29. Flexibility constants c_{γ_i} describing the tool carrier's displacements for an applied torque around the z-axis in the xy-plane at $z = 0$ for design variant I.

investigation of the other variants.

Even the minimum principal stiffnesses are encouragingly large, what will be seen in particular when compared to the other variants, discussed in the following sections. In order to indicate the global stiffness behavior in the complete workspace in Figure 28 (right) exemplary also the minimum principal stiffnesses $\min(\hat{k}_i^*)$ in xy- and yz-planes are plotted for $x = 0.1\,\mathrm{m}$ and $z = 0.1\,\mathrm{m}$, respectively, where again substantially large values are obtained.

Further, the flexibility behavior of the machine in the xy-plane at $z = 0$ for an applied torque around the z-axis l_z is illustrated in Figure 29, where the tool carrier's displacement is shown for each coordinate of $\boldsymbol{y}_{\mathrm{tc}}$. This configuration corresponds to the sample trajectory and the loading case applied in Section 3.3. It can be seen that low flexibilities appear in all directions which is consistent with the results found in Section 3.3.

Design Variant II For design variant II, the workspace contains a plane of kinematic singular positions located at $z = 0$, which divides the workspace into two usable domains, see Figure 30 (lower left). Within these domains, the prinicipal stiffnesses are relatively low, as can be seen by comparison of Figure 30 with the respective plots of variant I in Figure 28.

The sample trajectory used during the investigation of design II in Section 3.3 is completely located in the singular plane, which explains the very bad behavior of this design. However, the flexibilities are also disappointingly high in the usable domains ($z \neq 0$) and the observed coupling of the translation and the rotation with respect to the z-axis remains. This is illustrated in Figure 31 by the corresponding flexibility constants $c_{\gamma\gamma}$ and $c_{\gamma z}$ at $z = 0.1\,\mathrm{m}$, which are still much higher than the analogous flexibilities for design I at $z = 0$, see Figure 29.

Design Variant III For this design variant the workspace contains three planes of kinematic singular positions located at $x = 0$, $y = 0$ and a vertical one crossing the workspace diagonally. These planes divide the workspace into six domains. In Figure 32 (top), the minimum principal stiffnesses in yz-planes at $x = 0$ and $0.1\,\mathrm{m}$ are shown. Further, in Figure 32 (bottom) the minimum principal stiffnesses in xy-planes at $z = 0$ and $0.1\,\mathrm{m}$ are given which are also divided by the singular planes (lines with zero stiffness). The separated, relative small domains without singular configurations are surrounded by three singular planes each, which leads to disappointingly low minimum prinicipal stiffnesses. This can in particular be emphasized by comparison of the plots in Figure 32 with those of variant I, Figure 28.

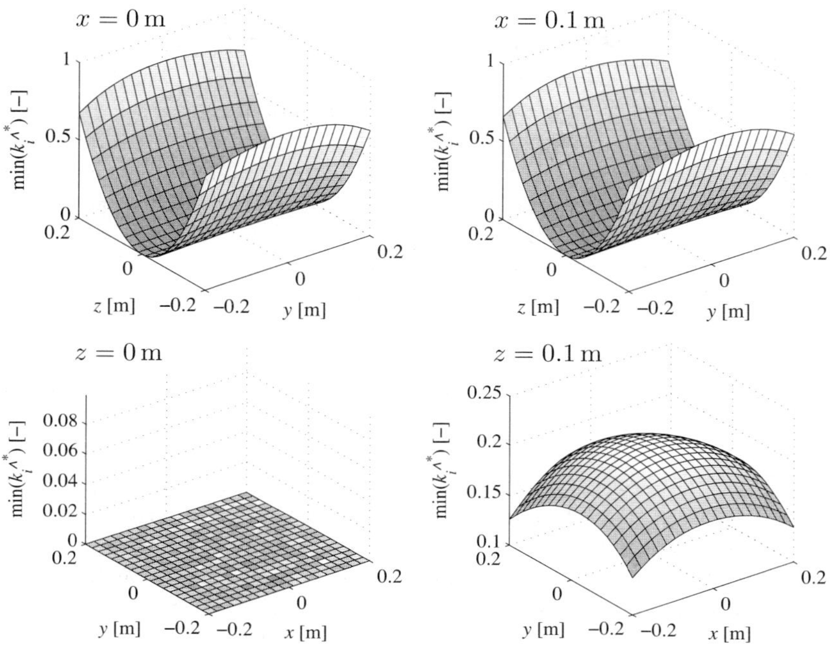

Figure 30. Minimum principal stiffnesses $\min(\hat{k}_i^*)$ in yz-planes at $x = 0$ and 0.1 m (top) and xy-planes at $z = 0$ and 0.1 m (bottom) for design variant II with $\zeta = 82.5°$.

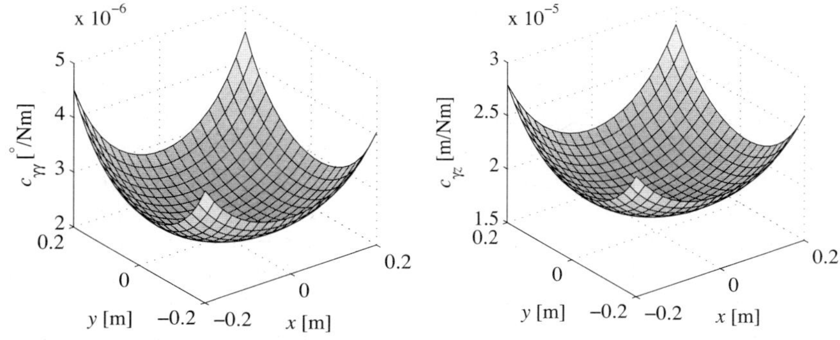

Figure 31. Flexibility constants $c_{\gamma\gamma}$ and $c_{\gamma z}$ in xy-plane at $z = 0.1$ m for design variant II.

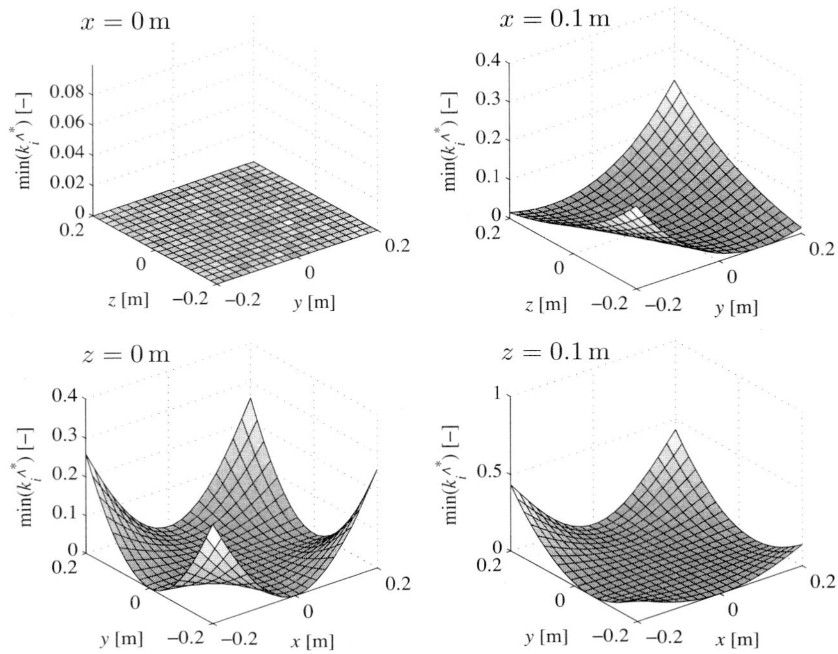

Figure 32. Minimum principal stiffnesses $\min(\hat{k}_i^*)$ in yz-planes at $x = 0$ and $0.1\,\mathrm{m}$ (top) and xy-planes at $z = 0$ and $0.1\,\mathrm{m}$ (bottom) for design variant III with $\zeta = 34.1°$.

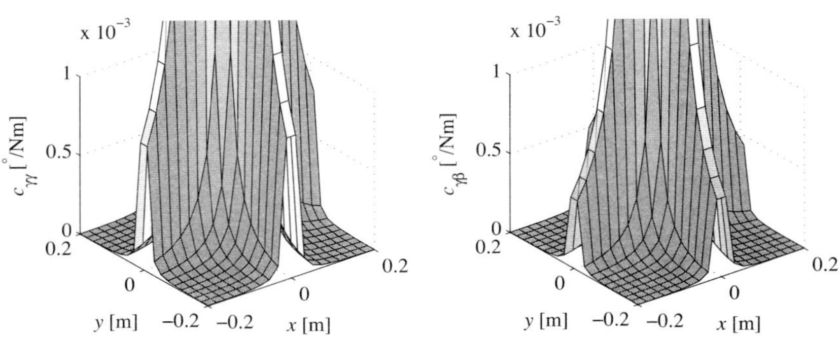

Figure 33. Flexibility constants $c_{\gamma\gamma}$ and $c_{\gamma\beta}$ in xy-plane at $z = 0$ for design variant III.

In Section 3.3 large deflections of the z- and y-rotation appeared for the applied torque around the z-axis for design variant III. This behavior can also be seen when regarding the flexibilities $c_{\gamma\gamma}$ and $c_{\gamma\beta}$ in Figure 33. High values of these flexibilities clearly show this strong coupling.

3.5 Summary

In order to increase the stiffness of the hexapod machine HEXACT and to eliminate singular configurations three variants of alternative designs were discussed. Variant I with equally rotated joints leads to a substantial improvement in both translational and rotational flexibility in comparison to the current design. A global investigation of the tangential flexibilities in the entire workspace shows the absence of any singular configuration which confirms the results received during optimization. A drawback of this design is that pre-stressing, used for compensation of bearing clearance, is no more possible.

Contrary to the expected behavior the flexibility cannot be improved by design variant II with counter-rotated joint positions, which is due to a singular plane in the workspace that also affects the flexibility behavior in the adjacent domains. Further, coupling of the translation along and the rotation around the tool axis causes a non-negligible deviation of the position in z-direction when torques are applied. Both results disqualify design variant II. Analogous results are found for variant III, now with three singular planes dividing the workspace into eight domains and a coupling of the z- and y-rotation.

From the three investigated designs, only variant I promises a substantial improvement of the static compliance. This is especially appreciated as for this design variant, no modification of the hardware is necessary. The initial rotation of the tool carrier can be achieved by a modification of the coordinate transformation routines inside the controller.

4 Simulation of Impacts in Geartrains Using Different Approaches

Mid- and large-scale Diesel engines often demand gear drives for durability and maintenance reasons to drive the camshaft through the crankshaft. Usually, these gear trains are further used to drive auxiliary devices like high-pressure pumps, etc. that may introduce high dynamic loads. Compared to the rated torque needed to drive a gear train, the dynamic forces caused by gas forces and by auxiliary devices are normally comparatively high and cause the flanks of the teeth to lift off and reestablish contact in the form of impacts. These impacts cause performance and emission problems and also noise. This phenomenon is called gear hammering. Furthermore, design problems arise, concerning the involved gears due to the highly nonlinear character of the impacts. Certainly, the most influencing quantity for the design process is the magnitude of the contact force. Many commercial multibody packages provide tools to simulate contacts between meshing gears, which are most often based on rigid body contact models. However, gearwheels in gear trains are often very thin to reduce inertia effects and the elastic compliance of the gear body must not be neglected. Furthermore, commonly used input parameters like the tooth or the single stiffness are very often chosen according to static methods as described in the standard ISO 6336-1 or Niemann and Winter (1989) that are only valid for cylindrical gear bodies. For a closer insight into gear body effects during impact, three different models are used. First, a basic multibody model, similar to those used in many commercial program systems is used. Further, a nonlinear finite element model is used for static and dynamic simulations. The third model is a modally reduced model with a local force law. The impact investigations are carried out for just one single impact between one pair of teeth because of the large simulation times needed for the transient finite element analyses. Simulations are carried out for different initial and boundary conditions, including fixed and pivoted gearwheels and impacts in different contact points along the flanks.

4.1 Investigated system

The chosen system is a typical gear pairing of a mid-scale Diesel engine. It consists of an idler gear and a crankshaft gear, both shown in Figure 34. In the following, the wheels are denoted by the subscripts i and c, respectively. Both gears have $z_i = z_c = 97$ teeth of modulus $m = 6\,\text{mm}$ without addendum modification, the normal angle is $\alpha = 20°$. The face widths of the gears are $b_i = 50\,\text{mm}$ and $b_c = 57\,\text{mm}$. The mass moments of inertia follow as $I_i = 2.27\,\text{kgm}^2$ and $I_c = 4.08\,\text{kgm}^2$, see Ziegler (2005).

The pairing is chosen because of the particular idler gear body, which is very thin and additionally weakened by several axial holes, see Figure 34. Therefore, strong gear body effects are expected. For the simulations, a material with Young's modulus $E = 210000\,\text{MPa}$, density $\rho = 7850\,\text{kg/m}^3$ and Poisson ratio $\nu = 0.3$ is assumed. Further, fully elastic material behaviour is assumed and material damping is neglected. For the dynamic analyses two different model setups are investigated. In the first setup, the crankshaft wheel is pivoted in its centre, in the second setup the crankshaft wheel is fixed in its centre. Both gears are setup such that both flanks touch in the

Figure 34. Idler gear (left) and crankshaft gear (right) of a typical gear train of a mid-scale Diesel engine.

impact point and the impact is established by an initial rotational velocity ω_0 prescribed to the idler gear.

Multibody Model Multibody models have been used in many investigations of impacting gears, due to their simplicity, see e.g. Blankenship and Kahraman (1995), Kahraman and Blankenship (1997), Kücükay (1984), Kücükay (1987) Pfeiffer (1984). Here, a model similar to the one proposed in Prestl (1991) is used, which is very similar to those used in many commercial multibody program packages. It consists of two rigid bodies, with one rotational degree of freedom each, denoted by φ_i and φ_c, as shown in Figure 35.

Both wheels are connected by a linear spring-damper combination including backlash e, acting on the base radii. The spring with stiffness c

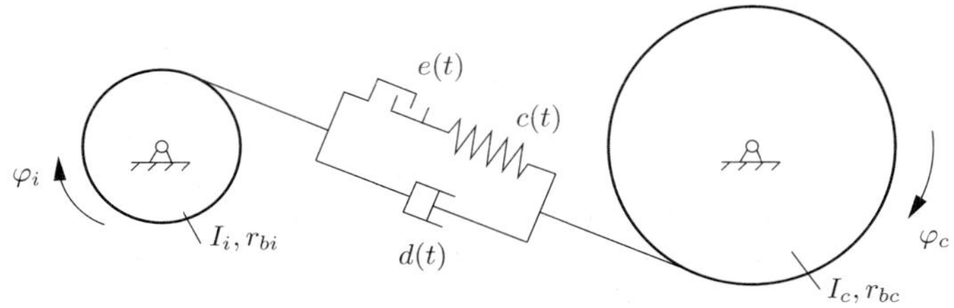

Figure 35. Multibody model for gear pairing.

accounts for the elasticity of the gear body, the bending of the teeth in contact and the contact stiffness. The damper with damper ratio d accounts for material damping and a possible oil film. Since only one contact between a single tooth pair is investigated the stiffness is assumed to be constant, which is a good approximation as shown later in the results of the static finite element analyses. The backlash function e is chosen to be

$$e = \begin{cases} 0 & \text{for } \varphi_i r_{bi} + \varphi_c r_{bc} \geq 0 \\ \varphi_i r_{bi} + \varphi_c r_{bc} & \text{for } \varphi_i r_{bi} + \varphi_c r_{bc} < 0 \end{cases} \tag{52}$$

to account for the unilateral contact. In industrial practise the tooth stiffness c' is often chosen for the stiffness c. In the industrial standard ISO 6336-1 several methods are proposed to determine the tooth stiffness. Here, two different values are used. The first one is chosen according to method B. In this method the tooth stiffness is determined by comprehensive analysis based on measurements, theory of elasticity, or finite element analyses as carried out later. The second value is chosen according to method A. There, results from theoretical as well as experimental studies of the elastic behaviour of solid spur gears are approximated in a series expansion from which a characteristic diagram is developed. From this diagram the single stiffness can easily be determined from geometric parameters of the gear pair. For the implementation of the multibody model and all simulations MATLAB is used.

Finite Element Model The use of transient, nonlinear finite element analyses is a widely used and well developed method for contact simulations as shown in Eberhard (2000), and the results can be regarded to be very precise. However, the spatial discretization and the modelling of the contact itself has to be chosen very carefully to ensure this accuracy. Here, emphasis is placed on a precise description of the contact geometry, namely the tooth profile. Therefore, the actual production process is mimicked theoretically by describing the geometry of milling and grinding tools used to produce the teeth. Generally, this process can be reduced to describing the profile of the tool rolling on the pitch circle, according to Padieth (1978). This procedure allows not only to obtain the genuine profile subject to addendum modifications, but also the exact root diameter, root fillet radius and profile modifications due to special purpose gear hobs often used in practise.

Impact problems often show wave propagation phenomena up to over 50 kHz. Hence, for the spatial discretization a maximum overall global element length of 7 mm is chosen, to be able to evaluate frequencies up to about 80 kHz with the used material. For the contact region with a width of about 1 mm, clearly a much finer discretization is necessary. Therefore, an element length of 0.1 mm is chosen in the contact region. It is necessary that the mesh density changes rapidly with increasing distance to the contact region to avoid an unnecessary large number of nodes. Unfortunately, it is very difficult to create continuously changing element sizes using automatic meshing algorithms provided with the used finite element software ABAQUS. In order to still obtain a rapid mesh transition with a reasonable number of nodes, the contacting teeth are amply cut from the gear body and both parts are meshed independently. During the simulation both parts are put back together by constraining the nodal degrees of freedom of the finer mesh to the motion of the nearest node of the coarse mesh. If the stress gradient and the change in the mesh density over the intersection are small, the results can be regarded as meaningful, see ABAQUS (2004b), while allowing efficient simulation. Both wheels are meshed with linear hexahedral elements with hourglass control.

Figure 36 shows the fine mesh of the contacting teeth of both gears with the discontinuous mesh transition. This meshing strategy results in a finite element model with about 80000 nodes and 65000 elements for each gear. For the contact node-to-segment elements are used in a two pass algorithm. For the static analyses a Lagrangean contact formulation is chosen, for the dynamic analyses a penalty, as well as the 'kinematic

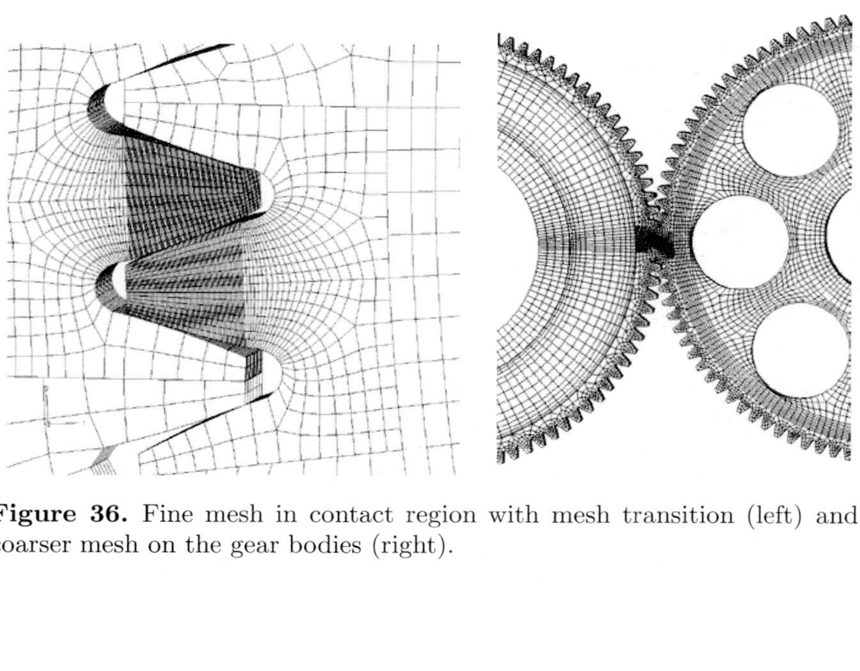

Figure 36. Fine mesh in contact region with mesh transition (left) and coarser mesh on the gear bodies (right).

contact' formulation, proposed in ABAQUS (2004a) are used. For efficient application of boundary conditions, the motion of the nodes lying on the fit bores are constrained to reference points defined in the wheel centres by a kinematic constraint. Then, the boundary conditions can easily be defined by applying a rigid body motion to the reference points.

Modal model To overcome the extremely large simulation times for the transient, nonlinear finite element model, a method proposed in Seifried (2005) is used. Since in impact problems the nonlinear behaviour is often only limited to a small contact region the system is decomposed in two submodels. The first submodel accounts for the locally limited nonlinear contact behaviour, the second submodel describes the global response of the impacting bodies and is assumed to be linear. Under the assumption of known contact forces the global behaviour can be simulated with the linear submodel, derived e.g. from a finite element model, reading

$$\boldsymbol{M} \cdot \ddot{\boldsymbol{U}} + \boldsymbol{K} \cdot \boldsymbol{U} = \boldsymbol{F} \tag{53}$$

with the mass matrix $\boldsymbol{M}$, the stiffness matrix $\boldsymbol{K}$, the vector of nodal displacements $\boldsymbol{U} \in \mathbf{R}^n$ and the vector of applied forces $\boldsymbol{F}$ which includes the contact forces. For a further reduction a modal description is chosen and modal condensation is applied. Therefore, in the transformation to modal coordinates

$$\boldsymbol{U} = \boldsymbol{\Phi} \cdot \boldsymbol{X} \text{ with } \boldsymbol{\Phi} = [\varphi_1, \ldots, \varphi_m] \tag{54}$$

only the first $m \ll n$ eigenmodes are considered, see Bathe (1996), and the reduced equations of motion read as

$$\ddot{\boldsymbol{X}} + \hat{\boldsymbol{K}} \cdot \boldsymbol{X} = \boldsymbol{\Phi}^T \cdot \boldsymbol{F}, \tag{55}$$

$$\hat{\boldsymbol{K}} = \boldsymbol{\Phi}^T \cdot \boldsymbol{K} \cdot \boldsymbol{\Phi} = \text{diag}\{\omega_1^2, \ldots, \omega_m^2\} \tag{56}$$

with the corresponding eigenfrequencies ω_j and eigenmodes φ_j. Due to the diagonal structure of $\hat{\boldsymbol{K}}$, the Equations (55) are fully decoupled and of much lower order compared to Equation (53). Hence, they allow a much faster solution, see Cook, Malkus and Plesha (1989) or Maia and Silva (1998), compared to the full model.

To determine the contact force needed in Equation (55) a local force law is used that provides a relationship between displacements of nodes outside the contact region and the contact force. This can be done by using a nonlinear finite element model of the contact region and coupling both submodels in a co-simulation, see Seifried (2005). However, if quasi-static contact behaviour is assumed, the co-simulation can be avoided by using a

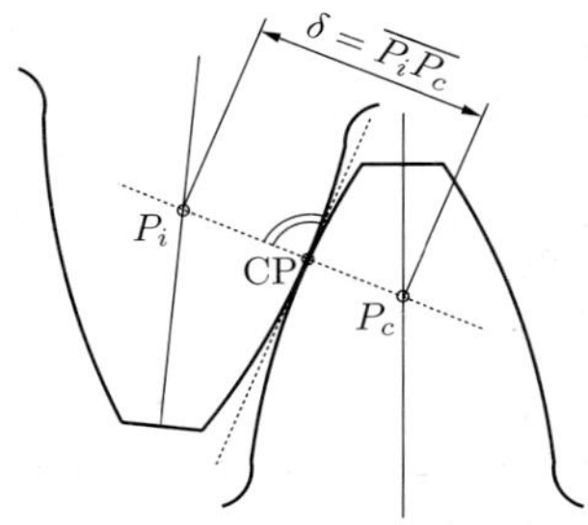

Figure 37. Points P_i and P_c used in the definition of the approach δ for the determination of the displacement-force relationships used as local contact law.

precalculated displacement-force relationship. The local displacement-force relationships used here can be derived from static finite element analyses. Here, for every contact point two points P_i and P_c are defined on the contacting teeth. These points are defined by the intersection of the symmetry plane of these teeth and the normal to the flank through the contact point, as shown in Figure 37. The distance between these two points is called approach δ and the resulting normal contact force is related to this approach. The relationships are determined for different contact points and approximated in polynomials of order three and are further stored in diagrams to achieve a very efficient evaluation during time integration.

The displacement-force relationships, however only yield a scalar contact force, which has to be distributed to nodal forces in the contact area. Here, the contact area is simplified as a contact line and the contact force is equally distributed over the corresponding nodes. Further, the location of the contact point, as well as the direction of the contact forces are assumed to be constant during one impact. For the calculation of the eigenmodes as well as for the static finite element analyses ABAQUS is used, but any finite element program can be used instead. For the implementation of the modal model and the determination of polynomial approximation of the displacement-force relationships and all simulations MATLAB is used.

4.2 Simulation results

In the following results of simulations carried out with the three different models are presented and the results are compared. Of major interest is the precision of the contact forces and the rotational angles and velocities, respectively.

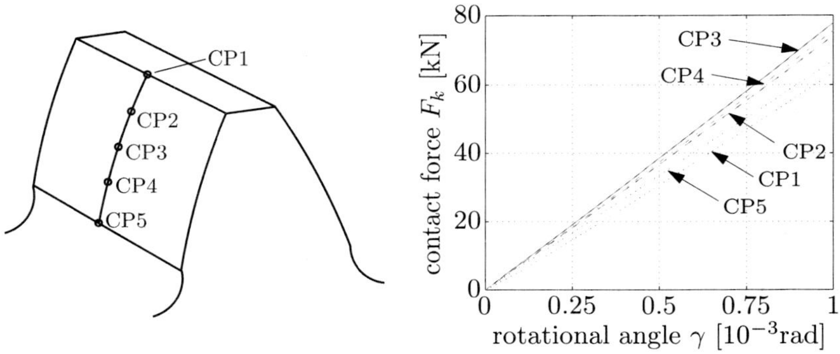

Figure 38. Different contact points (left) and contact force over rotational angle γ for static finite element analyses (right).

Static analyses One model parameter needed for the dynamical simulations using multibody models is the stiffness c. Obviously, there is a great demand for a physically motivated choice of this stiffness. In industrial practise, most often the tooth stiffness c_γ is used, which is a static quantity by definition from ISO 6336-1. The values given there are only valid for cylindrical gear bodies. Since the investigated gear bodies differ significantly from simple cylindrical bodies, a more sophisticated investigation of the tooth stiffness is carried out with a static finite element model. Furthermore, the results from the static finite element analyses can be used to determine the local force law for the contact force.

Both gears are oriented such that they touch each other in the contact point. For the idler gear a linear increasing rotation angle γ around its axis is prescribed, all other degrees of freedom of the reference points are constrained to zero. The maximum angle γ_{max} is chosen such that a maximal contact pressure of about 70% of the maximal durably endured contact pressure of the used material is obtained. These simulations are carried out for 5 different contact points called CP1, ..., CP5, shown in Figure 38 (left) for the idler gear. They are distributed over the heights of the flanks. In the following, CP1 means contact between the addendum of the idler gear and the dedendum of the crankshaft gear. On a Pentium 4 computer with 2.8 Ghz a typical simulation takes 2 hours.

From the normal contact forces for the contact points CP1,..., CP5 shown in Figure 38 (right), an almost linear characteristic between the contact force and the angle γ can be observed. This justifies the choice of a constant stiffness c in the multibody model. To be able to compare this stiffness with the tooth stiffness defined in the industry standard ISO 6336-1 as

> "... *the requisite load over* 1 *mm face width, directed along the line of action to produce in line with the load, the deformation amounting to* 1 μm*, of one or more pairs of deviation-free teeth in contact*...,"

which is the deformation

> "... *equal to the base circle length of arc, corresponding to the load-induced rotation angle of one gear of the pair when the mating gear is held fast.*"

It is related to the face width and the deformation on the line of action. The maximum of the tooth stiffness is called single stiffness c' and normally occurs for contact in the pitch point. From Figure 38 (right) it can be seen that this is not the case here. Due to the particular gear body design and the asymmetry of the gear pair with different compliances for both gear bodies the maximum stiffness occurs in CP4. The single stiffness follows as

$$c'_A = \frac{F_{k,max,CP4}}{\gamma_{max}\frac{m}{2}b_i z\cos\alpha} = 5.68\frac{\mathrm{N}}{\mu\mathrm{m\ mm}}. \tag{57}$$

The subscript A denotes that the determination of the single stiffness is according to method A described in ISO 6336-1. However, in industrial practise most often method B is used due to its simplicity. According to Ziegler (2005), method B yields a single stiffness of

$$c'_B = 15.66\frac{\mathrm{N}}{\mu\mathrm{m\ mm}}. \tag{58}$$

This tremendous difference can only be explained by gear body effects due to the very compliable idler gear body compared to a cylindrical gear. Particularly, for the idler gear this can very well be seen in a deformation plot scaled by factor 300 as shown in Figure 39. Due to the radial holes of the idler body a polygonisation takes place. This phenomenon is not seen in cylindrical gears.

The results from the static finite element model can further be used to compute the local displacement-force relationship diagrams for the local contact force law needed for the modal model. Therefore, the points according to Figure 37 are determined and their approach δ is evaluated.

In Figure 40 the diagrams for the different contact points are plotted. In contrast to the plots in Figure 38, a clearly nonlinear behaviour similar to the Hertzian relationship is observed. This is due to the description of a local contact behaviour which is bound to a small region, whereas in Figure 38 the relationship of an overall behaviour is presented.

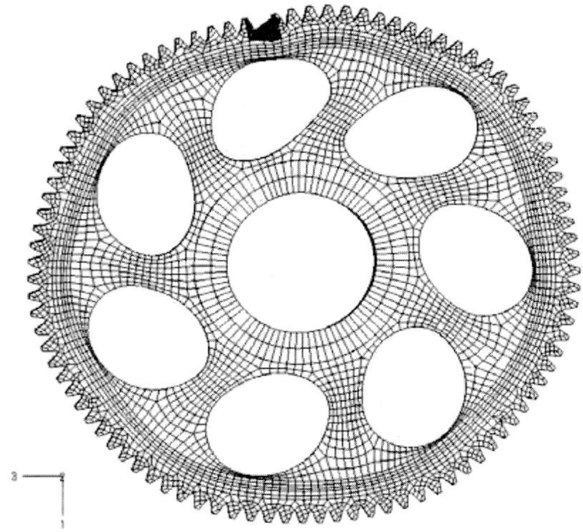

Figure 39. Polygonisation of the idler gear due to the highly compliable gear body (deformation plot scaled by factor 300).

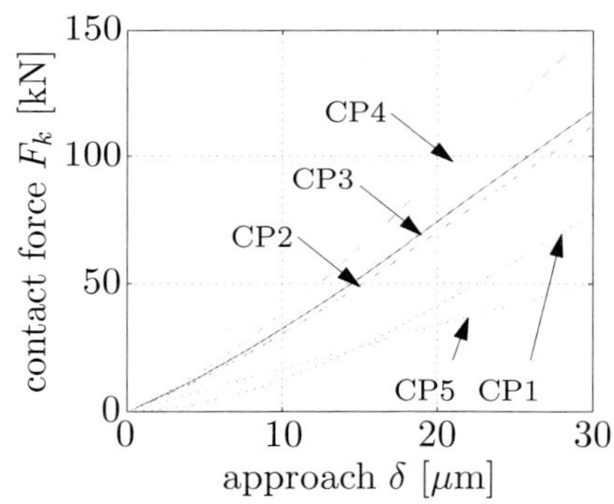

Figure 40. Displacement-force relationship for different contact points.

Dynamic analyses Figure 41 shows the contact forces calculated by a transient finite element analysis for an initial angular velocity of $\omega_0 = 100°/\mathrm{s}$ of the idler gear for both setups, considering contact point CP3.

The contact time is between 0.7 and 0.8 ms, the maximum contact forces are 52 kN and 38 kN, respectively. Very noticeable is a high frequency oscillation of about 60 kHz at the beginning of the contact, fading with increasing

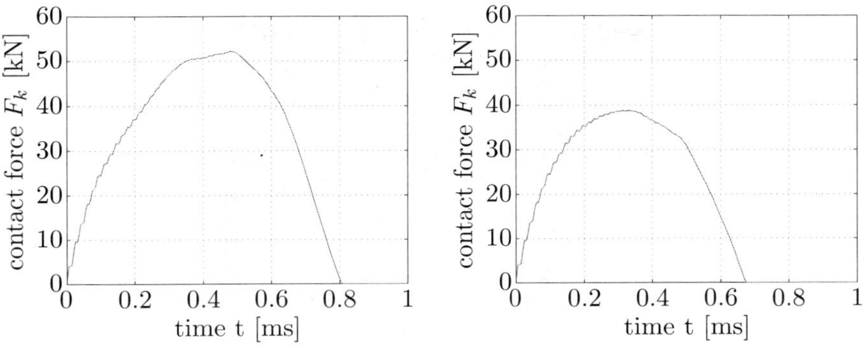

Figure 41. Contact forces of the finite element analyses for fully constrained crankshaft wheel (left) and pivoted crankshaft wheel (right).

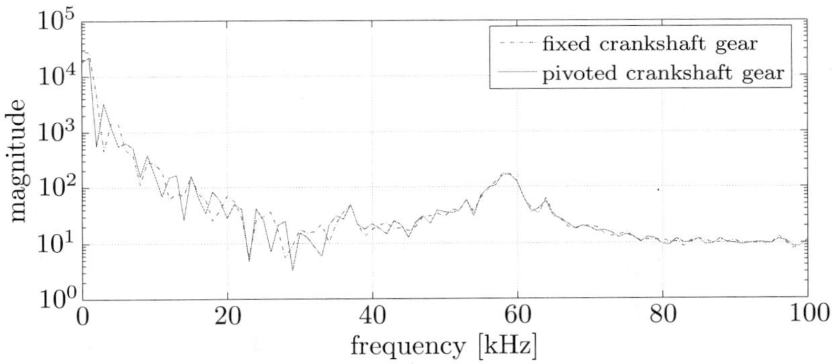

Figure 42. Power density spectrum for the contact forces.

contact duration. Further investigations reveal that this vibration originates in a motion dominated by tooth bending, resembling to the first bending mode of a single sided fixed cantilever beam. This motion decays, as energy from this motion is transferred to other motions by wave propagation through refraction and diffraction within the structure. This frequency can also be found in a spectrum derived by means of a Fourier transformation of the contact forces, as shown in Figure 42.

Furthermore, a slight indentation in the force progression near the maximum can be recognised, which is due to a radial deformation of the outer idler gear rim, similar to the polygonisation effect seen in Figure 39. In the compression phase the rim is radially deformed, the contact force lowers

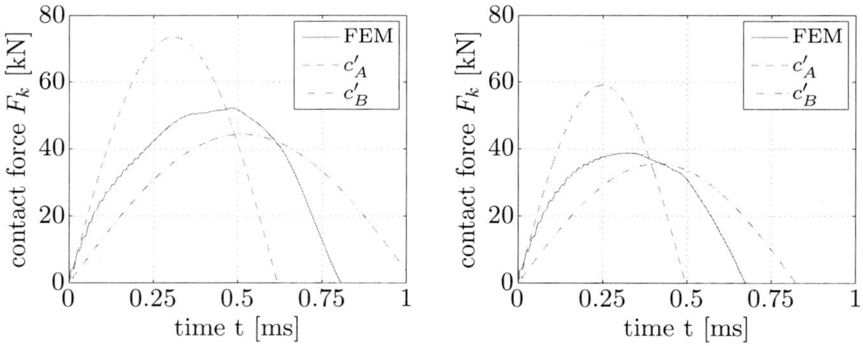

Figure 43. Contact forces for the fixed crankshaft gear (left) and the pivoted crankshaft gear (right) for the multibody model and different stiffnesses.

and energy is stored in the structure. During decompression this energy is released resulting in an increase of the contact force. For both setups this de- and increase can be seen at $t \approx 0.4\,\text{ms}$ and $t \approx 0.5\,\text{ms}$, respectively.

The results for impacts considering CP3 using the multibody model are shown in Figure 43 for both setups. For comparable results, the damping coefficient d of the multibody model is set to zero. It shows that the pragmatic choice according to Equation (58) yields an unrealistically high contact force and short contact duration for both setups. Even though the more physically motivated choice using a static finite element preanalysis shows fairly good results for the maximum contact force, the contact duration is apparently too long. Of course, the multibody model is not able to describe gear body effects and radial deformations as mentioned above, and can therefore not be observed in the force plots.

Figure 44 shows the rotational angles and angular velocities for both multibody models and the finite element model. Although the rotational angles show considerable differences to the finite element solution for both stiffnesses, at least the mean value of the angular velocities after the contact are met very good by the multibody models. Here, the finite element solution shows a very noticeable vibration after the impact. This vibration is characterised by an out of phase torsional vibration of the inner rim of the idler gear body against the outer rim with the teeth. Again, the multibody model is not capable to account gear body effects alike. One way to account for such gear body effects is the addition of degrees of freedoms to the rigid bodies. One approach to improve the results for gear stages with compliable gear bodies is to introduce tangentially movable teeth bound to the gear body by elastic elements, proposed in Ebrahimi and Eberhard (2006). As shown there, the precision of the results improve significantly due to the introduction of additional degrees of freedom.

In Figure 45 the simulation results for the contact forces using the modal model are shown for both setups. The forces are plotted for different dimensions m of the reduced model, taking eigenfrequencies up to different frequencies into account.

With increasing m a very good convergence of the solutions to the finite element solution is achieved. For higher m the same characteristic oscillation

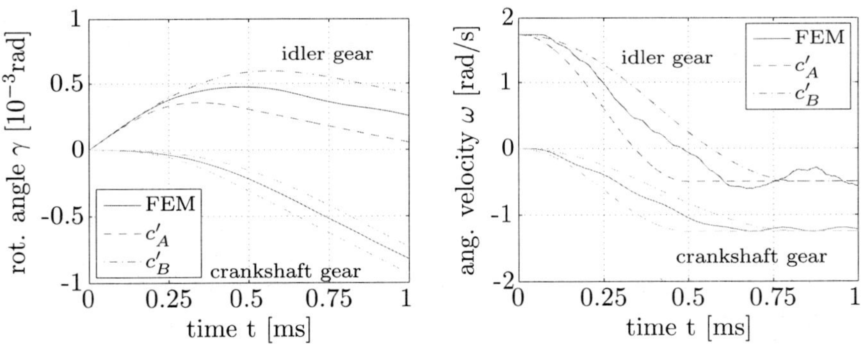

Figure 44. Rotational angles (left) and angular velocities (right) for the multibody model and different stiffnesses.

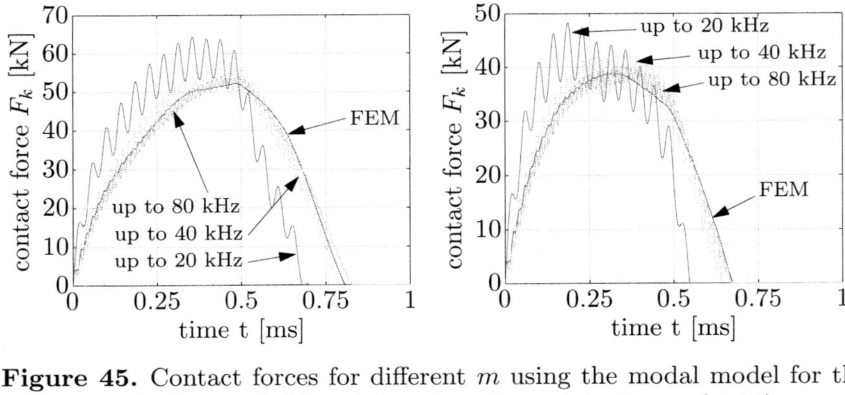

Figure 45. Contact forces for different m using the modal model for the fixed crankshaft gear (left) and the pivoted crankshaft gear (right).

of approximately 60 kHz, as seen in Figure 41 can be observed. Apparently, the energy from the bending motion of the contacting tooth is transferred to high frequency modes by refraction and diffraction. It is further remarkable that eigenmodes with an eigenfrequency of up to about 80 kHz are needed for very good results, which coincides with the highest dominant frequencies in the Fourier spectrum, shown in Figure 42. A comparison of the rotational angles and velocities of the gears for the modal model also shows a very good agreement between the finite element analyses and the modal model, as shown in Figure 46 for the pivoted setup only.

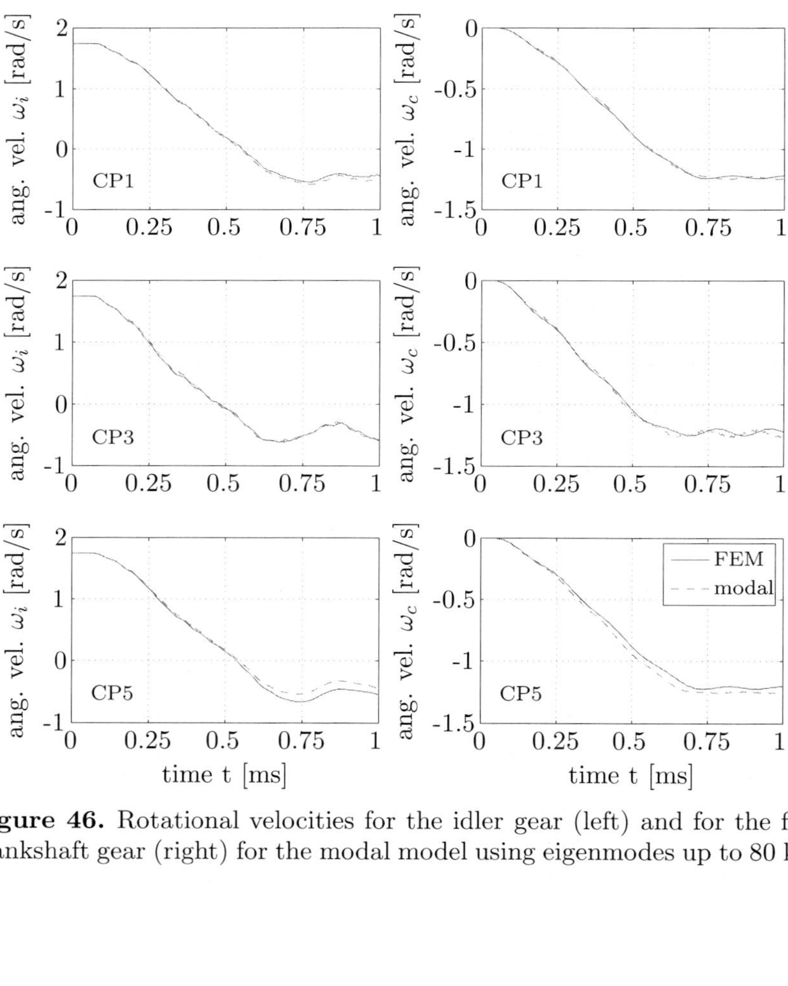

Figure 46. Rotational velocities for the idler gear (left) and for the fixed crankshaft gear (right) for the modal model using eigenmodes up to 80 kHz.

4.3 Numerical effort

Regarding the finite element solution as a reference solution, the results using the multibody model are only of limited precision, whereas the modal models show very good agreements for all contact points if a sufficiently high number of eigenmodes is considered. However, regarding the integration time of about 10 h on a Pentium 4 computer with 2.8 Ghz for one single impact using the transient finite element model, clearly shows that a trade off between precision and computational effort is necessary. Also, the three approaches differ in the time needed to setup the models. For the finite element model, for example, the generation of the exact geometry is very time-consuming, even if the geometry is read in from a CAD system. Further the generation of a useful mesh can take a considerable amount of time, as described before. In contrast, the setup of the multibody model is rather simple, since one has to provide only a few geometrical parameters. But the different approaches also distinguish in the time needed for integration. Therefore, the numerical effort in terms of model setup, computation and required disk space is summarised in Table 3.

Table 3. Comparison of modelling and numerical effort for different models.

model	model setup	int. time 1 impact	total time 1st impact	int. time 10^3 impacts	required disk space
FEM	8 h	8 h	16 h	$\gg$ 8000 h	$< 1\,GB$
MBS	10 min	0.21 s	$\approx$ 10 min	$\approx$ 0.25 h	$\ll 100\,MB$
modal	2 h	0.64 s	$\approx$ 22 h	$\approx$ 30 h	$\approx 8\,GB$

Considering only the computational effort, the multibody model is far best. However, the simulation results show that in some cases elastic effects of the gear body can have a significant influence on the contact force and thus on the dynamic behaviour of the entire gear train. Very precise results in short integration time can be achieved using the proposed modal model, even though a considerable amount of precalculations have to be carried out. For this system a total preprocessing time of 28 hours is needed to build finite element models to carry out the eigenmode analyses and determine the displacement-force relationships for the local contact law. However, all the precalculated data may be reused for consecutive contacts or parametric studies.

Of course, this precomputed data has to be stored either in the computer memory or on the harddisk. For the spatial finite element models used for the eigenmode analysis with each having about 80000 nodes, one can easily compute the memory request for storing one eigenmode as

$$M = 3 \cdot 80000 \cdot 8\,\text{byte} \approx 2MB \tag{59}$$

if double precision variables are used. For $m = 2100$ eigenmodes as used in some simulations and two gears, a total memory of about 8 GB is needed. Consequently, the modal matrix does not fit completely into the memory of a standard desktop computer anymore and the modal transformations have to be carried out piecewise. However, the memory request strongly depends on the size of the finite element model used for the eigenform analysis. Thus, the memory request for smaller gears is significantly smaller.

4.4 Summary

Contacts between a typical gear pair, consisting of an idler gear and a crankshaft gear, have been investigated using a finite element model, a multibody model and a modal model in conjunction with a local force law. Due to the particular gear body design typical for idler gears significant gear body effects could be observed already in the static precomputations for the multibody model. These show large differences in the tooth stiffnesses compared to values given in industry norms for cylindrical gears.

In the impact investigations using a fully nonlinear finite element model also very noticeable gear body effects are observed. Even after contact, gear body effects are observed in out of phase vibrations of the outer rim against the inner rim. Due to the model assumption of rigid bodies the multibody model is not capable to describe these effects and thus the results differ considerably to the finite element solutions.

The results of the modal model with a precomputed force law show a very good agreement to the finite element solutions for a sufficiently high number of considered eigenmodes. It is also possible to simulate the dynamical gear body effects after contact. However, a sigificant numerical effort is necessary for the needed precomputations like the eigenmode analyses an the determination of the displacement-force relationships used as local contact laws. But since this data can be reused for consecutive contacts and the integration time itself is very small, the modal approach is most advantageous for investigations of many contacts.

Bibliography

ABAQUS. *ABAQUS – Theory Manual.* ABAQUS Inc., Providence, 2004a.

ABAQUS. *ABAQUS – Verification Manual.* ABAQUS Inc., Providence, 2004b.

A. Ast and P. Eberhard. Flatness-based Control of Parallel Kinematics Using Multibody Systems - Simulation and Experimental Results. *Archive of Applied Mechanics*, 76:181–197, 2006.

A. Ast, S. Braun, P. Eberhard, and U. Heisel. Adaptronic Vibration Damping for Machine Tools. *Annals of the CIRP*, 56(1):379–382, 2007.

K.J. Bathe. *Finite Element Procedures.* Prentice Hall, Upper Saddle River, 1996.

J. Berkemer, R. Altenburger, T. Koch, and W.D. Lehner. Effektive Nutzung des Leistungspotentials von Direktantrieben. *wt Werkstattstechnik online*, 94(6):244–241, 2004. (in German).

D. Bestle. *Analyse und Optimierung von Mehrkörpersystemen.* Springer, Berlin, 1994. (in German).

D. Bestle and P. Eberhard. Analyzing and Optimizing Multibody Systems. *Mechanics of Structures and Machines*, 20:67–92, 1992.

D. Bestle and P. Eberhard. NEWOPT/AIMS 2.2. Ein Programmsystem zur Analyse und Optimierung von mechanischen Systemen. Manual AN–35, Institute B of Mechanics, University of Stuttgart, 1994. (in German).

C. Bischof, A. Carle, P. Khademi, and A. Mauer. ADIFOR 2.0: Automatic Differentiation of Fortran 77 Programs. *IEEE Computational Science & Engineering*, 3:18–32, 1996.

G.W. Blankenship and A. Kahraman. Steady State Forced Response of a Mechanical Oscillator with Combined Parametric Excitation and Clearance Type Non-Linearity. *Journal of Sound and Vibration*, 185(5):743–765, 1995.

C.R. Boër, L. Molinari-Tosatti, and K.S. Smith, editors. *Parallel Kinematic Machines – Theoretical Aspects and Industrial Requirements.* Advanced Manufacturing Series. Springer, London, 1999.

U. Brahms. *Regelung von Lineardirektantrieben für Werkzeugmaschinen.* Fortschritt-Berichte VDI, Reihe 8, Nr. 978. VDI-Verlag, Düsseldorf, 1998. (in German).

I. Bronstein, K. Semendjajew, G. Musiol, and H. Mühlig. *Taschenbuch der Mathematik.* Verlag Harri Deutsch, Thun, Frankfurt am Main, 1997. (in German).

R.D. Cook, D.S. Malkus, and M.E. Plesha. *Concepts and Applications of Finite Element Analysis.* John Wiley & Sons, New York, 1989.

J.J. Craig. *Introduction to Robotics.* Addison-Wesley, Reading, 1989.

B. Dasgupta and T. Mruthyunjaya. Force Redundancy in Parallel Manipulators: Theoretical and Practical Issues. *Mechanisms and Machine Theory*, 33(6):727–742, 1998.

F. Dignath. *Zur Optimierung mechatronischer Systeme mit nichtdifferenzierbaren Kriterien.* Fortschritt-Berichte VDI, Reihe 8, Nr. 1031. VDI-Verlag, Düsseldorf, 2004. (in German).

F. Dignath and D. Hempelmann. Grundlagenuntersuchungen zum thermischen Einfluss – Bericht 2002. Report ZB–131, Institute B of Mechanics, University of Stuttgart, 2002. (in German).

P. Eberhard. *Zur Mehrkriterienoptimierung von Mehrkörpersystemen.* Fortschritt-Berichte VDI, Reihe 11, Nr. 227. VDI-Verlag, Düsseldorf, 1996. (in German).

P. Eberhard. *Kontaktuntersuchungen durch hybride Mehrkörpersystem/Finite Elemente Simulation.* Shaker Verlag, Aachen, 2000. (in German).

P. Eberhard and D. Bestle. Integrated Modeling, Simulation and Optimization of Multibody Systems. In G. Johannsen, editor, *Integrated Systems Engineering*, pages 35–40. Pergamon, Oxford, 1994.

P. Eberhard, F. Dignath, and L. Kübler. Parallel Evolutionary Optimization of Multibody Systems with Application to Railway Dynamics. *Multibody System Dynamics*, 9:143–164, 2003.

S. Ebrahimi and P. Eberhard. Rigid-Elastic Modeling of Meshing Gear Wheels in Multibody Systems. *Multibody System Dynamics*, 16(1):55–71, 2006.

R. Fletcher. *Practical Methods of Optimization.* John Wiley & Sons, Chichester, 1987.

O. Föllinger. *Regelungstechnik.* Hüthig Buch Verlag, Heidelberg, 1992. (in German).

C. Gosselin and J. Angeles. Singularity Analysis of Closed–Loop Kinematic Chains. *IEEE Transactions on Robotics and Automation*, 6(3):281–290, 1990.

V. Hagenmeyer and E. Delaleau. Exact Feedforward Linearization Based on Differential Flatness. *International Journal of Control*, 76(6):537–556, 2003.

U. Heisel and W. Maier. Fachwerkstrukturen für hochdynamische Werkzeugmaschinen. *wt Werkstattstechnik online*, 7(8):333–339, 2002. (in German).

U. Heisel, V. Maier, and E. Lunz. Auslegung von Maschinenkonstruktionen mit Gelenkstab-Kinematik-Grundaufbau, Tools, Komponentenauswahl, Methoden und Erfahrungen. *wt – Werkstattstechnik*, 88(4):75–78, 1998. (in German).

INTEC GmbH. *SIMPACK v8.8.* For contact information and informative literature see `www.simpack.de`.

A. Isidori. *Nonlinear Control Systems.* Springer-Verlag, London, 1995.

ISO 6336-1. Calculation of Load Capcity of Spur and Helical Gears – Part 1. International Organisation for Standardisation.

A. Kahraman and G.W. Blankenship. Experiments on Nonlinear Dynamics Behavior of an Oscillator with Clearance and Periodically Time-Varying Parameters. *Journal of Applied Mechanics*, 64:217–226, 1997.

J. Kirchner. *Mehrkriterielle Optimierung von Parallelkinematiken.* PhD thesis, Fraunhofer Institut Werkzeugmaschinen und Umformtechnik IWU, Chemnitz, 2000. (in German).

E. Kreuzer and G. Leister. Programmsystem NEWEUL'90. Manual AN–24, Institute B of Mechanics, University of Stuttgart, 1991. (in German).

L. Kübler, C. Henninger, and P. Eberhard. Multi-Criteria Optimization of a Hexapod Machine. *Multibody System Dynamics*, 14:225–250, 2005.

F. Kücükay. Rheonichtlineare Zahnradschwingungen. *Journal of Applied Mathematics and Mechanics*, 64:T58–T61, 1984. (in German).

F. Kücükay. *Dynamik der Zahnradgetriebe.* Springer Verlag, Berlin, 1987. (in German).

T. Kurz, C. Henninger, and P. Eberhard. SYMBS - Symbolical Analysis and Optimization of Multibody Systems in Matlab. In *Proceedings of the European Nonlinear Oscillations Conference ENOC*, St. Petersburg, June 30 – July 4 2008.

J. Liu. Kinematics and Dynamics of Spatial Hexapod Motions with Thermal Disturbances. Report IB–38, Institute B of Mechanics, University of Stuttgart, 2001.

N.M.M. Maia and J.M.M. Silva, editors. *Theoretical and Experimental Modal Analysis.* Research Studies Press, Baldock, 1998.

J.-P. Merlet. Singular Configurations of Parallel Manipulators and Grassmann Geometry. *Journal of Robotics Research*, 8(5):45–56, 1989.

J.-P. Merlet. *Parallel Robots.* Kluwer, Dordrecht, 2000.

G. Niemann and H. Winter. *Maschinenelemente – Band 2.* Springer Verlag, Berlin, 1989. (in German).

J. O'Brien and J. Wen. Redundant Actuation for Improving Kinematic Manipulability. In *Proceedings of the 1999 IEEE International Conference on Robotics and Automation*, volume 2, pages 1520–1525, Detroit, 1999.

R. Padieth. Exakte Ermittlung der Zahnform. *Antriebstechnik*, 17(10): 434–436, 1978. (in German).

F. Pfeiffer. Mechanische Systeme mit unstetigen Übergängen. *Ingenieur-Archiv*, 54:232–240, 1984. (in German).

W. Prestl. *Zahnhämmern in Rädertrieben von Dieselmotoren.* Fortschritt-Berichte VDI, Reihe 11, Nr. 145. VDI-Verlag, Düsseldorf, 1991. (in German).

A. Ratering and P. Eberhard. Flatness-based Control of a Machine Tool with Lambda-Kinematics. In F. Allgöwer and M. Zeitz, editors, *Proc. of the NOLCOS 2004, 6th IFAC-Symposium on Nonlinear Control Systems, Stuttgart, Germany*, volume 1, pages 197–142, 2005a.

A. Ratering and P. Eberhard. Nonlinear Control of a Machine Tool with Parallel Kinematics. In J. Goicolea, J. Cuadrado, and J.G. Orden, editors, *Proc. of the ECCOMAS Thematic Conference on Multibody Dynamics, Madrid, Spain*, 2005b.

S. Riebe and H. Ulbrich. Modelling and Online Computation of the Dynamics of a Parallel Kinematic with Six Degrees-of-Freedom. *Archive of Applied Mechanics*, 72:817–829, 2003.

R. Rothfuß. *Anwendung der flachheitsbasierten Analyse und Regelung nichtlinearer Mehrgrößensysteme.* Fortschritt-Berichte VDI, Reihe 8, Nr. 664. VDI-Verlag, Düsseldorf, 1997. (in German).

R. Rothfuß, J. Rudolph, and M. Zeitz. Flachheit: Ein neuer Zugang zur Steuerung und Regelung nichtlinearer Systeme. *at – Automatisierungstechnik*, 11:517–525, 1997. (in German).

J. Rudolph. *Flatness Based Control of Distributed Parameter Systems.* Berichte aus der Steuerungs- und Regelungstechnik. Shaker-Verlag, Aachen, 2003.

W. Schiehlen and P. Eberhard. *Technische Dynamik.* Teubner, Wiesbaden, 2004. (in German).

L. Sciavicco and B. Siciliano. *Modeling and Control of Robot Manipulators.* McGraw-Hill, New York, 1996.

K. Sedlaczek and P. Eberhard. Using Augmented Lagrangian Particle Swarm Optimization for Unconstrained Problems in Engineering. *Structural and Multidisciplinary Optimization*, 32(4):277–286, 2006.

K. Sedlaczek and P. Eberhard. Augmented Lagrangian Particle Swarm Optimization in Mechanism Design. *Journal of System Design and Dynamics*, 1(3):410–421, 2007.

R. Seifried. *Numerische und experimentelle Stoßanalyse für Mehrkörpersysteme.* Schriften aus dem Institut für Technische und Numerische Mechanik der Universität Stuttgart, Band 2. Shaker Verlag, Aachen, 2005. (in German).

Y. Takeda and H. Funabashi. Kinematic and Static Characteristics of In–Parallel Actuated Manipulators at Singular Points and in Their Neighborhood. *JSME International Journal, Series C*, 39(1):85–93, 1996.

Y. Takeda and H. Funabashi. Kinematic Synthesis of In–Parallel Actuated Mechanisms Based on the Global Isotropy Index. *Journal of Robotics and Mechatronics*, 11(5):404–410, 1999.

L.-W. Tsai. *Robot Analysis: The Mechanics of Serial and Parallel Manipulators.* John Wiley, New York, 1999.

M. Valášek, Z. Šika, V. Bauma, and T. Vampola. Design Methodology for Redundant Parallel Robots. In *Proc. of AED 2001, 2nd Int. Conf. on Advanced Engineering Design*, pages 243–248, Glasgow, 2001.

M. Weck. *Werkzeugmaschinen, Fertigungssysteme. Volume 3: Automatisierung und Steuerungstechnik.* VDI-Verlag, Düsseldorf, 1995. (in German).

H.H. Winkler, D. Prust, W. Maier, R. Kirchberger, C. Kramer, G. Rogers, K.H. Wurst, W.D. Lehner, R. Altenburger, J. Berkemer, and C. Eppler. Konzeptionen für hochdynamische Maschinen. In *Stuttgarter Impulse, Tagungsband zum Fertigungstechnischen Kolloquium Stuttgart*, pages 89–137, 2003. (in German).

C. Woernle. Dynamik von Mehrkörpersystemen. Lecture Notes. University of Rostock, 2001. (in German).

X. Zhengyi, X. Fengfeng, and C.K. Mechefske. Kinetostatic Analysis and Optimization of a Tripod Attachment. In C. Gosselin and I. Ebert-Uphoff, editors, *Proc. of the Workshop on Fundamental Issues and Future Research Directions for Parallel Mechanisms and Manipulators*, pages 294–303, Quebec City, Canada, October 3–4 2002.

P. Ziegler. Statische und dynamische Zahnkontaktuntersuchungen. DIPL-104, Institute B of Mechanics, University of Stuttgart, 2005. (in German).

P. Ziegler, P. Eberhard, and B. Schweizer. Simulation of Impacts in Geartrains Using Different Approaches. *Archive of Applied Mechanics*, 76(9-10):537–548, 2006.